Nanocarbon Allotropes Beyond Graphene

Synthesis, properties and applications

Online at: https://doi.org/10.1088/978-0-7503-5177-5

Nanocarbon Allotropes Beyond Graphene

Synthesis, properties and applications

Edited by
Arpan Kumar Nayak
Department of Energy Engineering, Konkuk University, 120 Neungdong-ro, Seoul, Republic of Korea

Santosh K Tiwari
Department of Chemistry, NMAM Institute of Technology, Nitte (Deemed to be University), Nitte-547110, Karnataka, India

IOP Publishing, Bristol, UK

ISBN 978-0-7503-5177-5 (ebook)
ISBN 978-0-7503-5175-1 (print)
ISBN 978-0-7503-5178-2 (myPrint)
ISBN 978-0-7503-5176-8 (mobi)

DOI 10.1088/978-0-7503-5177-5

Version: 20230601

IOP ebooks

British Library Cataloguing-in-Publication Data: A catalogue record for this book is available from the British Library.

Published by IOP Publishing, wholly owned by The Institute of Physics, London

IOP Publishing, No.2 The Distillery, Glassfields, Avon Street, Bristol, BS2 0GR, UK

US Office: IOP Publishing, Inc., 190 North Independence Mall West, Suite 601, Philadelphia, PA 19106, USA

Dedicated to Lord Jagannath

Contents

Preface

Carbon is unique, and it surprises the scientific community time and time again, from the early benzene rings and complex polymers to the all-carbon caged fullerenes and tubular nanotubes, grapheme nanoribbons, and many more. At present, it is nearly impossible to imagine commercial aspects of the Internet, mobile phone, optical fiber, pharma–biomedical, space, defense, automobile, and textile industries without carbon materials. According to a recent report, carbon-based materials will replace almost 55%–60% of non-carbonaceous semiconductor materials used in various high technologies, including supercomputers and next-generation communications devices, by 2040. Such tremendous growth and demanding applications of carbon materials are due to their tunable physical properties and exceptional bonding capabilities. That is why carbon as an element is has a significant lead in in terms of functionality compared to the other elements in the periodic table, and it is currently the most prominent candidate for nanoscience and advanced technologies. According to the Scopus database, almost 30%–35% of all scientific publications are directly or indirectly associated with carbon-based materials. Thus, carbon materials research is dominating the entire scientific research field for various industrial applications. The year 1985 was an early year for nanotechnology and materials engineering, when nanocarbons received huge scientific attention owing to their exceptional properties. In this regard, fullerene was discovered by Kroto *et al* in 1985, which created excitement over the possibility of nano-allotropes of carbon. The discovery of fullerene was a sensation and within ten years of its discovery, reports on the applications of fullerene, especially for biomedical and bioimaging, drug delivery, HIV treatment, and enhancement of the thermo-mechanical properties of polymers, were developed and utilized accordingly. The extensive research around the globe on fullerene and fullerene-like (carbon onion and higher fullerene) nanomaterials led to the discovery of carbon nanotubes, which was first reported by Ijima *et al* in 1992. This was another unique carbon atom-based nanomaterial with outstanding thermo-mechanical strength, very high chemical inertness, great flexibility, light weight, high surface area, and very high electronic conductivity. Interestingly, different kinds of nanotubes have been fabricated including several inorganic nanotubes, and many of them are part of the modern technologies of the twenty-first century. Similarly, in recent years graphene (the discoverers of which won the Nobel Prize in 2010) and its derivatives have caught great attention and within a short period of time thousands of articles have been published.

Overall, this book deals with various aspects of graphene, graphyne, graphdiyne, graphene nanoribbon, etc, synthesis and their advanced applications. We concentrate on progress in new nanocarbons, considering the problems with cost, optimization, and commercialization.

Santosh K Tiwari and Arpan Kumar Nayak

Acknowledgement

We are very grateful to all the contributors (authors and co-authors of the chapters), in particular for their timely completion and revision as per the reviewers' comments, during the difficult working conditions of the COVID-19 pandemic. We are also very grateful to IOP Publishing, in particular Dr Caroline Mitchell, for allowing enough time to prepare the manuscript. We very sincerely thank the researchers whose work is directly or indirectly cited in this book in the form of text, research data, figures, and tables.

Editor biographies

Dr Arpan Kumar Nayak

Arpan Kumar Nayak earned his MSc from Jadavpur University, India. He then completed his PhD degree at the Indian Institute of Technology Kharagpur (IIT-KGP), India. He was a post-doctoral fellow at Hanyang University, South Korea. Currently he is working as Research Professor at the Department of Energy Engineering, Konkuk University (120 Neungdong-ro, Seoul 05029, Republic of Korea). His current research mainly focuses on the synthesis of various nanostructured materials and carbon-based materials towards environmental and energy applications. He has published more than 56 articles in various international journals (Google scholar citation count: 1436; h-index: 22; i10-index-37) and edited four books for Elsevier, IOP Publishing, and IOR International Press. He is Editor-in-Chief of the *International Research Journal of Multidisciplinary Technovation* and *Bulletin of Scientific Research*, and Associate Editor of *Nanoscale Reports*.

Dr Santosh K Tiwari

Dr. Santosh K. Tiwari is Assistant Professor at NMAMIT, Nitte, India (Visiting Scientist, Guangxi Institute Fullerene Technology, GU, Nanning, China). Dr. Tiwari received a Ph.D. in 2D materials-based nanocomposite from IIT Dhanbad, India. He worked more than five years with several high ranked research labs and universities around the globe, including HSCL & Hanyang University (Seoul Korea), GIFT, GU, Nanning (China), and University of Warsaw, Poland. He has published over 70 articles including 6 books which are cited over 2100 times with an h-index of 24. His research interests are in the areas of new materials, 2D nanomaterials, polymer composites and mechanical properties.

List of contributors

Hossem Belhamdi
M'Hamed Bougara University, Boumerdes, Algeria

Paritosh Chaudhuri
Institute for Plasma Research, Bhat, Gandhinagar, India

Maya Devi
Department of Physics, School of Applied Sciences, KIIT Deemed to be University, Bhubaneswar, Odisha, India

Asma Nour El Houda
M'Hamed Bougara University, Boumerdes, Algeria

Tapas Ghatak
Department of Chemistry, Vellore Institute of Technology, Vellore, India

Amina Hachaichi
M'Hamed Bougara University, Boumerdes, Algeria

Bhim Prasad Kafle
Department of Chemical Science and Engineering, School of Engineering, Kathmandu University, Dhulikel, Kavre, Nepal

Dhulikhel, Kavre
Department of Chemistry, University of Warsaw, Warsaw, Poland

Benalia Kouini
Laboratory of Coatings, Materials and Environment, M'Hamed Bougara University, Boumerdes, Algeria

Siladitya Laha
Department of Chemistry, Indian Institute of Technology Kanpur, Kanpur, India

Sutripto Majumder
Department of Physics, Yeungnam University, Gyeongsan, Republic of Korea

Umer Mehmood
Polymer and Process Engineering (PPE) Department, University of Engineering and Technology (UET) Lahore, Pakistan

Srikanta Moharana
School of Applied Sciences, Centurion University of Technology and Management, Odisha, India

Arpan Kumar Nayak
Department of Energy Engineering, Konkuk University, 120 Neungdong-ro, Seoul, Republic of Korea

Rabia Nazar
Polymer and Process Engineering (PPE) Department, University of Engineering and Technology (UET), Lahore, Pakistan

Raunak Pandey
Department of Chemical Science and Engineering, Kathmandu University, Dhulikhel, Nepal

Swetapadma Praharaj
Department of Physics, School of Applied Sciences, KIIT Deemed to be University, Bhubaneswar-24, Odisha, India

V Priyanka
Vellore Institute of Technology, Vellore, India

Dibyaranjan Rout
Department of Physics, School of Applied Sciences, KIIT Deemed to be University, Bhubaneswar, Odisha, India

Siddheswar Rudra
Department of Chemistry, National Institute of Technology Meghalaya, Bijni Complex, Laitumkhrah, Shillong, India

Rajashree Sahoo
Institute for Plasma Research, Bhat, Gandhinagar, India

Bibhuti B Sahu
Department of Physics, Veer Surendra Sai University of Technology, Odisha, India

Lakshmi Sajeev
Department of Physics, School of Advanced Sciences, Vellore Institute of Technology, Vellore, Tamil Nadu, India

Kuldeep Singh
Schulich Faculty of Chemistry, Technion—Israel Institute of Technology, Haifa City, Israel

Pamula Siva
Department of Physics, School of Advanced Sciences, Vellore Institute of Technology, Vellore, Tamil Nadu, India

C Sreelakshmi
Department of Physics, School of Advanced Sciences, Vellore Institute of Technology, Vellore, Tamil Nadu, India

A Thennarasi
Department of Physics, School of Advanced Sciences, Vellore Institute of Technology, Vellore, Tamil Nadu, India

Santosh K Tiwari
Department of Chemistry, University of Warsaw, Warsaw, Poland
Department of Chemistry, NMAM Institute of Technology, NITTE (Deemed to be University), Mangalore, Karnataka, India

Kuraganti Vasu
Department of Physics, School of Advanced Sciences, Vellore Institute of Technology, Vellore, Tamil Nadu, India

Nannan Wang
Key Laboratory of New Processing Technology for Nonferrous Metals and Materials, Ministry of Education, School of Resources, Environment and Materials, Guangxi University, Nanning, China

IOP Publishing

Nanocarbon Allotropes Beyond Graphene
Synthesis, properties and applications
Arpan Kumar Nayak and Santosh K Tiwari

Chapter 1

Recent advances in nanocarbons: status and prospect

Raunak Pandey and Santosh K Tiwari

The constant changes in the field of nanotechnology need to be analyzed to obtain an idea of the present state of these technologies. One branch of nanotechnology, nanocarbons, experiences constant changes and growth for effective, new, robust, and efficient applications that can fill the voids left by newer technologies. In this sense, the status of different nanocarbons should be assessed regularly and continuously so that the current state can be viewed and future outcomes can be predicted. We have found that chemical vapor deposition, laser ablation, and arc discharge are the most prominent methods for the synthesis of nanocarbons. In terms of properties, excellent electrical, mechanical, optical, and thermal properties have been shown by all the nanocarbons that are discussed here. Enhanced electrical and thermal conductivity, mechanical strength and transmissivity are some of the properties that are observed in these nanocarbons, which provide applications in electronics, energy storage, drug carriers, biosensors, biomedicine, aerospace, thermal management, etc. In this chapter we review five different types of nanocarbons, namely carbon nanotubes, fullerenes, graphene, diamane, and diamanoids. Analysis of their methods of synthesis and properties are provided, which can subsequently be used in the brilliant applications of nanotechnology.

1.1 Introduction

Even though carbon is extensively available in nature, the Earth's crust consists of only 0.025% carbon. Carbon is a non-metallic, tetravalent element of the periodic table that offers four electrons on the outer orbit to form covalent bonds with other elements. Carbon can be synthesized by igniting helium so that three nuclei of helium are fused to produce a carbon nucleus, thus producing carbon atoms. Although there is low availability in the crust, the formation of compounds of carbon is more plentiful than for any other element. An estimated number of more

doi:10.1088/978-0-7503-5177-5ch1

than one million carbon compounds are synthesized by chemists or available in nature and this number is still growing at a reasonable pace [1]. Tuning the parameters while synthesizing carbon compounds can help to fabricate new compounds with enhanced physical, chemical, environmental, and mechanical properties. Scientists and researchers are working to produce carbon compounds with maximum new and enhanced properties to fulfill the needs of the booming new technological advances.

The search for compounds exhibiting extraordinary properties, being lightweight but equally or more efficient than other bulky compounds, is vital in nanotechnology. Nanotechnology is not new to the twenty-first century, as several nanoproducts and their derivatives have already infiltrated the global market before then. Although the development of nanotechnology is expanding, research on this topic is limited to only some parts of the world. Figure 1.1(a) shows that the USA has the greatest number of research institutes for nanotechnology while others are far behind in this sector. However, data relating to the institutes in South America, Africa, and Oceania could not be retrieved by the authors. According to figure 1.1(b), the USA is leading in the revenue market as well and figure 1.1(c) shows the revenue generated by materials science in comparison to other fields [2]. These data show the conditions and future predictions of materials science and nanotechnology. The search for the right compounds to perform at the nanometer scale with extreme precision and efficiency in this growing market is compelling. Within this extensive research to find the right materials that encompass important qualities, carbon compounds are seen as the best fit compounds. This research

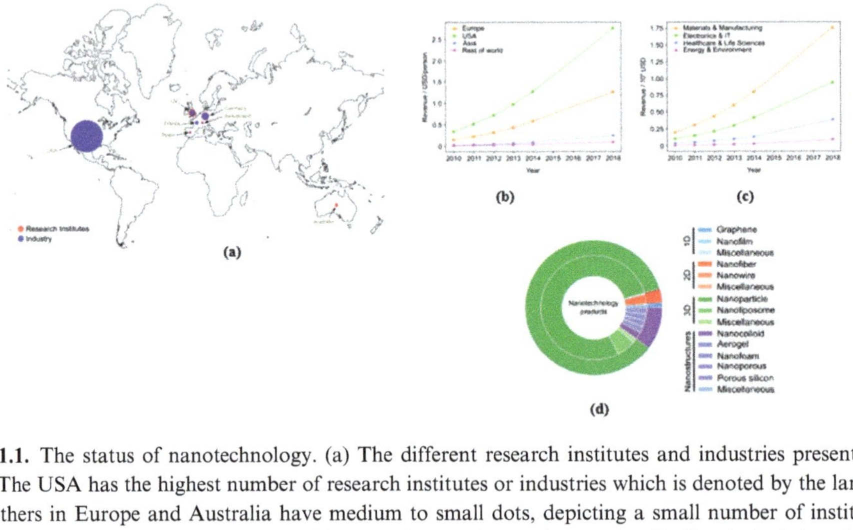

Figure 1.1. The status of nanotechnology. (a) The different research institutes and industries present in the world. The USA has the highest number of research institutes or industries which is denoted by the large dot, while others in Europe and Australia have medium to small dots, depicting a small number of institutes or industries. Data were not available for Africa, South America, and Oceania. (b) Graph depicting the gross revenue generated from nanotechnology for different continents (data available from 2010 to 2018). (c) Graph depicting the gross revenue generated by the various fields of nanotechnology from 2010 to 2018. (d) Pie chart depicting the different nanomaterials that are used in nanotechnological products. (Reproduced with permission from [2]. Copyright 2021 American Chemical Society.)

eventually led to the birth of nanocompounds of carbon, or nanocarbons. Nanocarbons can be described as carbon compounds that are synthesized structurally in the nanometer scale with the goal of obtaining enhanced and abundant physical and chemical properties in comparison to bulk compounds. The comparison of nanocarbons with bulk compounds shows that nanocarbons exhibit much more efficient and specific applications than the bulk compounds. For instance, the development of computer processors from bulky devices to minute ones occurred thanks to nanocarbons. Nanocarbons can be found structurally in one-dimensional, two-dimensional, and three-dimensional forms. The exploitation of hybridization in every way possible along with assembly processes will assist in obtaining these structures of nanocarbons [3]. Different types of nanocarbons have their own importance regarding their exhibited properties and the use of these compounds towards the applications of the modern technological field of nanotechnology (figure 1.2). Exceptional properties in the form of electrical, magnetic, optical, and mechanical properties, etc, can be found in nanocarbons, such that applications in the fields of electronics, transistors, optoelectronics, composites, aeronautics, mechanics, light emitting diodes, etc, can benefit significantly from them, which play a major role in making robust, efficient, and lightweight devices.

As mentioned earlier, the classification of nanocarbons is done mainly based on their dimensions. 'Dimension' refers to the measurement of the height, width, and length of their particles. Zero-dimensional nanocarbons are those carbon particles whose dimensions are all in the nanometer scale. Some examples of zero-dimensional nanocarbons are quantum dots, polymer dots, inorganic quantum dots, conversion nanoparticles, metallic nanoparticles, etc. For one-dimensional

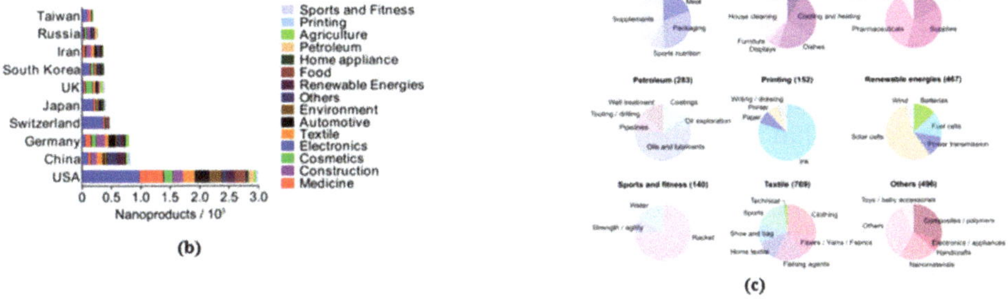

Figure 1.2. Various products that are the result of nanotechnology. (a) The different industries of nanotechnology. (b) The different nanotechnological industries in different countries. (c) The use of various nanoparticles in industrial products. (Reproduced with permission from [2]. Copyright 2021 American Chemical Society.)

nanocarbons, one of the dimensions of the nanocarbons is measured to be outside the nanometer scale. Some examples of one-dimensional nanocarbons are nanotubes, nanowires, nanorods, etc. For two-dimensional nanocarbons, the structure shows that two of the dimensions of the nanocarbons are measured to be outside the nanometer scale. Some examples of two-dimensional nanocarbons are graphene, nanocoatings, nanofilms, nanolayers, etc. For three-dimensional nanocarbons, all the dimensions of the nanocarbons are measured to be outside the nanometer scale. Some examples of three-dimensional nanocarbons are bundles of nanorods, nanotubes, nanowires, bulk powders, dispersions, etc [4]. Even though the synthesis of materials are performed in the nanometer scales, the properties that are exhibited by nanocarbons are enormous.

In this chapter we discuss the current state of some nanocarbons and their applications in the field of nanotechnology. We discuss in brief the current status and prospects of carbon nanotubes, fullerene, graphene, diamane, and diamanoids. The properties, methods of synthesis, and applications of these nanocarbons are the main focus of this chapter, with some future outlook as well, to allow scientists to gain some insight into how current gaps can be filled using nanocarbons.

1.2 Carbon nanotubes (CNTs)

Carbon nanotubes (CNTs), tubular structured nanoallotropes of carbon, are one-dimensional, cylindrically rolled, singled layers of carbon whose synthesis was first described by Iijima *et al* in 1991. According to their article, Iijima *et al* discovered the synthesis of 'needle-like tubes' of finite carbon structures through an arc discharge evaporation technique in which the formed carbon structures were later defined as 'multi-walled carbon nanotubes'. The formation of needle-like tubes was observed at the negative side of the electrode that was involved in the arc discharge [5]. CNTs are basically tubular structures that are synthesized by rolling up hexagonal sheets of carbon atoms or graphene. These structures depend upon sp^2 bonds and van der Waal's force to keep them in alignment and hold them together. They can also be covered with half fullerene molecules on each end but are stronger than the sp^3 bonds due to their unique structure [6]. There are two types of carbon nanotubes that have been discovered to date: single-walled carbon nanotubes (SWCNTs) and multi-walled carbon nanotubes (MWCNTs). Also, different morphologies such as helical, cup-shaped, bamboo-like, cotton-shaped, etc, are found. SWCNTs, also known as 'graphene nanotubes', are single layered carbon atoms or graphene layers that are rolled to form a tube. MWCNTs are SWCNTs that are arranged concentrically, or are multiple layers of carbon atoms or graphene that are concentrically rolled into tubular structures. The most important distinguishing feature between SWCNTs and MWCNTs is that an SWCNT contains one layer whereas an MWCNT contains multiple concentric layers [6–10]. The structure of carbon nanotubes including SWCNTs and MWCNTs can be seen in figure 1.4, where (a) depicts the structure of single layered graphene, (b) depicts the structure of SWCNTs, and (c) depicts the structure of MWCNTs. The crystal lattices of SWCNTs and MWCNTs do not experience any breakage throughout the length

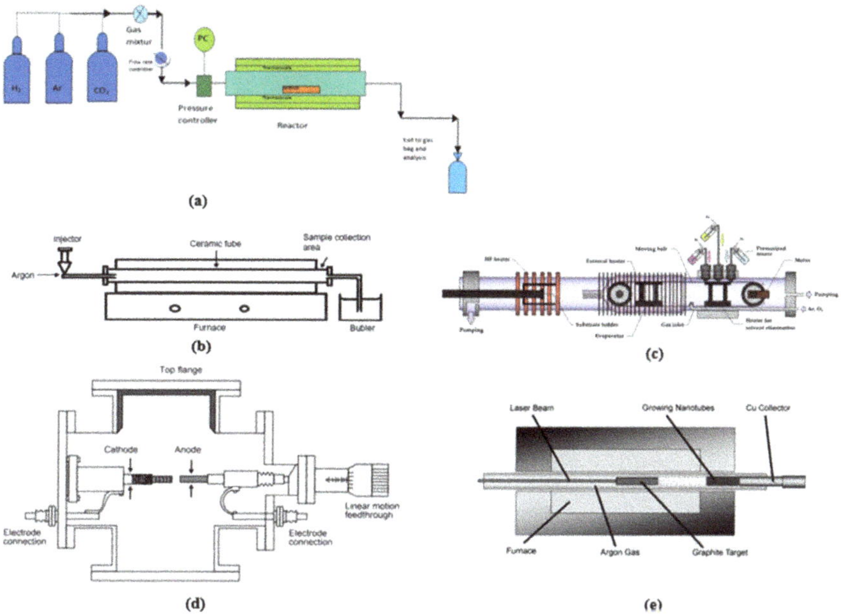

Figure 1.3. Different apparatus for the synthesis of nanocarbons. (a), (b), (c) Different chemical vapor deposition apparatus. (a) The general CVD apparatus. (Reproduced with permission from [12]. Copyright 2015 Elsevier.) (b) The floating catalyst CVD apparatus (Reproduced with permission from [6]. Copyright 2019 Elsevier). (c) The direct liquid injection CVD apparatus with solenoid electrovalves. (Reproduced with permission from [10]. CC BY 3.0.) (d) The arc discharge apparatus. (Reproduced with permission from [13]. Copyright 2007 Elsevier.) (e) The laser ablation apparatus. (Reproduced with permission from Springer from [14]. Copyright 2002 The Korean Fiber Society.)

of the tube making them perfect for applications [11]. The following sub-sections will describe the status of CNTs in terms of the methods of synthesis, properties, and applications that are demonstrated for nanotechnology.

1.2.1 Synthesis

The fabrication methods for CNTs consist of several techniques which have provided consistent synthesis of nanostructures with excellent efficiency. When Iijima *et al* first described the formation of CNTs, they used the arc discharge method to fabricate them. Several researchers have found other methods such as chemical vapor deposition (CVD), arc discharge, laser ablation, electrophoretic deposition, chemical grafting, electrospinning, forest spinning, solution spinning, aerogel spinning, etc, for the fabrication of CNTs (figure 1.3) [10, 16, 17]. The CVD method is the most common and prominent method for the fabrication of CNTs due to its qualities of being cost effective and easy to scale up, having a low synthesis temperature, the production of uniform structures on complex surfaces, etc. This technique employs catalyst and organic or inorganic supports for the fabrication of CNTs, which aid in decreasing agglomeration and achieve good dispersion of CNTs in the polymer matrix or nanocomposites when used in applications. Several metallic

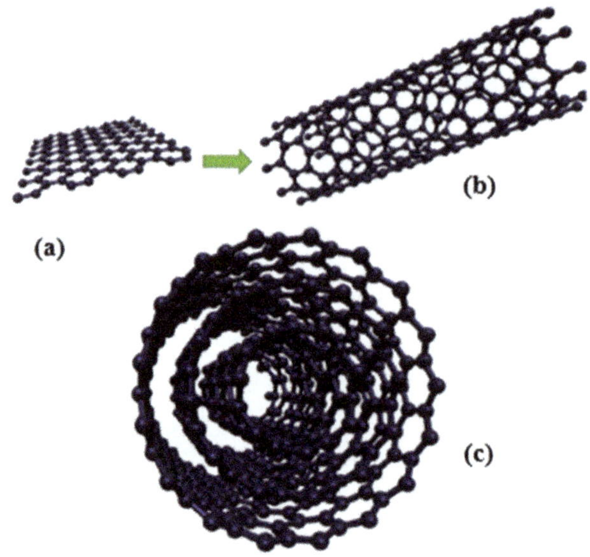

Figure 1.4. The process of formation of carbon nanotubes. (a) Graphene sheet. (b) Single-walled carbon nanotubes. (c) Multi-walled carbon nanotubes. Figures (a) and (b) show that single-walled carbon nanotubes are formed by rolling up graphene sheets. (Reproduced with permission from [15]. Copyright 2018 Elsevier.)

catalysts such as iron (Fe), nickel (Ni), molybdenum (Mo), cobalt (Co), ferrocene, etc, are used in CVD techniques for efficient production of CNTs. Along with this, new methods of the CVD process for synthesizing CNT have also been found, namely, plasma enhanced CVD, atomic layer CVD, microwave plasma CVD, etc [8, 9, 17–20].

According to the literature, the most common method to fabricate CNTs is the CVD approach. In an experiment conducted by Li *et al*, the CVD method was employed to deposit CNT on a fly ash surface. Here, nickel (Ni) was used as the catalyst, and acetylene and hydrogen were used as the hydrocarbon precursor and were were decomposed at 400 °C to fabricate CNTs [17]. In another experiment conducted by Pełech *et al*, MWCNTs were obtained via the CVD method by decomposing ethylene on a cobalt (Co) catalyst, iron (Fe) catalyst, and a mixture of Fe/Co catalysts [18]. Orbaek *et al* have also employed the CVD technique for the synthesis of CNTs using ferrocene as a catalyst, and hydrocarbons such as toluene, benzene, methylcyclopentane, cyclopentane, and cyclohexane inside a horizontal tubular reactor [11]. Analysis through an experiment and model fabrication for carbonaceous growth was done by Henao *et al* where they used CVD to catalytically decompose a methane precursor at 700 °C–800 °C. Cobalt (Co)–copper (Cu) was used as the catalyst and the carbon support was derived from cellulose [8]. Modification of CVD through microwave assistance uses microwaves to heat the substrate so that the deposition of CNTs can be done favorably [21]. Studies conducted by Burakova *et al* used microwave treatment of the metal catalyst so that carbonization is slowed down, and the specific surface area of the catalyst and the chemical activity were increased. Here, the nickel (Ni)/magnesium oxide (MgO) catalyst was treated with microwave radiation which then was used in fabricating CNTs instead of thermal heating [22]. In an experiment conducted by Sridhar *et al*, a microwave assisted CVD process was developed to produce 3D functional, mesoporous CNTs, where palladium and cobalt based zeolitic imidazolate frameworks (ZIFs) were used as the single precursors of CNTs, cobalt was used as a catalyst, and graphene was used as the anchor substrate to fabricate CNTs. Later, doping with nitrogen was also done to obtain nanostructures shaped like bamboo for applications as electrode materials [9]. Another method for the production of CNT-cotton was using the floating catalyst CVD (FCCVD) process where hydrocarbon was injected in the furnace along with catalytic precursors. Ismail *et al* have reported the fabrication of CNT superfiber cotton using FCCVD inside an FCCVD reactor (figure 1.3(b)). A quartz support, ethanol hydrocarbon precursor, and carbon or zeolite catalyst were used for the process [6]. The use of natural ingredients in fabricating CNTs through CVD has also been studied. Yuan *et al* used a natural catalyst (red soil) to fabricate MWCNTs using the CVD technique [19]. In another study conducted by Ismail *et al*, waste cooking oil was used in the FCCVD process for the fabrication of CNT-cotton. The authors discussed the synthesis of CNT-cotton from waste palm oil, which is the carbon source for the synthesis procedure. Here, thiphene was used as a promotor, hydrogen and argon were used as the carrier gas, and ferrocene was used as the catalyst precursor [6].

Apart from CVD, other techniques have also been used for the fabrication of CNTs. Sharma *et al* fabricated CNTs through high-resolution, real-time, *in situ* TEM synthesis using current based Joule heating and immediately performing an on-spot soldering process. Platinum was used as the catalyst which was embedded inside the amorphous carbon nanofiber by argon (Ar^+) ion irradiation. Y-junction CNTs were fabricated with the process along with broken CNTs. The soldering process, done immediately after, was able to reconstruct the broken CNTs and form a network of CNTs through the electromigration of metal by inducing Joule heating [23]. Another method for the synthesis of CNTs is the wet chemical process of the sol–gel method, which was used to fabricate carbon nanotube aerogels by removing the contained solvent using inert fluid or freeze-drying the wet gel [6]. Several new synthesis methods should also be studied to reach the pinnacle in terms of physical and mechanical properties of CNTs so that excellent applications can be realized.

1.2.2 Properties

Analysis of the properties of nanocarbons enables researchers to view the possibility of these structures towards applications. CNT nanostructures, through analysis, have also been observed to possess excellent electrical, mechanical, optical, and thermal properties. Enhanced thermal and electrical conductivity, excellent tensile strength and flexibility, toughness, good chemical and thermal stability, unique one-dimensional structure, light weight, minute size, oxidative and corrosion resistance, high aspect ratio and large surface area, low electrical resistivity, excellent adsorption capacity and rate, selectivity, etc, are some of the important features of CNTs [8, 10, 22, 24–26]. Among these, mechanical strength and electrical conductivity are the two most important properties of CNTs [26]. Enhanced mechanical strength helps to produce excellent super-strong materials, but the agglomeration of CNTs while in synthesis is a disadvantage to its mechanical properties. Without agglomeration, the mechanical properties of CNTs are enhanced at a high aspect ratio. There is an important need for research to remove the agglomeration of CNTs and realize these amazing properties [27]. Not only this, the morphology of the synthesized fibers also has an influence on the performance of CNTs. Alvarez *et al* have studied the changes in the physical properties of CNTs when they are made into threads. Changes in diameter, mechanical strength, and electrical resistivity were observed when changes in the thread assembly parameters were made. For instance, a decrease in the diameter of the CNT thread causes an increase in packing density, providing small improvements in the electrical and mechanical properties. Changes in the length of the thread directly affect the mechanical properties. An decrease in the diameter of the thread enhances the mechanical strength of the thread in comparison to larger ones. Changes in electrical resistivity with thread density and changes in the electrical conductivity while increasing the CNT-array width was also observed [26]. Assembling CNT wires and threads with full transfer of the exceptional properties is a challenge. Along with this, it has not been possible to control key parameters regarding the synthesis and properties. The length, diameter, and chirality are defined according to its synthesis but the full prediction of diameter and chirality is still uncertain. The number, density, and porosity of these CNTs need to be controlled to fabricate them into threads [26]. Thus, the morphological effects of CNTs influence property enhancement.

CNTs alone have exceptional properties but when they are added to nanocomposites and nanopolymers, further enhancement and modification of the properties of the structure are observed. The qualities of high aspect ratio, excellent electrical and thermal conductivity, high strength, high stiffness, enhanced interlaminar shear strength, enhanced hardness and toughness, flexible design, low thermal expansion, corrosion resistance, and excellent weight critical characteristics with enormous stiffness and strength are provided by carbon fiber (CNTs) when reinforced in polymer composite (figure 1.5). The interfacial interactions, polymer matrix, and fiber reinforcement determine the mechanical and physical properties of the matrix [16, 18, 27, 28]. In the case of nanocomposites fabricated using CNT additives, the growth of CNTs in carbon fiber fabrics enhances the resistance of impact damage and energy dissipation. These qualities improve the mechanical performance of the composite, such as the delamination of carbon fiber-reinforced polymer, improvement of fracture toughness, and resistance to crack propagation, which are useful for applications [28]. Also, factors such as matrix distributions patterns, orientations, dimensions, structure types, etc, influence the nanocomposite formation [27, 29]. According to computational studies to analyze the mechanical and percolating properties of CNT based nanocomposites, Zare *et al* have developed and analyzed the percolating and strengthening effects of CNTs on nanopolymer composites. The interfacial tensile modulus and strength of CNTs on general

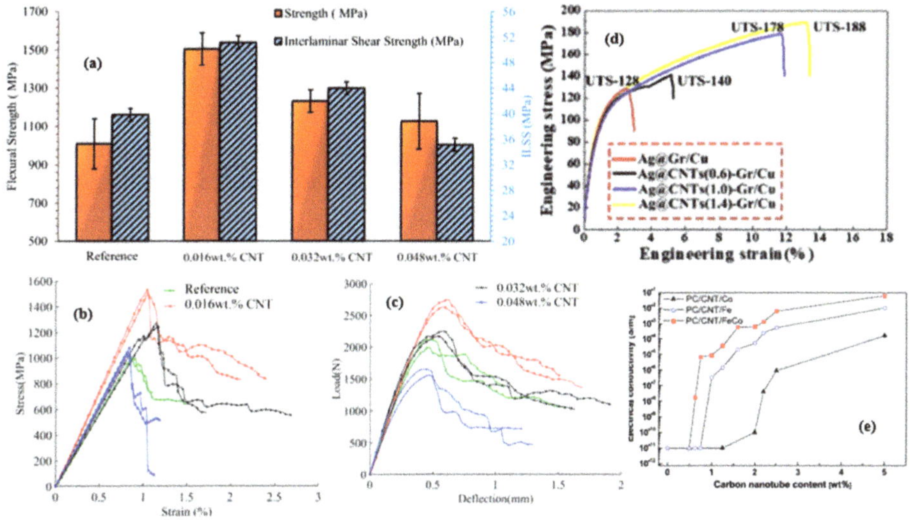

Figure 1.5. Mechanical and electrical properties of various reinforced CNT composites. (a) The interlaminar and flexural properties of carbon fiber-reinforced polymer composite (CFRPC) alone (reference) and CNT reinforced CFRPC. (b) A plot of the stress versus strain curve for composites of CFRPC alone (reference) and CNT reinforced CFRPC. (c) A plot of load–displacement short beam shear (SBS) testing for the composites of CFRPC alone (reference) and CNT reinforced CFRPC. (Reproduced with permission from [28]. Copyright 2021 Elsevier.) Commercial polyacrylonitrile (PAN)-derived, Aeropoxy PR2032 resin and PH3670 matrix were processed to form CFRPC where the reference one contained no CNT fibers. (d) A plot of tensile stress versus strain for the composites of Ag@Gr/Cu and Ag@CNTs-Gr/Cu. (Reproduced with permission from [35]. Copyright 2019 Elsevier.) (e) A plot of conductivity versus CNT content for composites of metal and CNTs. (Reproduced with permission from [18]. Copyright 2020 Elsevier.)

polymer nanocomposites were also studied. The group found that the improvement of interfacial interaction and thicker interphase enhanced the tensile modulus whereas stiffness was enhanced by the robust network of the nanocomposite. Extraordinary strength was also achieved by the strong interphase adhesion and high surface area of CNTs when added to the nanocomposite. Tensile properties, however, are enhanced through the effects on reinforcement of interphase and declination of the percolation points through the interphase area [30]. In another analysis, Parvin *et al* used a multi-scale method to construct models of nano-composites reinforced by CNTs. Enhancement in the elastic modulus and increased aspect ratios were observed in the nanocomposite once CNTs were added to them (figure 1.5(a)) [29].

In addition to computational studies, Schadler *et al* and Allaoui *et al* have discussed the mechanical properties of epoxy resin polymer matrix with added carbon nanostructures. The tension, compression yield strength and elastic modulus were found to be influenced by adding CNTs to the polymer. A small amount of CNTs provides a linear increase in elastic modulus and strength of the matrix but a large amount of CNTs demonstrated a nonlinear increase of strength and elastic modulus. An increase in the CNT aspect ratio, an increase in the efficiency factor of

CNT waviness, further improves the tensile strength and elastic modulus of the epoxy matrix [27, 31, 32]. Analyzing the properties of polystyrene nanocomposites with added CNTs, Qian *et al* have found that the Young's modulus and tensile strength of the polymer composite were increased after the addition of CNTs to the composites [33]. Another study conducted by Ekrem *et al* synthesized a reinforced nanocomposite mat of MWCNTs/PVA which was added to carbon fiber-reinforced epoxy (CFRE) polymers. An enhancement in the tensile properties, flexural elasticity modulus, stress, load, flexural strain, and toughness of nearly 50% was observed after the addition of the MWCNT/PVA to the CFRE (figure 1.5(c)) [34]. In another study performed by Yao *et al*, the group synthesized composites of polyacrylonitrile (PAN)-based composite fiber (CF)/epoxy where the CF consisted of carboxylic functionalized CNTs and GO that were deposited by an electrophoretic deposition (EPD) process. Enhancement in the interlaminar shear strength, interfacial adhesion, humidity resistance, wettability, and moisture absorption were observed which can be used in various applications [16]. Thus, CNTs provided enhancement in the properties when used in nanocomposites and nanopolymers.

1.2.3 Applications

The enhanced thermal and electrical conductivity, excellent tensile strength and flexibility, toughness, good chemical and thermal stability, and unique one-dimensional structures are some of the important features of CNTs that allow various applications in nanocomposites, wearables and smart materials, smart textiles, strain sensing fabrics, sensors and actuators, chemical and biological sensors, hydrogen storage, antennas, microprobes, electronics, biomedicines, artificial muscles, neural and brain electrodes, supercapacitors, batteries, energy storage devices, solar cells, aerospace, space, optoelectronics, catalysis, water treatment, construction materials, high energy springs, etc, and can be manufactured using CNTs and additives of CNTs [8, 10, 17, 22, 24–26, 36]. The various application of CNTs and CNT reinforced nanocomposites and nanopolymers are discussed below.

1.2.3.1 Electronics

Carbon nanotubes have provided their excellent properties most commonly to the electronics sector. Excellent electrical and thermal conductivity with enhanced mechanical properties of strength, hardness, and robustness are important for electronics. In the case of nanocomposites and nanopolymers, further adding CNTs to the structures has provided enhanced applications in electronics. Pełech *et al* used the technique of mixing MWCNTs with bisphenol A epoxy resin, where triethylenetetramine acted as the curing agent, to fabricate a nanocomposite. Enhancement in the electrical conductivity was obtained when (0%–5%) MWCNTs grown on Fe/Co catalyst were added to the epoxy but lower conductivity was seen with an Fe or Co catalyst. These nanocomposite provided applications in fabricating nanoelectronics devices. Also, signals associated with magnetic resonance were obtained with these composites, making them potential candidates in electromagnetic devices [18]. The interconnect materials in electronics can also use metallic CNTs because of their properties of excellent conductivity, high characteristics of frequency, and high ampacity. These interconnects are lightweight and provide a room to produce strong and tough fibers with enhanced strength and tensile modulus [9, 10, 25]. Zhao *et al* conducted experiments with an improved electroless plating method so that silver coatings/deposits with higher uniformity on CNT-graphite hybrid (Ag@CNT-graphite) surfaces can be obtained. The improved intrinsic strength of reinforcements, improved adhesion provided by the CNT bridge, overall bonding, and inhibition of nucleation and motion provide enhancement in the mechanical properties. Electrical conductivity was also found to be increased after the incorporation of CNTs (figures 1.5(b) and (d)). Increasing the CNT content provided a reduction in the wear rate and friction coefficient, but anti-friction properties were not improved. These characteristics have provided a framework for potential sliding electrical contact materials for applications in nanoelectronics for completing and interrupting electrical circuits [35]. Thus, CNT favor the development of electronic devices.

1.2.3.2 Energy storage

After electronics, CNTs have their most extensive applications in energy storage. The fabrication of supercapacitors, batteries, solar cells, etc, through CNTs is advantageous as it provides enhanced efficiency of power conversion, electrical conductivity, and electrical performance. Analyzing the situation of supercapacitors fabricated through carbon materials, it poses advantages of excellent corrosion resistance, high flexibility, enhanced electrical conductivity, and low density. Among them, carbon fiber possesses good toughness, superior power density, cost-effectiveness, light weight, fast charging and discharging, and enhanced electrical conductivity. Wang *et al* fabricated a flexible composite electrode for use in a supercapacitor comprising reduced graphene oxide (RGO)/manganese dioxide (MnO_2), CNTs, and carbon fiber (CF). A free standing electrode of CNT/CF/RGO/MnO_2 (CCRM) was obtained through a self-assembly method directed by filtration for applications as an electrode material. The CCRM provided enhanced electrical conductivity and rate performance, highly reversible and stable charging and discharging, enhanced double layered electrical capacitance, enhanced specific capacitance, effective charge transport, enhanced electrochemical performance, good conductance, high capacity and rate capability when used in supercapacitor applications [37]. Another experiment conducted by Mery *et al* featured the fabrication of a supercapacitor using CNT-manganese dioxide (MnO_2)/dimethylformamide (DMF) as the positive electrode and reduced graphene oxide as the negative electrode on potassium sulfate (K_2SO_4) electrolytic solution. CNT decreases the serial resistance of the capacitor device, thus improving the maximum power in the device. The presence of CNTs inside the composite enhances the electrochemical response, improving the specific surface and material conductivity of the composite. The device provided excellent pseudocapacitive behavior, storage capacity, and higher capacitance for outstanding performance in the areas of power, energy, and cyclability. These characteristics are ideal for supercapacitor applications [38]. In another experiment, conducting polymer was synthesized in a study conducted by Dai *et al.* The group have shown the vertical growth of MoS_2 nanosheets on CNT/PANI composites by 3D wrapping. These composites provide enhancement in the electrochemical properties of the composite, such as exceptional cycling stability, rate capability, high specific capacity, rapid and efficient conduction of charge carriers or electron transport, excellent capacitive performance, specific capacitance and rate capability, etc. These properties were suited to fabricating excellent supercapacitor devices which can be called 'energy stores' for their excellent capability to store and release charges when desired [39]. These excellent characteristics are helpful in the fabrication of supercapacitors using CNT based composites.

Energy storage through batteries is in increasing demand for newer technologies because of their excellent electrochemical properties, electrical conductivity, robustness, etc. These excellent features of batteries provide applications in electric vehicles, portable electronics, and energy storage. In the case of zinc–air batteries, the characteristics of environmental friendliness, economic feasibility, and technological maturity are observed, making them useful and prominent in applications of energy storage. Wu *et al* synthesized a catalyst of Co–N_x–C moieties and CoP_x which was embedded inside graphitic carbon shells. The support matrix was interlaced onto MWCNTs (CoP_x/Co–N_x–C@CNTs) with microwave assistance using phosphorization–carbonization reactions. A cobalt centered zeolite imidazole framework (ZIF-67) precursor was used for the battery as well. The prepared (CoP_x/Co–N_x–C@CNTs) showed excellent oxygen reduction reaction (ORR) catalytic activity and the MWCNTs helped to transfer electrons while in the ORR process. (CoP_x/Co–N_x–C@CNTs) also demonstrates excellent oxygen evolution reaction (OER) performances due to the influence of the MWCNTs. MWCNTs are highly graphitized, providing enhanced durability under strong oxidation and OER stability. The bifunctional catalyst based carbon nanostructures providing applications as flexible zinc–air batteries demonstrated enhanced power density, excellent mechanical flexibility, and cyclic stability. The fabricated liquid zinc–air battery provided applications in wearable electronics and miniature devices [40]. For applications as lithium-ion batteries, reduced graphene oxide cobalt decorated nitrogen doped carbon nanotubes (G-Co@NCNTs) can be applied as anode materials. The formed lithium-ion battery showed high energy storage capacity and enhanced synergy between CNT and porous graphene, but some disadvantages, such as increased volume changes and increased intrinsic resistivity during the lithiation and delithiation process persisted, causing truncated specific capacity and cyclic stability. However, formic acid electro-oxidation can be performed using reduced graphene oxide palladium decorated nitrogen doped carbon nanotubes (G-Pd@NCNTs) so that enhanced electrocatalytic properties in the application of the batteries can be observed. CNTs assisted in the healing of defects in graphene, which acts as an anchor substrate, due to processing and oxidation while in synthesis [9]. Thus, energy storage applications through CNTs have made it prominent for use due to its excellent electrocatalytic performance, stability, robustness etc.

1.2.3.3 UV protection, microwave absorption, and EMI shielding

Radiation that do not fall under the visible spectrum may pose some harmful effects to health and the environment, so it is necessary for us to avoid this radiation. CNT based nanocomposites and nanopolymers have the capability of shielding us from UV rays, electromagnetic interference, microwaves, etc. For applications in UV protection and electromagnetic interference (EMI) shielding, CNTs provide excellent electrical and thermal conductivity, and a high aspect ratio, permittivity, and dielectric loss, making them ideal candidates for applications as nanofillers to be used in blocking UV and microwave radiation [41]. The combination of carbon nanotubes with metal or metal oxides is advantageous to obtain exceptional thermal, electrical, and mechanical properties along with dielectric loss. Magnetic microwave absorption materials (MAMs), combined with CNTs, help to enhance the microwave absorption characteristics. Magnetic oxides have large densities making them confined in the field of microwave absorbing materials. As CNTs are light-absorbing materials possessing excellent capability of electrical dissipation, the combination of these materials is useful in this application. Experiments conducted by Ge *et al* have produced composites of CNTs/carbonyl iron (CI)@SiO_2 using a simple mechanical mixing methods to obtain microwave absorption capabilities. Electromagnetic properties are enhanced when this composite is formed where even a small amount of CNTs has demonstrated excellent electrical properties and permittivity. When CI@SiO_2 is added to make the composite, enhancement in the attenuation ability of the electromagnetic waves and a good property of oxidation–reduction is observed, providing improvement in microwave absorption. The permittivity of a composite depends upon the concentration of CNTs in the CNT/CI@SiO_2 as it increases with the concentration of CNTs and also displays frequency dependence [20]. In another study, Nasouri *et al* synthesized composites consisting of a conducting network that demonstrates excellent UV protection and EMI shielding properties. They synthesized a polymer composite of PAN/MWCNT using the electrospinning technique. The UV protection values increased exponentially when MWCNTs were increased, making them suitable for use as such shielders. Also, the possibility of radar stealth performance can also be seen with this composite [41]. In another optimization experiment conducted by Mingdong *et al*, the group fabricated CNT/nickel ferrite composites to observe the electromagnetic parameters and microwave absorbing properties. The results show that CNT content when increased will provide enhancement in the capability for electromagnetic dissipation. However, increment in the microwave frequency will cause the declination of electromagnetic dissipation. Thus, if the mass fraction of CNTs is large, the capacity of dissipation of electromagnetic energy is also strong. Optimization of the absorption properties of a suitable nanocomposite can be done by selecting an opposite component ratio. Along with this, a thicker sample coating will also provide enhanced properties of microwave absorption, whereas a thin coating discourages this property. The coated composite (nickel ferrite and 10 wt% CNTs) consists of a bigger microwave absorbing coefficient value that produces the best absorbing wave material. This composite also has a larger imaginary part of the complex permittivity, strong capability in dissipation of microwave energy, and a strong impedance matching capability. These excellent properties make the composite an excellent candidate for the absorption of microwaves. Increasing the real part of permeability provides greater absorption capability than in the imaginary part [42]. In another experiment conducted by Koirala *et al*, the group investigated an approach of integrating CNT sheets between the laminates of polyacrylonitrile (PAN)-based fiber composites so that EMI shielding applications could be realized. Mechanical properties concerning the flexural and interlaminar characteristics are enhanced along with the electrical conductivity. In the experiment, when two interlaminar CNT sheets were introduced, flexural strength, interlaminar shear strength, fracture toughness, and interlaminar load transfer were increased without influencing microstructure, thickness, or weight. However, further increase was not favored as agglomeration was observed. These sheets help to enhance the shear properties and fiber-matrix adhesion. However, when interlaminar CNT sheets are increased, the electrical conductivity is also increased. This implies that electromagnetic interference shielding can be increased when the conductivity is increased [28]. Thus, CNT based composites and polymers provide excellent properties for applications in UV and EMI shielding.

1.2.3.4 Other applications

CNT provides applications to improve the mechanical properties of cement based composites. Various studies have shown that adding CNT in cement based composites has enhanced the thermoelectric properties of the composites which later will have the capability to proliferate the electrical properties when heat is provided to the composite. Wei *et al* fabricated a cement based composite with the presence of MWCNTs at the surfaces. A high volume fraction and electrical conductivity were obtained when the composites had a high specific surface area. Mechanical strength and electrical conductivity were enhanced by the CNT bridge that connected the hydration products of the cement making them useful in the thermoelectric performance. Electrical conductivity was also seen to be influenced by temperature. Increased CNT content showed an increment in the electrical conductivity but thermal conductivity was decreased with an increase in CNT content [24]. CNT based nanocomposites possess adsorption capability as well. Electromechanical injected charges can be stored inside CNTs which helps to adsorb gas ions or solvents [10]. A green approach to recognize the applications of CNT based materials towards adsorption was observed through an experiment performed by Diel *et al*, where eucalyptus, pomegranate, and pecan extracts were used as reducing agents and MWCNTs with metallic NPs were used as a support for the synthesis of adsorbents. Copper and iron were the metallic nanoparticles used and MWCNTs were functionalized by carboxylic acid (COOH) or hydroxide (OH). Twelve adsorbents were developed which have the capacity to adsorb and remove glyphosate. Pecan based extracts with iron nanoparticle (COOH3-MWCNT) support have the maximum efficiency and are the most favorable particles for the removal of glyphosate. An increase in the concentration of MWCNTs provides a higher removal rate of glyphosate and other contaminations in a small contact time and without the need for pH change. Further optimization of the parameters will further help in the removal of these contaminants. Regeneration, optimization, and reuse were also analyzed with various models [36]. The applications provided by CNTs in these intriguing fields of nanotechnology excite researchers and scientists to further conduct experiments so that marvelous inventions and innovations are realized.

1.3 Fullerenes

The on-demand use of hollow carbon nanostructures in nanoreactors, catalysis, drug delivery, and energy storage has excited researchers to work on fullerenes [43]. Fullerenes are hollow sp^2 hybridized nanoallotropes of carbon molecules that form hollow closed cages in a shape of a sphere, ellipsoid, or cylindrical structure with carbons. Fullerenes (C_{60}) are also referred as 'buckminsterfullerene', named after the American architect R Buckminster Fuller, who is famous for designing the 'geodesic dome', similar to the structure of C_{60} molecules of fullerene [44–46]. C_{60} fullerenes were synthesized by vaporizing graphite using laser irradiation for the first time by Kroto *et al* in 1985. Vaporization of the carbon species of graphite was performed using a focused pulse laser with a high-density flow of helium. The understanding of the formation of spherical shells and inter-sphere spaces with long carbon chains led to the idea for the formation of long chain fullerenes (C_{60}). The authors suggested the morphology of a polygon with 32 faces and 60 vertices (a truncated icosahedron like a football) where 12 pentagons and 32 hexagons are observed to be formed [47]. The molecular nanostructure of fullerenes is symmetric, having the capacity to develop a crystalline superstructure. Fullerene comprises different dimensionalities and morphologies, for example nanowires, nanotubes, microcubes, nanodisks, nanorods, microhorns, etc, formed by the technique of solvent engineering [48]. It has been observed that superfullerenes or hyperfullerenes can also be produced by replacing the hexagonal or pentagonal rings inside the fullerene [49]. Although C_{60} is considered an ordinary fullerene, other types of fullerene with different amounts of carbon atoms are also found. For instance C_{28}, C_{60}, C_{70}, C_{80}, C_{90}, etc, are the fullerenes that have been used in research. In addition to carbon based fullerenes, boron nitride (BN) fullerenes are also gaining much interest from researchers as they provide exceptional physical and chemical properties. BN based fullerenes consist of four, six, and eight ringed members in the structures in contrast with the five and six ringed members of the carbon based fullerenes [50]. Figure 1.6 depicts the different structures of fullerene where (a)–(j) show the various shapes of fullerene structures and (k) shows a football which is imagined as being similar to Buckminster fullerene (C_{60}). The following subsections will describe the status of fullerenes and their prospects in terms of methods of synthesis, properties, and applications that are demonstrated for nanotechnology.

1.3.1 Synthesis

Fullerenes and their derivatives have been found to be prominent in terms of their applications and properties. Discussion of the various synthesis modes is a requirement so that researchers will be able to know about them and then analyze the changes that can be made in them. Various method of arc discharge, laser ablation,

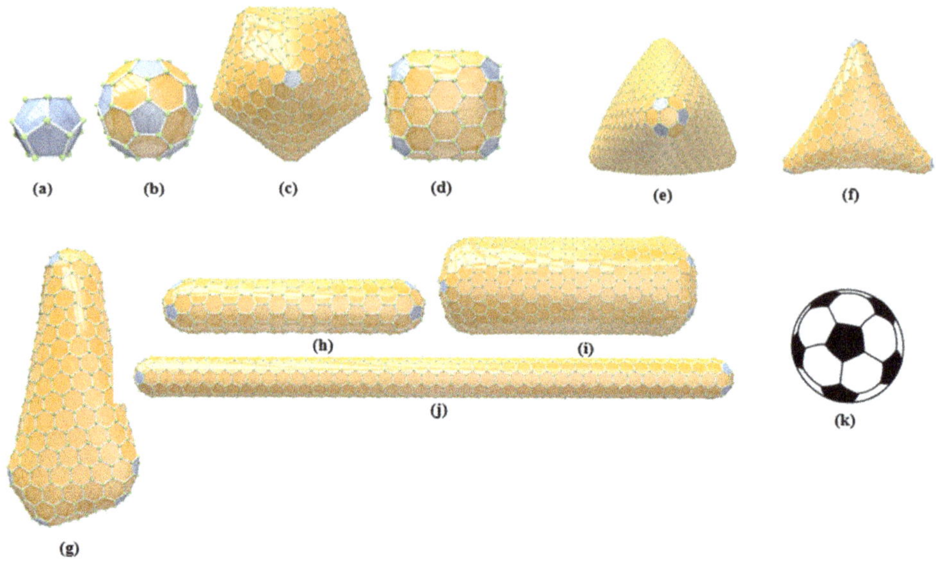

Figure 1.6. Different structures of fullerenes: (a) C20-Ih; (b) C60-Ih; (c) C960-Ih (DP = 12 × 1); (d) C140-D3h (DP = 6 × 2); (e) C1140-T d (DP = 4 × 3); (f) C440-D3 (DP = 3 × 4); (g) C524-C1 (DP = 5 + 7 × 1); (h) C360-D5h; (i) C1152-D6d; and (j) C840-D5d (DP = 2 × 6). Here, (a), (b), and (c) are spherically shaped (icosahedral), (d) is barrel shaped, (e) is pyramid shaped (tetrahedral structures), (f) is trihedrally shaped, (g) is nano-cone or menhir shaped, and (h),(i), and (j) are cylindrically shaped (nanotubes). (Reproduced with permission from [51]. Copyright 2015 Wiley.) (k) The shape of fullerene similar to the shape of a football as shown by Kroto *et al* in their papers where C_{60} molecules are depicted to be similar to this ball with a truncated icosahedral structure. (Reproduced with permission from [47]. Copyright 1985 Springer Nature.)

chemical vapor deposition, and ion beam irradiation are employed for the synthesis of fullerenes (figure 1.3). Similarly, in the case of fullerene derivatives, reactions such as cycloaddition, cyclodehydrogenation, sulfation, cyclopropanation, etc, are performed for their synthesis. Among the synthesis methods of C_{60} fullerene, the arc discharge method using helium atmosphere and a DC current is a prominent process [47, 52]. In another experiment conducted by Sachdeva *et al*, they used the CVD technique to synthesize fullerene C_{70} in the presence of a glass substrate [53]. Other techniques involve the solvent exchange technique with which aqueous fullerene C_{60} and fullerene C_{70} can be fabricated [54]. Analyzing the synthesis of different morphologies of fullerenes, rod-like one-dimensional fullerene structures are formed when fullerene C_{60} is mixed with a mixture of toluene/isopropyl alcohol (IPA) solvent. To obtain nanosheet style 2D structures of fullerenes, fullerene C_{60} is treated with a mixture of carbon tetrachloride (CCl_4)/isopropyl alcohol (IPA) solvent. Tang *et al* with the technique of the liquid–liquid interfacial precipitation (LLIP) method, have created fullerene microtubes from the blend of fullerenes C_{60} and fullerenes C_{70}. When fullerenes C_{60} and fullerenes C_{70} are subjected to inhomogeneous distribution, microtubes are changed to microhorns but the exposure of the microtubes to a mixture of mesitylene and alcohol is necessary [48].

In addition to the fullerene derivatives, several experiments for their synthesis have been conducted. An experiment conducted by Dorel *et al* employed a gas-phase cyclodehydrogenation process induced by laser and metal surfaced thermal cyclodehydrogenation for the synthesis of fullerene C_{60}. This method helped in the fabrication of polycyclic aromatic hydrocarbons (PAHs) and crushed fullerenes [55]. Another study conducted by Lin *et al* discussed the synthesis of cyclo [60] fullerenes using the mediation of fullerene-cation intermediates. These reactions are highly reactive, making the structure highly selective and can tolerate the functional group to a greater level. One group has fabricated fullerenes using a standard vacuum line method with a dry solvent in an argon atmosphere [56]. In another experiment, Vidal *et al* fabricated isomers of [70] methanofullerenes where the formed fullerene derivatives were obtained using a cyclopropanation technique. Alpha and beta isomers were synthesized and analyzed for their activity towards photovoltaic properties [57]. Another experiment conducted by Peng *et al* synthesized *N*-methyl-2-ferrocenyl-pyrrolidinofullerene C_{60} (denoted as Fc-C_{60}), a fullerene derivative, using a self-assembly technique through a template-free liquid–liquid interface precipitation (LLIP) method. Finally, the production of iron and nitrogen dual doped fullerene structures was done to obtain exceptional properties for zinc–air battery applications [58]. In another study, Yasuno *et al* fabricated various fullerene derivatives using cycloaddition reaction. Mono-adduct derivatives of fullerenes were observed to be functionalized with pyridinium groups [59]. Tokimaru *et al* synthesized heteroatoms of the fullerene derivatives buckybowls. A nitrogen consisting fullerene C_{80}-buckybowl with a hybrid of corannulene/azacorannulene was synthesized, demonstrating enhanced photophysical, structural, and electrochemical properties for energy storage and optical applications [60]. Another experiment conducted by Kinzyabaeva *et al* studied the synthesis route of a 1,4-oxathiane derivative of fullerene C_{60} based on a sonochemical reaction on inorganic bases between fullerene C_{60} and 1,2-hydroxythiols in toluene [61].

In another example of synthesizing fullerene derivatives, Hashikawa *et al* focused on the organic preparation of carbon nanoelbows from doubled holed fullerene C_{60} ([8:8]-C_{68} and [12:12]-C_{68} nanoelbows). Top-down approach of the Diels–Alder reaction of C_{60} and azine derivatives to provide a cage-opened derivative were performed where the spherical, highly strained fullerene structures were disrupted. Metallofullerenes and fullerene C_{70} can also be produced from this method [49]. In another study, Sergeeva *et al* employed fullerene precursors of chlorofullerene ($C_{70}Cl_8$ and $C_{60}Cl_6$) to fabricate the fullerene derivatives GI-761 and VI-419-P3K. The formed water soluble fullerene derivatives have been shown to influence the production of reactive oxygen species on human embryonic lung fibroblasts [62]. In another experiment studying a new type of fullerene C_{20} (a structure consisting of a dodecahedral cage), the structure was first synthesized by Rad *et al* using gas-phase production but currently laser ablation methods and ion beam irradiation techniques are also used for their synthesis [63]. In the case of inorganic fullerenes (Mo(W)S_2 particles with hollow or onion-like morphology), they were synthesized using metal–organic chemical vapor deposition, oxide sulfation, and a laser ablation process. Inorganic fullerene (IF-MoS_2) materials were synthesized by the topotactic solid–gas reaction of Scheelites ($AMoO_4$) nanoparticles (A = Ca, Sr, Ba) using a variety of temperatures and with a mixture of hydrogen sulfide (H_2S)–carbon tetrachloride (CCl_4)–argon (Ar) gases. The formed fullerenes possessed excellent catalytic activity for applications in energy [64]. In another experiment conducted by Han *et al*, the group have reported a new nanopottery technique for the synthesis of hollow fullerene structures by interconnecting flexible hollow single units to multi-compartment vessels. Inside a single opening, the liquid template was manipulated and regio-selectivity was allowed. The new synthesis approach of expansion, addition, and connection indicates a new facile route for the fabrication of nano-structures consisting of interconnected systems and hollow complex nanostructures [43]. Thus, various synthesis methods of fullerene and its derivatives indicate the excellent physical properties of the formed structured with intriguing applications.

1.3.2 Properties

The excellent properties of fullerene and its derivatives have provided intrigued applications in science and technology. The fullerene properties of enhanced electron transport, semiconducting properties, and the character of electron-deficiency provide a basis for fullerenes to become electron acceptors [65, 66]. Providing some insight on the photovoltaic properties, Umeyama *et al* in their analysis separated fullerene derivatives in accordance with their isomers. Endohedral or multiadduct fullerenes were separated which provided enhanced performance as electron acceptors. The difference in these isomers has provided enhancement in electrical properties, solubility, miscibility, and molecular arrangement. The steric orientations and positions of fullerene cages, optimization of composition ratios, structure, and purity of fullerenes influence their photovoltaic properties [66]. Also, the excellent characteristics of host–guest behavior, redox activities, oxygen reduction reaction (ORR) performance, and photophysical properties are also found in fullerene for optical and energy applications [49]. Enhanced electron-deficiency, electron mobility, and photosensitivity are some for the important optoelectronic properties in fullerenes that are useful in energy storage and applications along with sensing applications [48]. Also, mesofullerene in organic solar cells possesses excellent electrical and optical properties and shows ordinary dispersive characteristics. The optical constant of these materials has been observed to increase with decreasing wavelength and, consequently, the decreasing wavelength increases the refraction index. Increasing the interactions of incident photons provides an increase in the optical conductivity as well as the dissipation factor [53].

For biological applications, fullerene poses minimal toxicity when used as drug carrier or vehicle [54]. Also, research has shown that reactive oxygen species are neutralized by fullerene's chemical structures. Fullerene and its derivatives are anti-oxidants, i.e. they neutralize reactive oxygen species. It was found that the sizes and shapes of the cages of fullerene chemical structures and the attached chemical moieties on the fullerene cores influence the reactive oxygen species production and antioxidant properties [62]. For drug carrier applications, metallofullerenes were observed to have p-type semiconducting behavior. Charge transfer of the derivatives along with electrostatic interaction between two particles in drug carrier applications were observed. Along with electrical behavior with the drug carrier, increased adsorption behavior of the metallofullerene with drugs was also found [63].

In terms of mechanical properties, Turan *et al* analyzed the tribological, mechanical, and corrosion properties of composites of fullerenes consisting of a magnesium matrix. A semi-powder metallurgy process was employed for the successful production of fullerene based magnesium composite matrix for the first time. The specific surface area of fullerene is high making it useful in hardness performance. Improvement in the compressive strength and wear behavior (when load is applied) was observed, but the corrosion resistance of the composite was decreased after the addition of fullerene. Along with this, the increasing temperature of the composites reduces the yield strength of the specimen causing adhesive wear and softening behavior [46]. The lubricating behavior of fullerenes has also been observed due to the characteristics of robust intermolecular bonding and spherical morphology [46]. While analyzing the adsorptive behavior of fullerene derivatives, highly reactive structures of cyclo [60] fullerenes provide high selectivity along with increased tolerance towards the functional group [56]. Thus, the various intriguing properties of fullerene assist in its application in exciting fields of nanotechnology.

1.3.3 Applications

Fullerene and its derivatives have excellent physical properties for its applications in nanotechnology. The excellent magnetic, electrochemical, and photophysical properties of fullerenes makes them useful in organic and biological electronics and superconductors, photovoltaics, etc [65]. Also, fullerene provides applications in drug delivery, chemical sensors, artificial photosynthesis, bioimaging, sensing devices, molecular switches, radio therapy, spintronics, complementary metal-oxide semiconductor (CMOS)-based transistors, the conversion of light energy, etc [48, 67]. Further discussion on the applications of fullerenes and its derivatives is provided below.

1.3.3.1 Energy

Fullerene and its derivatives are providing increased applications in the fields of energy conversion and storage. Fullerenes are strong electron acceptors that possess enhanced optical properties, electron transport, semiconducting properties, electron-deficiency, and high surface area for application in solar cells, photocatalysis, lithium batteries, photoconductive devices, etc [43, 66]. Improvements in the stability, defect states, photoabsorption, and optical parameters have been found to be important for organic solar cells [53]. Also, energy conversion and light harvesting can be done effectively with fullerene due to its capability to absorb the strong UV–visible rays. They are large in size and symmetric so that transfer of electron and light-induced processes is achieved easily and effectively [68]. Fabrication of organic solar cells using C_{60} and C_{70} derivatives of fullerenes is of a great interest due to their excellent chemical and optoelectronic properties. The electron mobility of fullerene makes it useful in semiconductor applications. Inside photovoltaic systems, $PC_{71}BM$ ([6,6]-phenyl C_{71} butyric acid methyl ester) always acts as an electron acceptor even though a lot of non-fullerene acceptors are found due to their low reorganizational energy and suitable energy of molecular orbits. Perovskite solar cells (PSCs) are now receiving more attention than other cells due to their high efficiency of power conversion (around 24%) [57]. The research group of Lin *et al* explored the possibility of using indano [60]-fullerenes for the fabrication of perovskite solar cells. Power conversion efficiency in a perovskite solar cell is influenced by the solubility, passivation ability of the interface of the perovskite, hydrophobicity, fullerene's reorganization energy, etc. Thus, the balance of solubility, enhanced organizational energy, and capacity to passivate the active layer have applications in perovskite solar cells as electron-transport layers (ETLs). Overcoating (SnO_2) applications in perovskite solar cells was excellent for indano [60]-fullerene due to high power conversion efficiency through the ($NH_3CH_3PbI_3(MAPbI_3)$)-used perovskite solar cells. The enhancement in the charge dynamics of the device along with a dramatic reduction in hysteresis are some of the characteristics that have been achieved after cyclo [60] fullerenes were added to the metal-oxide ETLs as a coating [56]. The use of fullerenes in bulk heterojunction organic photovoltaics (BHJ OPVs) also provides enhanced efficiency of power conversion. Fullerene derivatives have been known to be used along with perovskites in ETL materials for their enhanced electron transport. In an experiment conducted by Umeyama *et al*, fullerene derivatives of [60] fullerene bisadducts, [70] fullerene bisadducts, and [70] fullerene monoadducts isomers were separated providing enhancement in the performances of OPVs. Mono or multiadducts of fullerenes have been applied as electron acceptors for ETL materials. Endohedral or multiadduct fullerenes provide enhanced performance as acceptors. The above isomers are the most suitable candidates to fabricate thin solar films having excellent performances for electricity generation [66]. Schottky organic solar cells based on fullerene C_{70} are also gaining much attention due to their enhancement of the efficiency of power conversion. Sachdeva *et al* fabricated a Schottky OSCs device where C_{70} was sandwiched between an anode buffer layer and a cathode buffer transport layer (molybdenum trioxide/C_{70}/lithium fluoride (MCL) ($MoO_3/C_{70}/LiF$)). The anode materials consisted of fluorine-doped tin oxide (FTO) coated on a glass substrate. The device showed excellent power conversion efficiency and electrical conductivity [53]. In the case of polymer solar cells, a blend of derivatives of C_{60} fullerene was found to be useful as an electron acceptor and a polymer was employed as the electron donor which were used as the cathode and anode, respectively, for solar cell applications. Barham *et al* fabricated a component of photovoltaic cell indene-C_{60} mono-adduct ($IC_{60}MA$) and indene-C_{60} bisadduct ($IC_{60}BA$) on a microwave flow reactor using a solvent of non-chlorinated nature. Enhanced solubility and efficiency regarding the power conversion is seen in them when used in solar cells [69]. Figure 1.7 depicts an example of the use of nanocomposites in solar cell applications where several graphs of optical and electrical properties are shown in the figure. In study conducted by Xu *et al*, the groups have discussed the fabrication method of shape persistent [2] rotaxanes containing a cycloparaphenylene macrocycle. Fullerene mono-adduct and [10] CPP interacted improving the reaction of stoppering so that potential

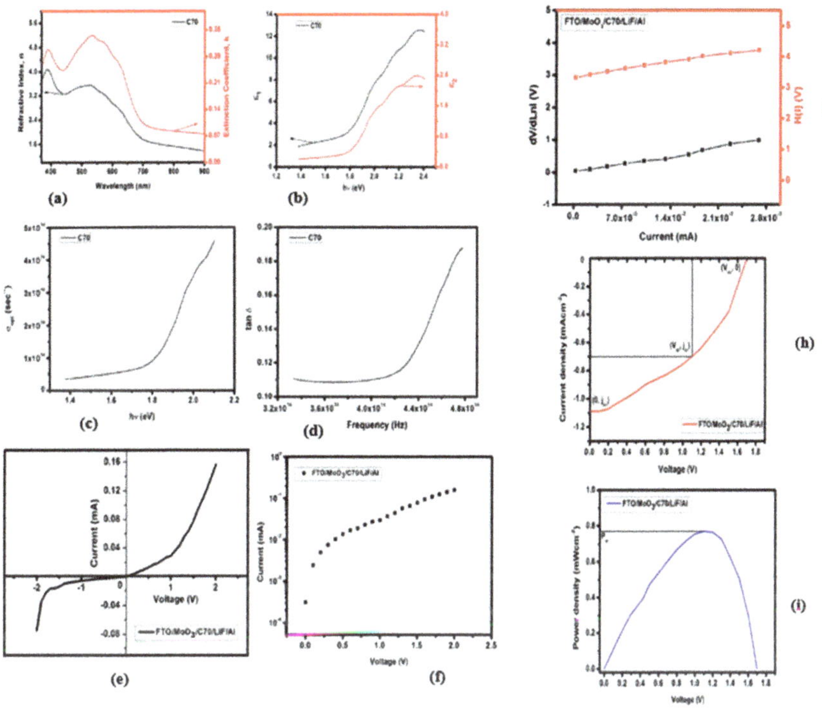

Figure 1.7. Optical and electrical properties of fullerene based nanocomposites for solar cell applications. (a) A plot of the variation of refractive index (n) and extinction coefficient (k) with the wavelength (λ) of fullerene (C_{70}) film. (b) A plot of the variation of the real ε_1 and imaginary ε_2 parts of the dielectric constant versus the photon energy ($h\nu$) of fullerene (C_{70}) film. (c) A plot of the optical conductivity (σ_{opt}) versus the photon energy ($h\nu$) of fullerene (C_{70}) film. (d) A plot of the dissipation factor ($\tan\delta$) versus frequency (ν) of fullerene (C_{70}) film. (e) Current and voltage characteristics of the FTO/MoO_3/C70/Al device. (f) Semi-logarithmic plot of the current and voltage characteristics of an FTO/MoO_3/C70/Al device. (g) Plots of $dV/d\ln I$ versus I and $H(I)$ versus I of the FTO/MoO_3/C70/Al device. (h) H, I, current density–voltage characteristics under illumination of the fabricated FTO/MoO_3/C70/LiF/Al device. (i) Power density–voltage curves of the fabricated FTO/MoO_3/C70/LiF/Al device. (Reproduced with permission from [53]. Copyright 2020 Elsevier.)

materials can be fabricated for applications in photovoltaics [70]. In applications towards zinc–air batteries, nitrogen, iron co-doped 3D porous carbon materials containing fullerene derivatives (named FMN600, FMN700, and FMN800) were used as an active ORR catalyst. The derivatives obtained a high surface area, boosted oxygen reduction, and enhanced the methanol tolerance, onset potential, and durability. These characteristics boosted the power density of zinc–air batteries providing enhanced performances in providing energy [58]. Hence, excellent ORR performance for energy storage applications by the aid of catalytic activity, electron transport, large hierarchical pore size distribution, and specific surface area was seen when fullerene was used as energy storage devices [58]. Along with this, enhanced power conversion and efficiency has also provided a platform for fullerene and their derivatives towards energy applications.

1.3.3.2 Pharmaceuticals

The ease of surface improvement and simple functionalization techniques have allowed fullerenes to be used as drug carriers and antioxidant reactors. Fullerene is a prominent drug carrier which has the capability to carry anti-cancer drugs, anti-inflammatory drugs, antiviral drugs, etc, due to its excellent properties of loading capability and enhanced protection. In the case of anti-cancer drugs, protection from poisonous chemotherapeutics that affect the essential organs such as the liver, heart, etc, is improved when fullerene is used. Fullerenes also possess the capability to instruct cell membranes to go towards tumor cells by concentrating in the cytoplasm, nucleus, and lysosomes. Controlled release of the drug Favipiravir can also be found by using fullerenes as their carriers. A simulation study conducted by Rad *et al* found metallofullerene as a possible carrier of Favipiravir for COVID 19 treatment. They used fullerene (C_{20}) doped with zinc (Zn), iron (Fe), titanium (Ti), nickel (Ni), and chromium (Cr). The DFT calculation was performed focusing on the molecular interactions and the best adsorbent for the job was analyzed. Metallofullerenes were observed to have p-type semiconducting behavior as charge transfer from the drug molecule to metallofullerene was observed. This characteristic showed an electrostatic interaction between them, thus confirming the binding of the drug inside fullerene. Among the listed metals, enhanced adsorption of the drug was seen when Cr, Fe, and Ni were doped to the fullerene structure due to their thermodynamic stability. The analysis of UV–visible spectra further confirmed the use of metallofullerenes in the adsorption of drugs and as carriers [63]. Amphiphiles have excellent biological applications such as in diagnostics, biomedical imaging, gene delivery systems, drug delivery systems, photodynamic therapies, etc. They are synthesized by conjugating hydrophilic polymers to fullerene surfaces. These amphiphilic fullerenes demonstrate high action against pathogens. For instance, mannose units when functionalized to amphiphilic fullerenes inhibit the infection of Ebola virus through blockage of target receptor of cells (DC-SIGN). Also, inhibition of influenza and HIV during entry to the cells are also blocked by these amphiphiles. In experiments conducted by Donskyi *et al*, the authors combined cellular and viral proteins, L-selectin and VSV-G (spike protein of vesicular stomatitis virus) with sulfated fullerenes (fullerene-polyglycerol sulfates (FPSs)) and non-sulfated fullerene-polyglycerol amphiphiles (FPAs) on the leukocyte surface. The adenocarcinoma human alveolar epithelial (A549) cells did not show cell toxicity in a broad range of concentrations with FPSs and FPAs. Both FPSs and FPAs interacted with the L-selectin and aided in inhibiting the infection in VSV. These FPSs and FPAs possess strong binding capability to leukocytes and viral proteins, enhanced biocompatibility, and water solubility for anti-inflammatory and antiviral applications [71].

Fullerenes, being hydrophobic, have some disadvantages for pharmaceutical applications. Hydrophilic (water soluble) constituents are thus added to the fullerene derivatives to remove this problem, which aids in the inhibition of various enzymes such as HIV protease, HCV NS5B polymerase, cysteinic/serinic protease, HCV NS3/4A protease, HIV reverse transcriptase, etc. The use of fullerene analogue

C_{60}-bis(N,N-dimethylpyrrolidinium iodide) provided anti-cancer activities by inducing apoptosis and enhancing the intracellular oxidative stress. Yasuno *et al* fabricated various fullerene derivatives (2′–18′) using a cycloaddition reaction. Mono-adduct derivatives of fullerenes were observed to be functionalized using pyridinium groups. The cytotoxicity test revealed that the formed fullerene derivatives showed induced apoptosis in HL60 cells. Also, when fullerene derivatives interacted with the cancer cells, anti-proliferative activities was observed and the effects were even observed in drug-resistant cells. Tumor growth was suppressed without significant toxicity, helping to fabricate anti-tumor agents [59]. An anti-retroviral drug, hydroxyurea (HU) has applications in chemotherapy for the treatment of breast, intestinal, and gastric cancers, etc. Toxicity, low dissolution and efficiency are some of the disadvantages that are reduced with the use of vehicles or drug carriers that bind with the designated drug. The drug carrier then transport drugs to suitable tissues with cancer for inhibition of the cancerous effects. BN based fullerenes (BNF) also have the capability to act as drug vehicles or carriers, and Xu *et al* have studied BNFs for carrying the HU drug for cancer treatment using theoretical analysis (DFT theory). The unchanged electronic properties of the BNF vehicle and drug suggests they are suitable for the transport of drugs. Also $B_{24}N_{24}$ becomes a better carrier or vehicle due to the fact that the distribution of electrons of HOMO and LOMO is situated near the HU and fullerene molecules, respectively. With this, the chemically active sites of HU are deactivated and these drugs go toward the cancer zone to be released there. The $B_{36}N_{36}$ structure and protonated HU are optimized to obtain a structure which acts as a vehicle for drug release toward cancerous cells where the pH is low. Optical sensors and drug monitoring in the human body can also be fabricated using fullerenes as they show red-shift in the UV–visible spectra [50]. Photosensitizers have also been fabricated using fullerenes (C_{60} and C_{70}) for generating singlet oxygen that can be used for the treatment of cancer cells. Experiments conducted by Martinez-Agramunt *et al* used host/guest complexes of fullerenes (C60@1 and C70@1) as photosensitizers to produce singlet O_2. They also facilitate C–O bond formation using various organic substrates. The fullerenes complexes of host/guest (C60@1 and C70@1) exhibit enhanced spin converting properties that are useful in photosensitizer applications in the generation of O_2 singlet. The reactions of peroxidation of cyclic and acyclic alkenes under air are performed. This performance of supramolecular systems can help as a guide for other host/guest complexes to be used in photocatalytic reactions of similar kinds and especially in the clinical photodynamic therapy treatment (PDT) of tumors [68].

Oxidative stress has been shown to promote various life-threatening diseases such as cancer, diabetes, Alzheimers, rheumatoid arthritis, cardiovascular diseases, schizophrenia, etc. Thus, anti-oxidants are crucial. Fullerene and its derivatives act as these anti-oxidants. Research has shown that reactive oxygen species are neutralized by fullerenes due to its chemical structures. Polyvinylpyrrolidone fullerene C_{60} had been found to provide protection for keratinocytes which are damaged by UV rays. The destruction of keratinocytes promote oxidative stress and reactive oxygen species production. In another analysis, the neutralization of TiO_2-photocatalized reactive oxygen species in skin tissues and keratinocytes was done using fullerene C_{60}. Also, human erythrocytes were protected by Fullerenol $C_{60}(OH)_{36}$ against the action of high energy electrons. In studies conducted by Sergeeva *et al*, the group investigated the water soluble fullerene derivatives C_{60} and C_{70} with solutions of aromatic acid and amino acid to see its influence on human embryonic lung fibroblasts (HELFs). The group employed fullerene precursors of chlorofullerenes ($C_{70}C_{18}$ and $C_{60}Cl_6$) to fabricate the fullerene derivatives GI-761 and VI-419-P3K. The formed fullerene derivatives showed antioxidant properties for reaction oxygen species for a short time. For an extended period of time, VI-419-P3K showed antioxidant properties while GI-761 enhanced the oxidation property which could damage the DNA cells of human embryonic lung fibroblasts [62]. Thus, pharmaceutical applications of fullerene and its derivatives have found a place in treating very sensitive and deadly diseases with excellent efficiency.

1.3.3.3 Other applications

The electronic, optical, and structural properties of fullerene will be influenced if foreign atoms are found in the structure. Paul *et al* have studied the optical properties of fullerene theoretically using a DFT study. The fact that the optical gap of fullerene is 1.1 eV suggests that the structure can act as an optical semiconductor. The onset absorption of fullerene falls entirely under the UV region of the electromagnetic spectrum and it is a potential candidate for applications in UV protection devices. The qualities of greater refractive index and low reflectivity can provide the application of fullerenes in solar cells and LEDs. The presence of a small static dielectric constant and the value of optical gap of the system can provide assistance in fabricating optoelectronic devices using fullerenes [67]. Fullerene and its derivatives possess magnetic properties when atoms or molecules were encapsulated inside their cages. Encapsulation techniques have been used to incorporate molecules and atoms that encourage interaction of the rotational state and nuclear spin. Various molecules such as HCN, CH_2O, CH_4, CH_3OH, NH_3, and N_2, etc, have been incorporated inside fullerenes but pristine closure of the structures has not been achieved. Nuclear spin quantum states and rotational states of a molecule can be achieved when fullerene is encapsulated with NH_3 and CH_4. These qualities can be used in the fabrication of spintronics. Bloodworth *et al* have discussed the encapsulation process of CH_4 and NH_3 in open fullerenes to obtain the rotational and nuclear spin states in them for spintronic applications [72]. While analyzing an antibacterial application, experiments conducted by Kinzyabaeva *et al* studied the synthesis route of a 1,4-oxathiane derivative of the C_{60} fullerene based on a sonochemical reaction on inorganic bases between C_{60} and 1,2-hydroxythiols in toluene. S and O heteroatom derivatives of fullerene provide enhanced biological activity for antibacterial, acricide, insecticide, and fungicide fabrication and use [61]. Fullerene has been used in space and aerospace applications, in particular as a stabilizer in propellant. Regioselective and efficient synthesis of unchained *cis*-1 and *cis*-2 bisadducts was performed using $C_{60}Cl_6$ precursors. These bisadducts were used as stabilizers in propellants due to their capability to forage nitroxide radicals [73]. In the case of using fullerenes in the adsorption of dyes, Elessawy *et al* synthesized functionalized nanocomposites of magnetic fullerenes using PET bottles where these bottles were thermally decomposed. After synthesis, these nanocomposites were analyzed for their application as a potential adsorbent for cationic and anionic dyes. PET bottles were subjected to catalytic thermal decomposition where ferrocene was used as the catalyst. Also, the magnetic nanoparticle precursor was ferrocene, which will be involved in the adsorption and separation of AB25 or MB dye [45].

1.4 Graphene

Graphene, the mother of all graphitic arrangements, is composed of hexagonal or honeycomb lattice structures bonded together by carbon atoms to form a single layer or sheet comprising of the thickness of one atom [74, 75]. Graphene is a high anisotropic, natural carbon nanostructure which is a 2D carbon network structure with sp^2 hybridization at the basin plane [76]. Through the mechanical exfoliation (Scotch tape) method, graphene was synthesized for the first time experimentally by A Geim and K Novoselov in 2004, even though the theoretical explanation was already provided in 1947 by P R Wallace [77–79]. The morphology of graphene is such that it can be found in the form of nanosheets, quantum dots, flakes, nanoribbons, platelets, etc [52, 80–82]. The structure of graphene is shown in figure 1.8. Tiwari *et al* have also discussed the properties, applications, and research opportunities of graphene and its related structures, and discussed the possibility of immense applications of graphene in the field of nanotechnology [83]. The following sub-sections will also describe the status of graphene in terms of the method of synthesis, properties, and applications that are demonstrated for nanotechnology.

1.4.1 Synthesis

A broad range of research has been conducted to understand and develop numerous synthesis methods for graphene and its use in applications of great interest for material science. Top-down and bottom-up approaches are used for the synthesis

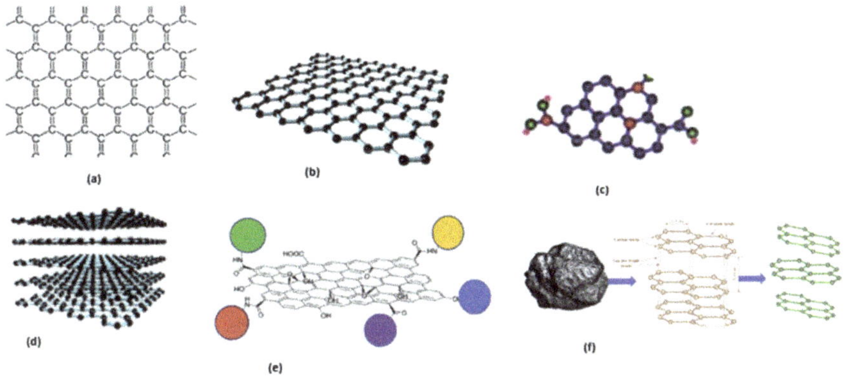

Figure 1.8. Different structures of graphene: (a), (b) pristine graphene, (c) graphene quantum dots, (d) few-layer graphene (e) functionalized graphene (f) origin of graphene from graphite. ((a), (c), (e), (f) Reproduced with permission from [83]. CC BY 4.0. (b), (d) Reproduced with permission from [74]. Copyright 2016 Elsevier.)

of graphene. Graphene has been synthesized using the arc discharge method, chemical vapor deposition method, mechanical exfoliation, chemical exfoliation, chemical oxidation, a solvothermal process using strong acid or oxidizers, electrochemical exfoliation, epitaxial growth, green one-pot hydrothermal processes, etc (figure 1.3), while still newer synthesis methods are being developed at an increasing rate [75, 80, 81, 84–86].

Analyzing the synthesis methods of graphene, Wang *et al* synthesized graphene using mechanical exfoliation techniques in an argon atmosphere with the presence of copper as matrix material [87]. In the case of the arc discharge method, experiments done by Zhang *et al* showed the DC arc discharge method for the synthesis of graphene flakes in the presence of hydrogen gas (H_2) as a buffer gas. Graphene synthesis was performed in an atmosphere of H_2 where excellent crystallanity was observed if the pressure of H_2 is increased. The dangling bonds of carbon were terminated by hydrogen leading to the formation of few-layer graphene [52]. In another study, Maritini *et al* synthesized armchair graphene nanoribbons using surface chemical vapor deposition on copper and in the presence of silicon dioxide (SiO_2)–silicon (Si) substrate [81]. In experiment conducted by Kim *et al* to fabricate gas sensors using graphene transistors, graphene was synthesized on copper foil using chemical vapor deposition techniques [88]. Similarly, in experiment conducted by He *et al*, the group synthesized graphene in copper foil using chemical vapor deposition [89]. In another experiment, Huang *et al* employed chemical vapor deposition to synthesize graphene on copper (Cu)/nickel (Ni) (111) alloy foil with methane (CH_4) being used as a precursor in an atmosphere of argon (Ar)/hydrogen (H_2) [90]. Another study conducted by Walker *et al* also synthesized graphene using plasma enhanced chemical vapor deposition and photolithography on copper foil. Passivation and patterning of graphene was achieved using titanium (Ti)/platinum (Pt) leads on 6′ silicon wafers [91]. According to an experiment conducted by Di Gaspare *et al*, the chemical vapor deposition technique was employed to fabricate graphene in a Ge(001) substrate using precursors of methane and hydrogen gas and a carrier gas of argon (Ar) [92].

In addition to chemical vapor deposition, epitaxial growth of graphene was also observed. Graphene synthesis through epitaxial growth was observed where the process of sublimation controlled by confinement was done. Silicon carbide (SiC) substrate was used for growing graphene through the process [93]. Exfoliation and oxidation of pure graphite also produces graphene. Graphene oxide was chemically reduced to form graphene using sodium borohydride, hydrazine hydrate, and hydrazine. Hou *et al* synthesized graphene by reducing graphene oxide (GO) to reduced graphene oxide (rGO) using artemisinin in ethanol [94]. Analyzing the electrochemical approach, graphene quantum dots were synthesized using an electrochemical approach where GO was treated with non-aqueous media of propylene carbonate with lithium perchlorate ($LiClO_4$) [95]. In another study conducted by Han *et al*, graphene platelets were synthesized by thermal shocking followed by ultrasonication [82]. Zhang *et al* conducted an experiment to fabricate graphene using benzene, where it is converted to graphene with the help of sulfur (triplet 3S2) as a precursor and commercial tetraphenyltin, or $Sn(Ph)_4$ at high temperatures [85]. Chen *et al* discussed a green, sustainable method of a one-pot hydrothermal process for the fabrication of graphene quantum dots using the mechanism of hydrolyzation and ring-closure condensation. Starch was used as the precursor for the first time, and is a long polymer consisting of high molecular weight and is also a natural polymer [80]. Another green, sustainable method for the synthesis of graphene is using a pyrolysis process involving a two-step mechanism. Waste plastics (polyethylene, polypropylene, and polystyrene) were pyrolyzed at 400 °C with nanoclay and then pyrolyzed again at 750 °C in a nitrogen atmosphere. The obtained charged black residue when characterized was found to be graphene. The process obtained 5.25 kg of graphene nanosheets from 35 kg of raw materials (plastics) [96]. Thus, the various synthesis methods of graphene have proven effective for excellent properties to be used in applications.

1.4.2 Properties

Graphene possesses excellent electrical, optical, mechanical, magnetic, self-healing, energy converting, lubricating, etc., properties for applications towards nanotechnology. All these celebrated properties suggest the fact that graphene is a 'supermaterial'. Various excellent properties of enhanced electron mobility and transport, high surface area, excellent thermal and electrical conductivity, high elastic modulus, excellent mechanical resistance, composite strength, stiffness, tensile strength, biocompatibility, low toxicity, tunable photoluminescence, photo-stability, hydrophilicity, chemical stability, low ductility, easier processing, cost-effectiveness, anti-oxidation, etc., are observed in graphene which will ultimately aid in excellent applications of nanotechnology [80, 87, 97–99]. The band gap of graphene is very low, so opening of the band gap is the main goal for realizing these excellent properties [88]. Modification of pristine graphene by doping to tailor its band gap is essential for tuning its electrical properties and chemical reactivity [100]. Analyzing the mechanical properties, the in-plane stiffness of graphene is maximum in comparison to other materials but the bending rigidity is very low. Wrinkles in graphene decrease the Young's modulus but increase the bending rigidity. Lopez-Polin *et al* have carried out three experiments to analyze the rippling effect and the characteristics of the membrane in graphene. Enhancement in elastic response and a negligible Young's modulus with a thermal expansion coefficient was observed when atomic defects and strain are introduced to graphene [101]. Figure 1.9(a) shows the rippling effects of graphene. Also, the self-healing properties discussed by Du *et al* in their review have shown that due to the excellent properties such as enormous specific surface area, good mechanical characteristics, enhanced thermal and electrical conductivity, enhanced antioxidant properties and thermal stability, graphene can provide a basis for the fabrication of self-healing composites when reinforced with graphene [102].

Alone, graphene produces excellent results of enhanced properties and when used as an additive, it has the capability to increase the properties of the materials. Graphene has been used to reinforce metal matrices, nanocomposites, and nanopolymers. Graphene has high mobility, and excellent thermal and electrical

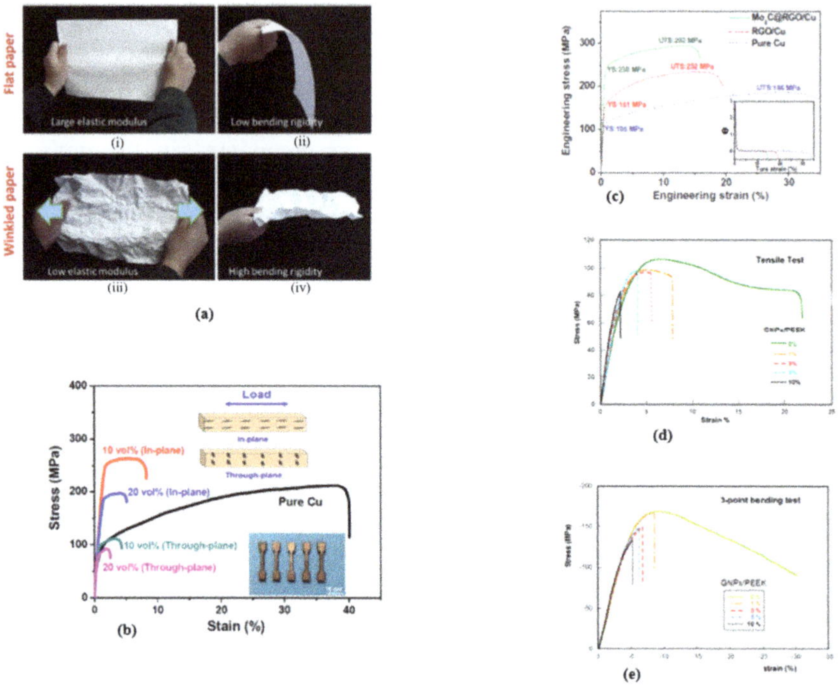

Figure 1.9. Mechanical properties of graphene based structures. (a) Illustration of rippling effects and the elastic properties. Paper acts as membrane where (i) and (ii) show flat paper without wrinkles with a large elastic modulus but low bending rigidity, and (iii) and (iv) show wrinkled paper with a low elastic modulus but high bending rigidity. (Reproduced with permission from [101]. Copyright 2022 Elsevier.) (b) A plot of tensile stress versus strain for pure Cu and 10 vol%/20 vol% GNP/Cu composites where the load is in parallel and perpendicular to the direction of alignment. (Reproduced with permission from [98]. Copyright 2018 Elsevier.) (c) A plot of stress versus strain of pure Cu, RGO/Cu, and Mo_2C@RGO/Cu composites with the corresponding curve of normalized strain hardening at the background. (Reproduced with permission from [112]. Copyright 2018 Elsevier.) (d) Tensile test for PEEK and PEEK/GNPx composite (stress versus strain curve), and (e) three point bending tests for PEEK and PEEK/GNPx composites (stress versus displacement curve). (Reproduced with permission from [97]. Copyright 2019 Elsevier.)

conductivity. Various nanocomposites reinforced by other carbon structures provide excellent mechanical properties but minimal development of electrical and thermal properties of the metal matrix nanocomposite has also been observed. However, when graphene was reinforced in a metal matrix composite, it enhanced all the electrical, thermal, and mechanical properties of the composite. Wang *et al* synthesized graphene using mechanical exfoliation techniques and a composite of graphene and copper metal matrix was synthesized using electric field-activated pressure-assisted synthesis (FAPAS). The hardness of the composite was enhanced, the electrical and thermal conductivities were improved, and the corrosion rate was slowed, providing improved resistance to corrosion [87]. In another experiment, bio-inspired synthesis of a graphene/copper (Cu) matrix was performed by Chu *et al* to provide enhancement in the mechanical properties and strength, electrical conductivity, and ductility.

Graphene nanoplatelets were added to the Cu matrix using vacuum filtration and then a spark plasma sintering process was followed. The in-plane tensile strength was enhanced after the addition of graphene but the through-plane tensile strength was decreased. Low ductility and strength improvement were also observed but all these mechanical properties can be decreased if graphene addition is increased continuously (figure 1.9(b)) [98]. Continuous and discontinuous fibers are used to enhance the properties while fabricating the composite, where strength is realized from the former and performance is given by the latter. Excellent specific strength and modulus, good damping, wear resistance, thermal conductivity, and a small coefficient of thermal expansion are found in an aluminum matrix partially reinforced by graphene, which is under speedy development [86]. In the case of a nanocomposite reinforced by graphene, adding carbon nanofillers such as graphene, CNTs, or their combination to multifunctional epoxy are some of the attempts to improve strength in epoxy. Enhanced strength, electrical conductivity, high aspect ratio and surface area, physical properties and toughness are provided by these nanofillers, giving superpowers to composites and adhesives [82]. In another study performed by Han *et al*, magnetic stirring and ultrasonic mixing were performed between graphene nanoplatelets and epoxy to obtain a reinforced nanocomposite. An expectation of high conductivity and excellent mechanical performance was observed while fabricating the nanocomposite. The results showed an increase in the performance of the epoxy after using nanofillers. Increments in tensile strength up to an acceptable concentration of nanofillers, enhancement in shear strength, Young's modulus, better performance in terms of mechanical properties when exposed to increased tensile forces, increased electrical properties, etc, were observed [82]. When conventional fiber-reinforced plastics (FRP) are fabricated, good environmental and chemical resistance, impact resistance, durability, excellent strength and stiffness, and low weight are sought so that applications in the automotive industry, construction materials, aeronautics, renewable energy, sports equipment, etc, can be realized. However, the disadvantages of interlaminar delamination still remain in the FRP. The use of nanoreinforcement, however, will be able to solve this problem. Graphene possesses excellent intrinsic properties which can also be used as reinforcements and has a some advantages over CNTs. Vázquez-Moreno *et al* analyzed the mechanical properties and morphology of epoxy composites containing graphene. A vacuum-assisted resin infusion molding process was used for the fabrication of the nanocomposite by adding graphene directly. Then, a spraying technique was used for calender mixing and adding graphene on the surface. Calender mixing provides increase in the viscosity of the resin with an increase in the mechanical properties of the resin. In the case of spraying, enhancement in the mechanical properties is achieved in low concentrations [99]. Thus, the excellent physical and mechanical properties of graphene inspires its use in nanotechnological applications.

1.4.3 Applications

The most prominent, extensive, and useful application of graphene is in nanocomposites. Different synthesis techniques allow the addition of graphene to polymers, metals, and metal oxides so that its superb properties can be passed on to those structures. Excellent conductivity, mechanical properties, and optical properties are some of the properties that are required by nanocomposites for their use in applications in space, aerospace, automobiles, construction, microelectronics, etc. Some important and prominent applications of graphene are presented below.

1.4.3.1 Energy

The field of energy is indebted to graphene as graphene has provided many useful modifications in this field. Batteries, supercapacitors, electronics, etc, are some of the fields that make advantage of the physical and mechanical properties of graphene to initiate excellent modifications to such devices. Rechargeable batteries, an important devices that the world will require forever, still needs to be more efficient for successful applications with low cost and duration of charging and high power density. Graphene has the capability to fulfill this need. Due to several defects and residues in graphene caused by exfoliation, the epitaxial method is used instead, which has the capability to produce high quality lithium-ion batteries. Potassium ion based batteries (PIBs) are also receiving attention as they possess extraordinary energy density, natural abundance of the mineral, and low voltage discharge, making them more compatible than the lithium-ion batteries. Graphite-like structures can be used in PIBs as anode materials. According to Zhang *et al*, their prepared sulfur assisted graphene possessed the properties of enhanced electron mobility and low contact resistance of the inter-sheet junction. Graphene is further layered to form few-layered graphene to form an anode so that potassium ion batteries can be prepared. The fabricated devices possess excellent capacity, enhanced mechanical properties without any defects, outstanding rate performance, low voltage discharge platforms, cyclic stability, and enhanced surface capacitive behavior [85]. In another experiment performed by Wu *et al*, the group fabricated a zinc ion battery using α-MnO_2 (manganese dioxide)/graphene scrolls (MGSs) where the MGSs are demonstrated to be a cathode material. High capacity, long cyclability, enhanced rate capability, and electrical conductivity are some characteristic of the formed Zn/MGS aqueous battery [103]. Another study conducted by Qiu *et al* stated that graphene/molybdenum disulfide (MoS_2) nanocomposites fabricated using the method of an L-cysteine-assisted solution-phase demonstrated enhanced electrochemical properties for its use in a lithium-ion battery as anode [104].

Graphene welded activated carbon materials were fabricated by Li *et al* which provided excellent electrical conductivity, optimized porosity, high energy and power density, and high surface area. The fabricated materials were then used as electrodes in an all-solid-state supercapacitor device with ionic gel of 1-ethyl-3-methylimidazolium tetrafluoroborate ($EMIMBF_4$)/polyvinylidene fluoride–hexafluoropropylene (PVDF–HFP) as the electrolyte. The fabricated device exhibited superior capacitance, high cyclic stability, excellent conductivity and electron transport, and enhanced stability, rate performance, electrochemical stability, power outburst, and energy storage. The high-voltage electrolyte that was used in the device also had excellent mechanical flexibility, a wide voltage window during operation and high safety characteristics [105]. For applications as a conductive electrode, graphene is a clear conductive electrode with excellent electrical, mechanical, and optical properties to be used in energy devices and touchscreens. The band gap of graphene is very low, thus expanding the band gap is the main goal for realizing these excellent properties [88]. In the case of electrodes for organic photovoltaic cells, a composite of graphene/glass substrate synthesized using spray coating was excellent in terms of applications. A conductive, transparent electrode in organic photovoltaic cells was fabricated which showed high transmittance, electrical conductivity, and excellent performance for energy conversion [106]. In another application of graphene as electrode, Martini *et al* fabricated a contact electrode using graphene nanoribbons (GNRs). The fabricated device and the electrode materials have shown that the electrical properties (conductance) of the GNRs can be altered by tuning the width and/or length of the GNR, thus controlling the electrical properties directly through fabrication. Optimization of the geometry of the electrode and GNR will be able to improve the conductance and output current when longer GNRs are fabricated from ultra-high vacuum. The formed materials have applications in optoelectronic devices, photosensors, and electrode materials in transistors [81]. In another study, Bronner *et al* synthesized hierarchical on-surface heterojunctions of GNRs using molecular precursors with awareness of their growth sequence. The structural control of the heterojunction is subtle which could be used in future nanoelectronic devices [107]. The energy sector has thus benefitted greatly from graphene based materials.

1.4.3.2 Biomedical

Due to the excellent electrical, thermal, mechanical, and optical properties of graphene, the use of these materials along with composites provide benefits in the field of biomedical applications. Reinforcing polyether–ether–ketone (PEEK) with fillers will tend to improve the frictional loss, osteointegration and, wear resistance when used as nanocomposites in biomedical applications. Among several nanofillers, graphene possesses an excellent elastic modulus and high surface area, mechanical resistance, composite strength, stiffness, tensile strength, and chemical stability to be used as a filler. Puértolas *et al* fabricated a solvent free GNP/PEEK composite using a melt blending process to analyze their mechanical and tribological properties to obtain high strength and stiffness, and low friction and wear in them. These composites can be used in cervical cages, temporomandibular joints, total facet arthroplasties, removable partial dentures, total joint replacements, etc. An increment in thermal conductivity and tensile strength, loss in ductility, enhancement of hardness at the highest concentration, improvement of frictional behavior, lower hardness but higher toughness, an increase in wear resistance associated with hardness, and a reduction in the coefficient of friction and wear rate are some of the properties of these composites applied for biomedical use (figures 1.9(d) and (e)) [97]. In another application, ultra-high molecular weight polyethylene (UHMWPE), a polymer used in total joint replacement (TJR), provided excellent mechanical and physical properties but the wear, efficiency, and failure of the material limits its lifetime in applications. The use of graphene provides a high surface area leading to an easier electrical conduction network, strong bonding, productive load transfer, superior lubrication capability, and decreased risk of wear. Alam *et al* and the group have synthesized UHMWPE nanocomposite reinforced with graphene nanoplatelets (GNP) by using the process of solution mixing followed by compression molding. An increment in yield strength, elastic modulus, tensile strength, fracture toughness, Young's modulus, yield strength, and elastic modulus was observed when these nanocomposites were used in applications [108]. In case of graphene quantum dots, they have properties of excellent biocompatibility, enhanced photoluminescent emission, excellent water solubility, high quantum yield, and low cytotoxicity useful in *in vitro* bioimaging and biolabelling to treat cervical cancer cells directly [80]. In the case of the treatment of virus infections, graphene sheets covered with polyglycerol sulfate have been observed to interact with African swine flu, herpes simplex, orthopox, etc. Vesicular stomatitis virus (VSV) has also been shown to interact with graphene sheets and sulfated gold particles. Thus, the interaction of these viruses with graphene has provided a framework for graphene encapsulated drugs to be used in the treatment of these diseases [71]. Hence, the biomedical application of graphene based materials is useful in this field to a great level.

1.4.3.3 Sensors

Graphene provides applications in terms of sensors as well. Chemical sensors and biological sensors have been fabricated using graphene. Field effect transistors (FET) can be used as gas sensors. In order to have this application, the device is fabricated using graphene. FET gas sensors shows responses towards H_2O, NO_2, and CO by a saturation test, detecting the increment in source–drain current and a decrement in electron mobility. Kim *et al* constructed a graphene based FET for effective gas sensing applications. A polystyrene polymer brush was treated on the surface of the substrate of silicon dioxide (SiO_2)/silicon (Si) to initiate and enhance the gas sensing properties of the FET. Synthesizing sensors also depend on doping, which in turn depends on the charged impurities present on the SiO_2/Si substrate used in the synthesis of the graphene based FET. Thus, due to the enhanced carrier mobility, carrier density, and increased adsorption, the FET has found applications in sensing ammonia (NH_3), carbon dioxide (CO_2), and relative humidity [88]. In another study conducted by Kalita *et al*, the group synthesized graphene quantum dots for application as moisture sensors in soil. A quick response, increased sensitivity, stability, and high accuracy towards the various contents of moisture was observed when the sensors were used to detect moisture in both red and black soil [95]. In terms of biosensing applications, graphene quantum dots provide an acceptable amount of attention for applications in optoelectronics, bioimaging, sensors, etc [80]. In a study conducted by Afashi *et al*, they used an encapsulation technique to capture the formed graphene in an epoxy matrix so that a biosensor surface of graphene could be fabricated based on field effect biosensing (FEB) technology. When anti-Zika NS1 was functionalized (covalent linkage) to an Agile R100 biosensor chip (data reader), ZIKV NS1 (a recombinant antigen) was detected using the biosensors. Portable and cost effective acquisition of binding data was obtained through the sensors. Thus, this device can help in the early detection of

Zika virus infection [91]. In another application of biosensors, a gold and platinum nanoparticle (AuPtNP)-Cys/graphene platform electrode (GPE) sensor was fabricated by electrodepositing nanoparticles of gold (Au) and platinum (Pt) on a graphene platform electrode (GPE) along with the addition of L-cysteine. The sensor detected epinephrine (EP) among samples of epinephrine (EP), ascorbic acid (AA), dopamine (DA), and uric acid (UA), showing extraordinary electrocatalytic activity, selectivity, reproducibility, stability, and sensitivity. This sensor can also be used to detect analytes in human serum [89]. Thus, graphene provides enhanced sensing applications.

1.4.3.4 Thermal management

In applications, sometimes heat provides disadvantages in terms of efficiency and operation in electronics and other devices. The sector of microelectronic industry needs a reliable, efficient, and miniaturized device with increased power density and efficient thermal management. Thermal management applications using metal matrix composites (MCCs) with carbon nanostructures such as graphene are exciting because of the enhanced thermal conductivity, thermal expansion coefficient, mechanical properties, etc. In this respect, when a considerable amount of graphene is added to the metal matrix, enhanced mechanical properties of the composite are observed. Also, a highly aligned structure of graphene will enhance the thermal conductivity of the composite. Thermally conductive fillers and metal matrix composites can be used in applications of thermal management devices as they provide enhanced thermal conductivity, outstanding mechanical properties, and a tunable thermal expansion coefficient. In studies conducted by Chu *et al*, they synthesized highly aligned graphene nanosheets in composites of graphene nanosheet (GNS)/copper (Cu) using the process of vacuum filtration followed by spark plasma sintering. This process achieved nanocomposites with excellent thermal conductivity, structural integrity, an enhanced elastic modulus, and laminated structure for applications in thermal management. In-plane thermal conductivity was enhanced while through-plane thermal conductivity was decreased when highly aligned graphene nanosheets were used in the composite to suppress interfacial thermal resistance [76]. These composites can be utilized in nanoelectronics where a high amount of heat is discouraged for applications. Also, space related applications which involve high heat can utilize this technology to generate electricity as well. In another experiment conducted by Li *et al*, they synthesized a nanocomposite of aluminum matrix that was reinforced by graphene nanoplatelets using the processes of ball milling, cold pressing, continuous casting, and subsequent rolling. The produced nanocomposites contained a high surface area, enhancement in the tensile strength, and a reduction of ductility, but a slight decrease in the electrical conductivity was also observed. These excellent properties was observed to be useful in fabricating electrical wires and cables and in thermal management of the finished cable products [109]. Due to its outstanding thermal conductivity, contact surface area, lower density, and greater specific heat capacity, graphene is a suitable material for the fabrication of nanofluids. Synthesized through dispersion or doping, graphene based nanofluids such as nitrogen doped graphene, copper/graphene, and silver decorated graphene are used as nanofluids due to their enhanced thermophysical properties, low bulk density, and excellent thermal conductivity. Further modification by adding metal oxides to graphene is also found, which will further provide enhanced properties. These nanofluids have applications in heat exchangers, heat pipes, photovoltaics, energy harvesters, water desalination, solar collectors, fuel additives for combustion, energy storage using PCM, etc [75]. Thus, efficient application in different areas of nanotechnology can be achieved when thermal management is done.

1.4.3.5 Other applications

In addition to these general applications, graphene possess excellent mechanical properties that can be used in vehicular and space applications. When using metal matrices, a flexible, lightweight material can be altered with graphene to realize excellent electrical and thermal properties, and enhanced hardness, density, strength, etc. Wang *et al* used a field-activated and pressure-assisted synthesis (FAPAS) method inside an argon atmosphere for the fabrication of a lightweight graphene/aluminum nanocomposite such that good wear resistance and interfacial bond strength, enhanced thermal conductivity, increment in the hardness and strength, and corrosion resistant properties could be attained. These properties provide applications in automobiles and trains for pistons and brake disks [86]. Along with this, a solid lubricant can also be formed by using graphene as an additive in liquid lubricant [97]. For cement applications, the mechanical strength and conductivity of the materials can be influenced by graphene based composites. Bai *et al* have discussed the disadvantages of electricity conduction in cement along with poor capacity and decreased tensile strength. The group suggested the addition of graphene to the cement materials so that excellent mechanical and electrical properties are provided. Silica fumes and graphene were added to the cement paste to observe its electrical and mechanical properties. An increment in the compressive strength was observed when graphene was added, which was further enhanced upon the addition of silica fumes to the nanostructures. The addition of graphene alone, however, reduced the electrical resistivity, but when silica fumes were used, the interfacial strength and electrical resistivity was increased with an increase in the concentration. Even at low graphene content, the addition of silica fumes enhanced the electrical and mechanical performance of the composite. Also, the addition of higher content of graphene and silica provided enhancement in the electrical conductivity [110]. In terms of corrosive resistant applications, the synthesis of bio-based hydroxyl epoxy phosphate monomer (PGHEP) from phloroglucinol, when functionalized with graphene sheets, provided applications of waterborne epoxy coatings. These structures demonstrated excellent physical barrier characteristics and anticorrosive properties due to the action of corrosion potential, pore resistance, and decreased corrosion current. Thus, the coatings provided applications in marine corrosion protection [111]. Its outstanding mechanical strength, electrical conductivity, and thermal conductivity along with high transmission of light in the visible-infrared region have made graphene suitable for applications in various fields of electronics, such as LEDs, touchscreens, solar cells, lighting, etc. However, the zero band gap of graphene limits its applications which are unhindered by heterostructures. The electrochemical and electrocatalysis characteristics of metal sulfide and oxides will be enhanced when graphene is added due to its excellent properties of high stability, chemical resistance, electrical conductivity, etc [104]. Thus, the applications of graphene are numerous and are increasing alongside the discovery of its newer properties.

1.5 Diamane

Diamane is a nano-allotrope of carbon whose structural configuration involves the presence of bilayer graphene that undergoes sp^3 hybridization. During the hybridization, the formation of covalent bonds between the carbon atoms and hydrogen atoms of the two sub-lattices is confirmed which leads to the synthesis of a C_2H layer on one side, while the other side of the carbon layer forms covalent bonds with the neighboring carbon layers forming the diamane structure [113]. Chernozatonskii *et al* observed the synthesis, properties, and structure of diamane for the first time in 2009 using theoretical simulation studies (Vienna *Ab Initio* Simulation Package (VASP)). They found that diamane, a diamond nanofilm, is synthesized through the hydrogenation of bilayer graphene and possesses excellent electrical and mechanical properties [113]. Although the name 'diamane' is commonly used by many researchers, other names such as bilayer graphene, hydrogenated bilayer graphene, hydrogenated few-layer graphene, and interlayer-bonded bilayer graphene are also used to describe the diamane structure [114, 115]. Observing the morphology of diamane, it is classified as Bernal stacking (the AB stacking sequence of diamane) and Lonsdaleite stacking (the AA stacking sequence of diamane). Here, Lonsdaleite stacking is the isomeric configuration of diamane which is the most stable structure, and under the application of pressure, 3D graphene crystals can also be synthesized from them [113, 115–117]. The structure of diamane is shown in figure 1.10. Without the attachment of chemical groups, hydrocarbons or hybridizations, or without the application of pressure, the stability of diamane cannot be attained due to the instability in the thermodynamic surface. Thus, various groups of nitrogen, hydrogen, fluorine, hydroxyl, etc, are attached to the carbon surface or a small amount of pressure is added to the carbon surface to obtain stability [118–120]. Diamane structure was found to be fabricated using defected graphene structure as well. When the two layers of graphene that are used to form diamane are rotated at certain angle of θ, the production of twisted bilayer graphene is attained. When the twisted bilayer graphene is functionalized, it forms the structure of diamane with a Moiré pattern [116, 121]. Diamane formation can be attained with both the configuration of bilayer graphene of Bernal stacking and isomeric Lonsdaelite stacking with the formation of a 0°–60° twisting angle [122, 123]. The various structures of diamane are depicted in figure 1.10. Excellent reviews describing the current status of diamane and diamanoids, their synthesis methods, properties, and applications,

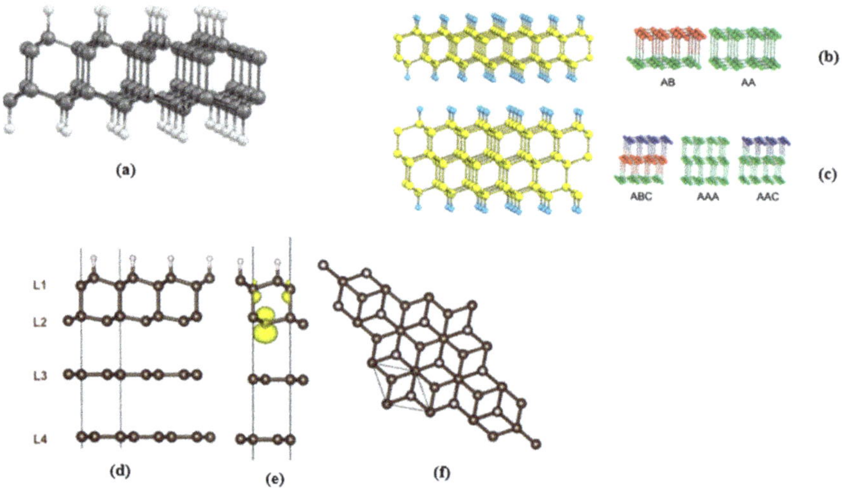

Figure 1.10. The structure of diamane and diamanoids. (a) Atomic configuration of diamane. The dark gray dots are carbon atoms and the light gray dots are hydrogen atoms. (Reproduced with permission from [121]. CC BY 4.0.) (b) and (c) Atomic structure of diamane with different stacking sequences, (b) shows the stacking sequences of two layer AB (Bernal) and AA (Lonsdaelite). (c) The stacking sequences of three layer ABC, AAA, and AAC, the yellow dots are carbon atoms and blue dots are hydrogen atoms. (Reproduced with permission from [126]. Copyright 2011 American Chemical Society.) (d) and (e) Side views of few-layer partially hydrogenated graphene (diamanoids) with ABBA stacking and (f) top view of few-layer partially hydrogenated graphene (diamanoids) with ABBA stacking. (Reproduced with permission from [127]. Copyright 2020 Elsevier.)

are presented by Sorokin *et al* and Tiwari *et al* [124, 125]. The following sub-sections describe the status of diamane in terms of the methods of synthesis, properties, and applications that are demonstrated for nanotechnology.

1.5.1 Synthesis

Diamane, a recently synthesized carbon nanostructure, was observed to possess excellent electrical, magnetic, optical, thermal, and mechanical properties. For this, careful synthesis methods to alter the properties of the diamane are essential. Thus, studying the various synthesis methods of computational approaches, CVD, pressure synthesis, atom substitution, and Birch reduction is important. Among them, the most prominent method is the CVD which involves the use of plasma or a hot filament procedure. Both the bilayer graphene and twisted bilayer graphene can be used to form diamane. Several scientists predicted the existence of diamane through a computational approach. Chernozatonskii *et al* discussed the existence of diamane through a simulative study using the VASP, where the group discussed the electrical and mechanical properties of diamane [113]. In addition, Momeni *et al* used a computational method to understand the formation, interlayer attractions, thermodynamic and kinetic stability, hybridization, breakage, and formation of bonds for the diamane structure using an atomistic model [128]. DFT calculation, molecular dynamics, and *ab initio* calculations are some other models that have been used to understand diamane [113, 129].

After the prediction of diamane, scientists were looking for feasible production methods for the experimental synthesis of diamane. CVD was one of the processes which could release and deposit chemical vapors of graphene on substrates. After the deposition, a functionalization technique is initiated which leads to the formation of diamane structures. Functional groups such as hydrogen, fluorine, hydroxyl, etc, were seen to be deposited at the bilayer graphene to from diamane. Hydrogenation and fluorination are the two processes that are used for the synthesis of diamane. In the case of hydrogenation, the hot filament process and plasma process are the two prominent processes for the synthesis of diamane. Discussing the hydrogenation of diamane using a hot filament process, Piazza *et al* used a low pressure and temperature reactor consisting of a hot filament to deposit bilayer graphene on gold Quantifoil TEM grids with AB stacking. After the deposition, subsequent hydrogenation to form hydrogenated diamane (H-diamane) was performed by dissociating hydrogen atoms. Heating process inside the reactor was performed by two tungsten wires while hydrogen (H_2) atoms for hydrogenation were provided by hydrogen gas instead of hydrocarbon [130, 131]. In another experiment conducted by Qin *et al*, a hot filament reactor was employed to deposit few-layer graphene on a platinum (Pt(111)) substrate and then deuterium gas was passed inside the reactor by a hot wire process to obtain H-diamane [117]. Another synthesis method for diamane formation is through hydrogenation using hydrocarbon. Piazza *et al* used a mixture of hydrogen gas and hydrocarbon. In this experiment, bilayer graphene was deposited on a copper substrate inside the hot filament reactor [132]. In terms of fluorination as well, CVD was used to fabricate fluorinated diamane (F-diamane) and Son *et al* have employed the process to deposit bilayer graphene on a silicon dioxide (SiO_2)/hexagonal boron nitride (hBN) substrate. Then, the formed bilayer graphene interacted with xenon fluoride (XeF_2) to obtain F-diamane. The top hBN was, however, etched away when in contact with the fluorinating agent [133]. In another experiment conducted by Bakharev *et al*, bilayer graphene was deposited on a crystal copper/nickel (Cu/Ni (111)) substrate and then fluorination through XeF_2 was chemisorbed to the deposited graphene to form F-diamane [119]. In another study conducted by Chen *et al*, fluorine gas was employed to heat the formed graphene flakes to synthesize $(C_2F)_n$ layers [134].

In the case of a plasma process, hydrogenation was performed by employing cold hydrogen plasma using aluminum electrodes with a mixture of argon and hydrogen gas. DC plasma was used where chemisorption of the bilayer graphene was observed with sp^3 hybridization on both sides but only the top and bottom layer were hybridized later [113, 135, 136]. Experiments conducted by Luo *et al* employed hydrogen plasma to perform the hydrogenation process to deposit bilayer graphene on a silicon dioxide (SiO_2)/silicon (Si) substrate to form H-diamane [137]. Another study conducted by Piazza *et al* used cold plasma to execute interlayer bonding between the hydrogen and bilayer graphene using a chemisorption process on metal [130]. In experiments to synthesize triamae, the irradiation of hydrogen plasma was done to obtain trilayer diamane when passed to the deposited trilayer graphene along with the physisorption of the water molecules [138]. An experiment using oxidative pit formation was also employed for the formation of diamane. Hydrogen plasma was involved in the hydrogenation of the deposited bilayer graphene to form diamane in the oxidative pit. The oxidative pits are etched away by the same plasma [121, 130, 139]. Plasma involved in ammonia adsorption can also be employed to form diamond films [136]. In the case of fluorination, different scientists have used the plasma process to obtain diamane. Various fluorinating agents, for example boron trifluoride (BF_3), tetrafluoromethane (CF_4), xenon fluoride (XeF_2), chlorine trifluoride (CIF_3), iodine pentafluoride (IF_5), etc, are involved in the process of fluorination [134, 140]. Qin *et al* and Feltan *et al* used CF_6, CF_4, and XeF_4 fluroinating agents to form F-diamane in a deposited graphene layer through interlayer bonding [117, 141]. In another experiment, bilayer graphene was fluorinated by xenon fluoride (XeF_2) at ambient conditions to obtain F-diamane on the bottom and top layers sequentially [119, 130, 142].

In addition to the chemical approach, using a physical approach such as pressure can also lead to the formation of diamane. Studies conducted by Pakornchote *et al* used the pressure method to fabricate diamane but a stabilization pressure of 5 GPa is mandatory to keep diamane from returning back to the structure of bilayer graphene [143]. In another experiment, Ke *et al* used trilayer graphene to form diamane using high pressure and temperature, but a stabilization pressure of limited GPa is mandatory [118, 144]. In another study conducted by Ke *et al*, they used 20 GPa pressure on trilayered graphene to form pristine state H-diamane, but 1 GPa stabilization pressure was mandatory to stabilize their structure [145]. In another study conducted by Qin *et al*, the group used 16 GPa pressure to fabricate pristine hexagonal diamane by employing three or more layers of graphene. Here too, the stabilization pressure of 1 GPa was mandatory [117]. In addition to these methods, diamane can also be synthesized using a substitution method by using atoms such as nitrogen, hydrogen, fluorine, etc, and the Birch reduction method. In the case of substitution by atoms, functionalization using various atoms of nitrogen, sulfur, lithium, fluorine, etc, was done to bilayer graphene to form diamane [143, 146]. Chemisorption or physisorption of atoms are mainly carried out, where Pakornchote *et al* used nitrogen atoms which form the configurations of CNCN, CNCC, NCCC, and NCCN, to bilayer graphenes to form diamane [143, 146]. In other studies, Fyta *et al*, Antipina *et al*, and Sahin *et al* discussed the substitution of boron and sulfur atoms in bilayer graphene to synthesize diamane [114, 122, 136]. Fyta *et al* have also discussed the possibility of hydrogen passivation on graphene to form diamane [114]. In another study, Li *et al* discussed an experiment to synthesize diamane using a phosphorous and lithium dopant vacancy complex [147]. In the case of using the Birch reduction method, a proton source and reductant in the form of water and lithium, respectively, was used to fabricate the diamane where bilayer graphene was fabricated using a mechanical exfoliation method at the silicon dioxide/silicon (SiO_2/Si) substrate [148]. In the case of the formation of diamane using twisted bilayer graphene, the formation of bigraphenes (twisted bilayer graphene) was observed to have taken place using a silicon dioxide (SiO_2)/hexagonal boron nitride (hBN) substrate. Then, the twisted bilayer graphene was fluorinated or hydrogenated to form a diamane structure [116, 121]. Thus, various synthesis methods for the formation of diamane can be observed such that excellent properties of the structures can be obtained.

1.5.2 Properties

Diamanes are structures that possess excellent electrical, magnetic, optical, thermal, and mechanical properties. The most prominent of these is the electrical properties. Graphene does not possess excellent electrical properties due to its zero band gap, but the case for diamane is the opposite. Due to the presence of functional groups and stacking sequences, the modification of electrical properties becomes easy. However, when the layer of diamane increases, the band gap decreases as an indirect band gap is formed [117, 121]. Graphene is semiconducting in nature, however, these properties are changed to semi-metallic when functionalized [114, 149]. A direct band gap along with dielectric properties was observed in diamane for stability and to obtain enhanced properties [126]. Functionalization and modification techniques open the narrow band gap in diamane to realize its excellent electrical properties [138, 150]. Functional groups such as hydrogen and fluorine that are used in diamane synthesis help to influence the electrical properties [142, 151].

Hydrogenation influences diamane when hydrogen atoms are adsorbed at carbon sites due to their configuration and concentration [152]. Hydrogenation initiates the opening of a band gap in diamane as the electronic structure is altered by the distance of the hydrogen regions. Birch reduction also influences the band gap of graphene [148]. Ammonia, water, and hydrogen are some agents for hydrogenation [136]. Fluorination also influences the electrical properties, modifying the band gap of diamane to initiate semi-conduction [123]. The substitution of atoms closes the band gap of diamane but passivation allows the band gap to open [143]. Ultra-thin hydrogenated diamond (UTHD) films possess a narrow gap which is opened by substituting phosphorous and lithium atoms onto the surface according to the studies conducted by Li *et al*, in which a semi-conducting nature with n-type doping is seen [147]. In the case of twisted bilayer graphenes forming diamane, semi-metallic or semiconducting properties are observed [153]. Hydrogenation of twisted bilayer graphene forming a Moiré pattern widens the band gap, shifts the frequency, provides active phonons and peculiarities to the diamane which provides more band gap than in ordinary diamane [116, 121, 153, 154]. Semi-metallic, insulative, and semi-conductive properties are shown by twisted bilayer graphene [154]. Superconductivity and insulative behavior are also provided by diamane. Superconductivity is attained if the interlayer interaction is done at 1.1° along with the electrical phonon coupling [117, 155]. In addition, dual properties, for instance Janus and non-Janus nature, are also exhibited by diamane. When fluorinated, bilayer F-diamane demonstrates insulating behavior. Functionalization with fluorine from both sides initiates insulative behavior, while functionalization on one side demonstrates conduction behavior on the non-functionalized side and insulative behavior on the functionalized side [133, 144].

Electrocatalytical and electrochemical behavior is also enhanced when diamane is fluorinated [153].

In the case of magnetic properties, doping or substitution of atoms is performed on the bilayer graphene so that diamane can be formed. With the formation of this type of diamane, magnetic properties are also opened for exciting applications. Boron, sulfur, nitrogen, etc, are doped or a nitrogen vacancy is created which enables the band gap to close and open the magnetic characteristics [143, 151]. When doped or substituted with nitrogen, paramagnetic and ferromagnetic properties are shown where nitrogen substitution enables ferromagnetic behavior and nitrogen vacancy enables paramagnetic behavior [146, 156]. In another case, the addition of an electric field will reduce the band gap, showing ferromagnetic characteristics [121]. For the case of semi-hydrogenation of diamond films, ferromagnetic properties are enabled [157]. Hydroxylated diamane also exhibits ferromagnetic behavior, which is discussed in a study conducted by Qin *et al* [117].

The optical properties in diamane can be enhanced by optical peculiarities, direct energy gaps, and a wide band gap [113, 121, 128, 142]. Functional groups that are present in bilayer graphene, such as hydrogen, fluorine, and chlorine, to form diamane, improve the optical properties of diamane. Absorption of light in the visible region by chlorinated diamane shows the optical properties of diamane [142]. Moiré diamane and hydrogenated graphene heterostructures provide improved optical properties through the thin mini-bands that are present inside the extended band gap [116]. UTHD films, when substituted with atoms such as lithium and phosphorous, also provided enhanced optical properties [147]. For thermal properties, Lonsdaleite diamane provides higher thermal conductivity than Bernal stacking because of horizontal reflection symmetry. Conductivity in diamane decreases when the temperature is increased and also with the increase in strain [153, 158–160]. Exposure of functional groups to bilayer graphene to fabricate diamane provides enhanced thermal properties of excellent thermal conductivity [158]. Thermal insulation in diamane is observed when hydrogenated and increases with an increase in temperature [135]. In the case of fluorinated diamane thermal insulation is also observed. The insulative property is greater in fluorinated diamane than in hydrogenated diamane because fluorinated diamane demonstrates an increase in the optical phonon mode in comparison with hydrogenated diamane [158, 161]. Along with fluorinated diamane, chlorinated diamanes are also strong insulators which have high thermal resistance [142].

Insight into the mechanical properties of diamanes shows that they are hard materials which have extraordinary strength but are brittle due to their high stiffness coefficient [113, 114, 117, 138]. Hybridization influences the mechanical properties, which require strong covalent bonding to realize maximum strength [114, 162]. Nanosheets of diamane demonstrate isotropic elastic characteristics with low mechanical strength [142]. While diamane is synthesized, functional groups such as hydrogen and fluorine are added to it, which improves the mechanical properties. Hydrogenation induces brittleness while fluorinated diamane contributes to an increase in the elastic modulus [140, 149]. Nanoribbons of diamane show enhanced mechanical and vibrational properties with high natural frequency and Q factor, decreased energy dissipation, along with increased tensile strength and figure of merit [163]. Twisted bilayer graphene will also have enhanced mechanical properties when functionalized. Improvement in the shear modulus and Young's modulus was observed in the structure which consisted of twisted bilayer graphene [149]. Instead of deposition, if substitution of atoms is done, it also provides enhancement of mechanical strength. When nitrogen atoms are substituted in bilayer graphene, an enhancement of tensile strength and elastic modulus was observed [143, 146]. Piezoelectric response, transition from metal to insulator, changeable wettability, and friction can be provided by diamane as well [133]. Hence, the mesmerizing properties of diamane contributes to the excellent applications of the structure in nanotechnology.

1.5.3 Applications

The functionalization of bilayer graphene synthesizes diamane to pave the way for enhanced electrical, optical, thermal, magnetic, and mechanical properties. The band gap is open in diamane, which is the most important property for maximum application in nanotechnology. This favors other important features for enhancing optoelectrical applications, thermoelectrical applications, mechanical applications, etc. Some of the applications of diamane following these properties are discussed below.

1.5.3.1 Nanoelectronics

The possibility of using diamane in electronics comes when its band gap is open [164, 165]. Supercapacitors, field effect transistors, batteries, miniature devices, and laser media can be fabricated using diamane when the band gap is opened during the synthesis [118, 130, 132]. Hydrogenated, fluorinated, and chlorinated diamane possess excellent hole and electron mobility which are useful in semi-conducting applications, spintronics, and nanoelectronics applications [121, 133, 142, 166]. Fluorinated diamane also provides a framework for applications in tunnel contact which need insulative behavior and electron beam irradiation [119, 133]. Electronic packaging of a fan-out wafer was also performed using diamane [140]. Semi-hydrogenated diamane can provide applications in ferromagnetic semiconductors [121]. Not only nanoelectronics but also thermoelectronics takes advantage of diamane. The properties of high thermal conductivity and wide band gap are shown by diamane when added with functional groups during synthesis [136, 160]. Simultaneous power generation along with cooling applications can be favored by diamane [122]. The heat transfer characteristics of fluorinated carbon on a copper surface are also excellent, favoring applications in refrigeration [140]. In the case of optoelectrical applications, diamanes can be used as flexible and lucid optoelectrical devices with excellent properties [128, 142]. Its wide band gap allows its application in nanophotonics and nano-optics [121]. Huang *et al* fabricated a photonic device using Bernal stacked bilayer graphene on SiO_2/hBN, where hBN is the top gate dielectrics and SiO_2 is the back gate dielectrics [167]. Photocatalysis applications can be achieved by employing fluorinated or chlorinated diamanes which possess thermal and dynamic stability [142]. Moiré graphene also has the capability to be used in optoelectronics [116]. Fluorinated diamane has been used in nanopatterning techniques, and provides applications as an irradiation device, important in nano-optics and in nanophotonics [119, 168]. Janus diamanes which possess monolayers of C_4FCl and C_4HCl are useful in the fabrication of optoelectronics [150]. Application of a laser using diamane can be achieved where the diamane can act as an active laser medium [126, 130, 132, 152, 162, 169]. Optical waveguides and electron waveguides can be formed using diamane and hydrogenated twisted bilayer graphene possessing dielectric nanostrips, integral waveguides, and superlattices [113, 126, 152, 154, 169]. When substituted in bilayer graphene, nitrogen and boron form diamane and the dopant added ultra-thin hydrogenated diamond (UTHD) films possess applications as active laser media in optical waveguides and photoelectronics [114, 147]. Optoelectrical sensors, linear waveguides, and tunnel devices are fabricated using diamane [130, 132]. For electromechanical applications, the mechanical and thermal properties needs to be tuned for nanosensor, electromechanical, and nano-electrical applications [153]. Electromechanical vibrators can be fabricated using diamane due to its important properties of high hardness, wide band gap, and enhanced thermal conductivity [121]. Piezoelectricity can also be found in diamane, which is useful in electromechanical systems [113]. High mechanical stiffness, good mobility, and large band gaps can be attained using functional groups and when these functional groups are used for diamane formation (H-diamane, F-diamane) they provide applications in electromechanical systems [121, 166]. Thus, the different modification and synthesis steps provide enhanced applications of diamane in the field of nanoelectronics.

1.5.3.2 Sensors

Due to its excellent vibrational properties, selectivity, and electrical properties, diamane enables excellent sensory applications. Sensors based on ultra-sensitive resonators and mechanical resonators were fabricated using diamane as it possesses an increased quality factor, plane stiffness, figure of merit, less energy dissipation, and a natural frequency [117, 130, 132, 140, 163]. H-diamane also has the capability to form mechanical resonators [150]. Adding a dopant to bilayer graphene paves the way for resonating devices. Nitrogen doped diamane has a low band gap but increased magnetic and vibrational properties useful for nano-magnetic resonance imaging (nano-MRI) by creating nitrogen vacancies [114, 151]. Moiré diamane which is formed with nitrogen vacancies and vacancy centers provides applications in temperature, electric, and magnetic sensing devices [116]. Electrochemical sensing was also realized by diamane where hydrogenation, oxygenation, and fluorination of diamane provided chemical and biosensors [141]. Explosives, dopamine, uric acid, and ascorbic acid were sensed using diamane due to its excellent electrochemical properties. Oxygen reduction and hydrogen evolution sensing can also be achieved using diamane. In addition formaldehyde, ethanol, methanol, and CO_2/N_2 sensitivity is also seen. Thus, the sensing field benefits to a great extent when diamane is used in the fabrication of the devices.

1.5.3.3 Quantum devices

Quantum information processing and computing is the process which involves the processing of information based on quantum mechanics. Diamane also provides enhanced applications in this field. In particular modified diamane, such as diamane formed of twisted bilayer graphene or the vacancy created diamane, provides excellent quantum computing applications. Quantum information processing was achieved using Moiré diamane formed by substitution of nitrogen vacancies and vacancy centers [116]. Diamanes provide hosts and predictions to optimum two level quantum systems which can be used for sourcing single photons. These are later important for quantum information processing [156]. Quantum computing can be done through nitrogen vacancies created in diamane which produce quibit configurations [132]. Thus, diamane can be an ideal material for use in quantum information processing.

1.5.3.4 Coatings and lubricants

Diamane can be used in applications in the coatings and defense industries [129]. Ultra-thin coatings produced from low frictional hydrogenated diamanes can help to improve wearability in mechanical parts. The enhanced stiffness and strength provided by hydrogenated diamanes can be applied in composite and aerospace applications [117, 130, 132, 170]. Diamane formed by nitrogen substitution provides super-hard materials for coatings which can be used in waterproofing and wear resistance [146]. In the case of lubrication, these materials decrease the coefficient of friction for its use in mechanical parts and other mechanical based applications [140, 159]. Thus, the excellent mechanical properties of enhanced strength, toughness, and hardness provided by diamane are important for lubrication and coating applications.

1.6 Diamanoid

The formation of diamane occurs when two layer of graphene are hydrogenated and bonded. When the number of layers of hydrogenated graphene increases, a new carbon allotrope called 'diamanoid' will be synthesized. Chemisorption of hydrogen radicals produced by low temperature and pressure in a hot filament process on base planes of few-layer graphene, where the top layer is hydrogenated and the bottom layer is covalently bonded with neighboring graphene, can lead to the formation of diamanoids [127, 130]. Diamanoids can also be synthesized by sequencing bilayer graphene and Lonsdaleite structures. The method of high pressure for few-layer graphene or functionalization of few-layer graphene by fluorine, hydrogen, etc, can also lead to the formation of diamanoids. Diamanoids can obtain various stacking sequences, such as (ABB) diamanoids, (ABBA) diamanoids, and (AABBCC) diamanoids but the formation of Bernal stacked diamanoid due to the disadvantage of thermodynamic instability cannot take place. The nomenclature of few-layer graphene does not specify which number of layers is considered to be 'few', so the maximum number of layers used in diamanoid synthesis is yet to be finalized by proper research [132]. The structure of diamanoids is depicted in figure 1.10.

1.6.1 Synthesis

When a higher number of layers are functionalized to obtain stability, diamanoids are thought to be synthesized [171]. The synthesis of diamanoids is not limited to stacking many layers of graphene, as the quality and lateral resolution of the graphene are the features that provide the basis for the formation of diamanoids [131]. Piazza *et al* used chemical vapor deposition to deposit few-layer graphene on a nickel substrate where the precursor methane was used for the synthesis. When formed, the deposited graphene was transported to TEM grids so that contamination was minimized [127, 130]. In another experiment, Subrahmanyam *et al* employed the process of exfoliation where graphite oxide is exfoliated and then the graphite is evaporated under hydrogen to obtain few-layer graphene. Chemical exfoliation was employed to obtain hydrogenated graphene and, along with this, the Birch method can also be used as an alternative [172]. In another experiment to form diamanoids, Piazza *et al* used the hot filament process involving a hot filament reactor with tungsten wires used for heating purposes. A maximum pressure of 10 torr was used for the process. Elements were heated by wires and hydrogen gas with high purity was introduced, instead of hydrocarbon (to minimize contamination), at the top of the chamber in low temperature and pressure conditions. The top and bottom layers of few-layer graphene are hydrogenated, leading to the formation of diamanoids [127]. ABBA, ABAB, ABA, ABCA, and ABBB are some of the stacking sequences that are used and are most stable [130]. Different synthesis methods, even though they have not been researched extensively, have provided diamanoids with good efficiency.

1.6.2 Properties

When the number of layers of graphene is increased followed by functionalization, diamanoids are synthesized [121]. A mixture of diamond and lonsdaelite diamane can also result in the formation of diamanoids. Bernal stacking in diamanoids is not thermodynamically stable so stacking sequences other than Bernal are used, such as AAA…, ABA…, ABC…, etc, but the number of layers should be greater than two [127]. The electrical properties of diamanoids can be tuned by changing the direct band gap of the structure through the crystalline structure and film thickness [121]. Research on diamanoids is in a very early stage, so only expectations of enhanced strength, biocompatibility, and decreased coefficient of friction can be suggested [127]. Hydrogenation of few-layer graphene to form diamanoids provides enhancement in the thermal and electrical conductivity which increase on dehydrogenation [172]. Also, negative electron affinity, inertness towards chemicals, and high bond energies are found in diamanoids. Even though a lot of research has not taken place for diamanoids, these materials have shown their excellent capabilities. Thus, diamanoids have the capability to replace current carbon nanostructures due to their potential electrical, optical, thermal, and mechanical properties.

1.6.3 Applications

The exploration of diamanoids and their properties and applications has not been extensive enough to accumulate enhanced applications in nanotechnology. However, the small progress in finding their electrical, thermal, and mechanical properties has provided some help towards applications. Diamanoids posses high electrical and thermal conductivity, a wide band gap, chemical inertness, and high bond energies, providing applications in harsh environmental conditions. Cathodes that have applications in space satellites can be fabricated using cold-resistant, robust diamanoids providing increased wear resistance and with high conductivity and catalytic performance. Data storage applications using graphene based materials that are doped with SI, NO_2, or NO radicals can be used for nanodevices [173]. Wear resistance applications for coated materials along with anti-adhesion properties are provided by diamanoids. Active laser media, lubrication, gas sensing, thermal management, band gap engineering, electromechanical systems, protective coatings, composites in aerospace, supercapacitors, tunnel devices, optoelectronic sensors, optical waveguides, and lithium batteries are some of the possible applications that can be provided by diamanoids [120]. Doped diamanoids provided quantum computing applications where nitrogen vacancy centers in diamanoids have been shown to be used in quantum applications. Their biocompatibility, excellent strength, and low frictional coefficient makes diamanoids suitable in biomedical devices and miniature low power devices [127]. Thus, diamanoids possesses potential in all these applications of nanotechnology in which they could be a replacement for the currently used carbon nanostructures (figure 1.11).

1.7 Summary and outlook

After the introduction of nanotechnology by Richard Feynman in 1959, considerable research and its results in nanotechnology have been undertaken by numerous scientists and researchers. Among the different fields, one branch of research has moved toward nanocarbons. Nanocarbons have the reputation of providing exceptional properties and applications. The recent trends in these nanocarbons have shown that the method of catalyst chemical vapor deposition needs more research as it has the ability to decompose hydrocarbons on transition metals and also provides enhanced electrical properties [17, 22]. Research has shown that waste products can fulfill the need for materials in the fabrication of nanocarbons. Up-cycling methods can be utilized for obtaining raw materials that are useful in the synthesis of nanocarbon as well [45]. In this chapter, we have addressed the topic of nanocarbons and its types. We have discussed recent trends in the methods of synthesis, properties, and applications of carbon nanotubes, fullerene, graphene, diamane, and diamanoids. Carbon nanotubes, fullerene, and graphene have found significant applications when added to nano-composites, nanopolymers, and metal matrices. These nanocarbons have provided enhanced electrical, mechanical, and thermal properties in all of the types mentioned. However, diamane and diamanoids are still a new areas of research in nanocarbons, so they are yet to be used in nanocomposites, nanopolymers, and metal matrices. Analyzing their capabilities and the available properties, these structures have allowed

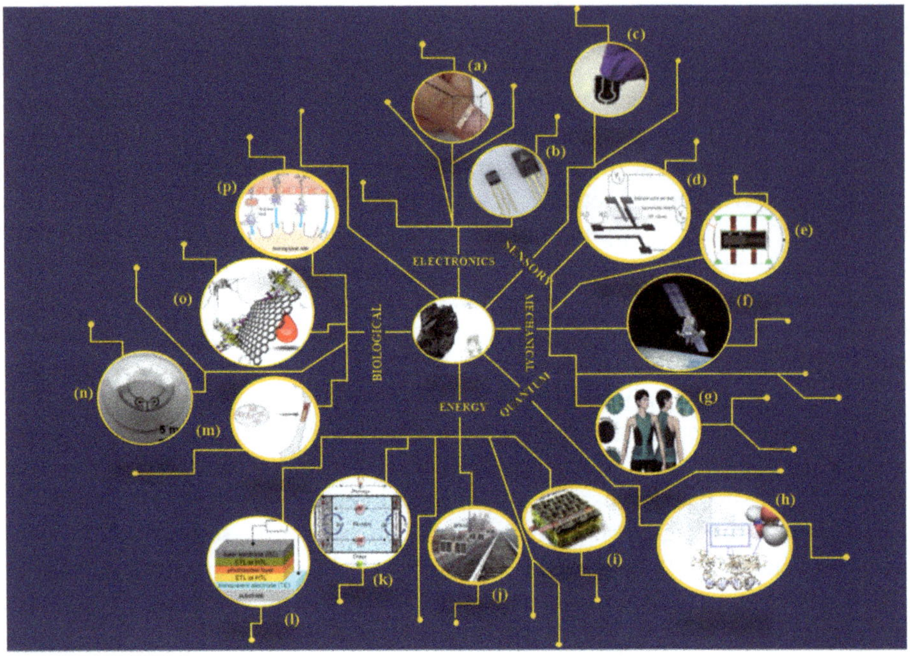

Figure 1.11. Applications of nanocarbon. (a) and (b) Applications as nanoelectronics: (a) nano-electrical devices and (b) transistors. (c) Applications as sensors. (d), (e), (f) Applications as mechanical devices: (d) strain based devices, (e) piezoelectric devices, and (f) satellite devices. (g) Smart textiles fabricated using carbon compounds. (h) Applications in the biomedical field using quantum computing. (i), (j), (k) and (l) Applications of carbon compounds in energy storage and conversion devices: (i) supercapacitors, (j) solar energy harvesting units, (k) zinc–air battery, and (l) the schematic diagram of organic perovskite based ETL materials. (m), (n), (o), and (p) Applications of carbon compounds in the field of biology: (m) tissue engineering, (n) a contact lens that can sense lactate in human tears, (o) Zika virus biosensors, and (p) nanomedicines. ((a) Reproduced with permission from [174]. Copyright 2019 Springer. (c) Reproduced with permission from [15]. Copyright 2018 Elsevier. (d) Reproduced with permission from [175]. Copyright 2021 Springer Nature. (e) Reproduced with permission from [176]. Copyright 2021 Elsevier. (g) Reproduced with permission from [177]. Copyright 2014 Wiley. (h) Reproduced with permission from [178]. Copyright 2020 Wiley. (i) Reproduced with permission from [179]. Copyright 2020 Elsevier. (j) Reproduced with permission from [180]. Copyright 2021 Wiley. (k) Reproduced with permission from [181]. Copyright 2014 Elsevier. (l) Reproduced with permission from [66]. Copyright 2019 American Chemical Society. (m) Reproduced with permission from [182]. Copyright 2021 Wiley. (n) Reproduced with permission from [183]. Copyright 2012 Elsevier. (o) Reproduced with permission from [91]. CC BY 4.0. (p) Reproduced with permission from [184]. Copyright 2016 Elsevier.)

us to determine that diamane and diamanoids will soon replace all the additives that are being used currently. This chapter will allow scientists and researchers to obtain an overview of different nanocarbons and will provide the path to follow from this point onwards to obtain materials for the applications of nanotechnology.

References

[1] Encyclopedia Britannica 2022 Carbon: Facts, Uses, & Properties [online] https://britannica.com/science/carbon-chemical-element (Accessed: 9 June 2022)

[2] Talebian S, Rodrigues T, Das Neves J, Sarmento B, Langer R and Conde J 2021 Facts and figures on materials science and nanotechnology progress and investment *ACS Nano* **15** 15940–52

[3] Perathoner S and Centi G 2018 Advanced nanocarbon materials for future energy applications *Emerging Materials for Energy Conversion and Storage* (Amsterdam: Elsevier) pp 305–25

[4] Wang Z, Hu T, Liang R and Wei M 2020 Application of zero-dimensional nanomaterials in biosensing *Front. Chem.* **8** 320

[5] Iijima S 1991 Helical microtubules of graphitic carbon *Nature* **354** 56–8

[6] Ismail I, Yusof J M, Nong M A M and Adnan N L 2019 Synthesis of carbon nanotube-cotton superfiber materials *Synthesis, Technology and Applications of Carbon Nanomaterials* (Amsterdam: Elsevier) pp 61–76

[7] Tuball.com 2022 Single-walled carbon nanotubes: structure, properties, applications, and health & safety [online] https://tuball.com/articles/single-walled-carbon-nanotubes (Accessed: 9 June 2022)

[8] Henao W, Cazaña F, Tarifa P, Romeo E, Latorre N, Sebastian V, Delgado J J and Monzón A 2021 Selective synthesis of carbon nanotubes by catalytic decomposition of methane using Co–Cu/cellulose derived carbon catalysts: a comprehensive kinetic study *Chem. Eng. J.* **404** 126103

[9] Sridhar V and Park H 2019 Zeolitic imidazolate frameworks as novel precursors for microwave synthesis of carbon nanotubes *J. Alloys Compd.* **781** 166–73

[10] Astié V, Millon C, Decams J M and Bartasyte A 2018 Direct liquid injection chemical vapor deposition *Chemical Vapor Deposition for Nanotechnology* (London: IntechOpen)

[11] Orbaek A W, Aggarwal N and Barron A R 2013 The development of a 'process map' for the growth of carbon nanomaterials from ferrocene by injection CVD *J. Mater. Chem.* A **1** 14122–32

[12] Allaedini G, Tasirin S M and Aminayi P 2015 Synthesis of CNTs via chemical vapor deposition of carbon dioxide as a carbon source in the presence of NiMgO *J. Alloys Compd.* **647** 809–14

[13] Paradise M and Goswami T 2007 Carbon nanotubes—production and industrial applications *Mater. Des.* **28** 1477–89

[14] Chatterjee A and Deopura B L 2002 Carbon nanotubes and nanofibre: an overview *Fibers Polym.* **3** 134–9

[15] Malhotra B D and Ali M A 2018 Functionalized carbon nanomaterials for biosensors in nanomaterials in biosensors: fundamentals and applications *Nanomaterials for Biosensors* William Andrew Publishing pp 75–103

[16] Yao X, Gao X, Jiang J, Xu C, Deng C and Wang J 2018 Comparison of carbon nanotubes and graphene oxide coated carbon fiber for improving the interfacial properties of carbon fiber/epoxy composites *Composites* B **132** 170–7

[17] Li F, Zhou C, Yang P, Wang B, Hu J, Wei J and Yu Q 2019 Direct synthesis of carbon nanotubes on fly ash particles to produce carbon nanotubes/fly ash composites *Front. Struct. Civ. Eng.* **13** 1405–14

[18] Pełech I, Pełech R, Narkiewicz U, Kaczmarek A, Guskos N, Żołnierkiewicz G, Typek J and Berczyński P 2020 Magnetic and electrical properties of carbon nanotube/epoxy composites *Mater. Sci. Eng.* B **254** 114507

[19] Yuan X, Liang C, Ruan C, Chang Y, Xu L, Huang H, Chen M and Yong Z 2021 Low-cost synthesis of multi-walled carbon nanotubes using red soil as catalyst *Diam. Relat. Mater.* **112** 108241

[20] Ge C, Wang L, Liu G and Wang T 2018 Enhanced electromagnetic properties of carbon nanotubes and SiO_2-coated carbonyl iron microwave absorber *J. Alloys Compd.* **767** 173–80

[21] Kuo T F, Juang Z Y, Tsai C H, Tsau Y M, Cheng H F and Lin I N 2001 Microwave-assisted chemical vapor deposition process for synthesizing carbon nanotubes *J. Vac. Sci. Technol.* B **19** 1030–3

[22] Burakova E A, Dyachkova T P, Rukhov A V, Tugolukov E N, Galunin E V, Tkachev A G and Ali I 2018 Novel and economic method of carbon nanotubes synthesis on a nickel magnesium oxide catalyst using microwave radiation *J. Mol. Liq.* **253** 340–6

[23] Sharma S, Rosmi M S, Yaakob Y, Yusop M Z M, Kalita G, Kitazawa M and Tanemura M 2018 *In situ* TEM synthesis of carbon nanotube Y-junctions by electromigration induced soldering *Carbon* **132** 165–71

[24] Wei J, Fan Y, Zhao L, Xue F, Hao L and Zhang Q 2018 Thermoelectric properties of carbon nanotube reinforced cement-based composites fabricated by compression shear *Ceram. Int.* **44** 5829–33

[25] Barnett C J, Gowenlock C E, Welsby K, Orbaek White A and Barron A R 2018 Spatial and contamination-dependent electrical properties of carbon nanotubes *Nano Lett.* **18** 695–700

[26] Alvarez N T, Miller P, Haase M R, Lobo R, Malik R and Shanov V 2019 Tailoring physical properties of carbon nanotube threads during assembly *Carbon* **144** 55–62

[27] Hassanzadeh-Aghdam M K and Jamali J 2019 A new form of a Halpin–Tsai micro-mechanical model for characterizing the mechanical properties of carbon nanotube-reinforced polymer nanocomposites *Bull. Mater. Sci.* **42** 117

[28] Koirala P, van de Werken N, Lu H, Baughman R H, Ovalle-Robles R and Tehrani M 2021 Using ultra-thin interlaminar carbon nanotube sheets to enhance the mechanical and electrical properties of carbon fiber reinforced polymer composites *Composites* B **216** 108842

[29] Alemi Parvin S, Ahmed N A and Fattahi A M 2021 Numerical prediction of elastic properties for carbon nanotubes reinforced composites using a multi-scale method *Eng. Comput.* **37** 1961–72

[30] Zare Y and Rhee K Y 2020 Analysis of the connecting effectiveness of the interphase zone on the tensile properties of carbon nanotubes (CNT) reinforced nanocomposite *Polymers* **12** 896

[31] Schadler L S, Giannaris S A and Ajayan P M 1998 Load transfer in carbon nanotube epoxy composites *Appl. Phys. Lett.* **73** 3842–4

[32] Allaoui A, Bai S, Cheng H M and Bai J B 2002 Mechanical and electrical properties of a MWNT/epoxy composite *Compos. Sci. Technol.* **62** 1993–8

[33] Qian D, Dickey E C, Andrews R and Rantell T 2000 Load transfer and deformation mechanisms in carbon nanotube–polystyrene composites *Appl. Phys. Lett.* **76** 2868–70

[34] Ekrem M 2019 The effects of carbon nanotubes added polyvinyl alcohol nanofibers on mechanical properties of carbon reinforced composite laminates *Sādhanā* **44** 179

[35] Zhao Q, Gan X and Zhou K 2019 Enhanced properties of carbon nanotube–graphite hybrid-reinforced Cu matrix composites via optimization of the preparation technology and interface structure *Powder Technol.* **355** 408–16

[36] Diel J C *et al* 2021 Green synthesis of carbon nanotubes impregnated with metallic nanoparticles: characterization and application in glyphosate adsorption *Chemosphere* **283** 131193

[37] Wang M, Yan Q, Xue F, Zhang J and Wang J 2018 Design and synthesis of carbon nanotubes/carbon fiber/reduced graphene oxide/MnO_2 flexible electrode material for supercapacitors *J. Phys. Chem. Solids* **119** 29–35

[38] Mery A, Cagnon B, Van F T, Autret C and Ghamouss F 2018 Solvent-assisted synthesis of carbon nanotubes–manganese oxide hybrid materials for high voltage aqueous supercapacitor *J. Alloys Compd.* **763** 62–70

[39] Dai J *et al* 2021 Effect of morphology and phase engineering of MoS_2 on electrochemical properties of carbon nanotube/polyaniline@ MoS_2 composites *J. Colloid Interface Sci.* **590** 591–600

[40] Wu X, Chen S, Feng Y, Yuan Q, Gao J, Chen Y, Huang Y, He Y B and Gan W 2019 Microwave-assisted synthesis of carbon nanotubes threaded core–shell CoP_x/Co–N_x–C@ CNT and its performance as an efficient bifunctional oxygen catalyst for the rechargeable zinc–air battery *Mater. Today Phys.* **9** 100132

[41] Nasouri K, Shoushtari A M, Mirzaei J and Merati A A 2020 Synthesis of carbon nanotubes composite nanofibers for ultrahigh performance UV protection and microwave absorption applications *Diam. Relat. Mater.* **107** 107896

[42] Mingdong C, Huangzhong Y, Xiaohua J and Yigang L 2018 Optimization on microwave absorbing properties of carbon nanotubes and magnetic oxide composite materials *Appl. Surf. Sci.* **434** 1321–6

[43] Han F, Wang R, Feng Y, Wang S, Liu L, Li X, Han Y and Chen H 2019 On demand synthesis of hollow fullerene nanostructures *Nat. Commun.* **10** 1548

[44] Encyclopedia Britannica 2022 Richard E Smalley: American chemist and physicist [online] https://britannica.com/biography/Richard-Smalley (Accessed: 9 June 2022)

[45] Elessawy N A, El-Sayed E M, Ali S, Elkady M F, Elnouby M and Hamad H A 2020 One-pot green synthesis of magnetic fullerene nanocomposite for adsorption characteristics *J. Water Process Eng.* **34** 101047

[46] Turan M E, Sun Y and Akgul Y 2018 Mechanical, tribological and corrosion properties of fullerene reinforced magnesium matrix composites fabricated by semi powder metallurgy *J. Alloys Compd.* **740** 1149–58

[47] Kroto H W, Heath J R, O'Brien S C, Curl R F and Smalley R E 1985 C_{60}: buckminsterfullerene *Nature* **318** 162–3

[48] Tang Q, Maji S, Jiang B, Sun J, Zhao W, Hill J P, Ariga K, Fuchs H, Ji Q and Shrestha L K 2019 Manipulating the structural transformation of fullerene microtubes to fullerene micro-horns having microscopic recognition properties *ACS Nano* **13** 14005–12

[49] Hashikawa Y, Fushino T and Murata Y 2020 Double-holed fullerenes *J. Am. Chem. Soc.* **142** 20572–6

[50] Xu H, Tu X, Fan G, Wang Q, Wang X and Chu X 2020 Adsorption properties study of boron nitride fullerene for the application as smart drug delivery agent of anti-cancer drug hydroxyurea by density functional theory *J. Mol. Liq.* **318** 114315

[51] Schwerdtfeger P, Wirz L N and Avery J 2015 The topology of fullerenes *Wiley Interdiscip. Rev. Comput. Mol. Sci.* **5** 96–145

[52] Zhang D, Ye K, Yao Y, Liang F, Qu T, Ma W, Yang B, Dai Y and Watanabe T 2019 Controllable synthesis of carbon nanomaterials by direct current arc discharge from the inner wall of the chamber *Carbon* **142** 278–84

[53] Sachdeva S, Singh D and Tripathi S K 2020 Optical and electrical properties of fullerene C70 for solar cell applications *Opt. Mater.* **101** 109717

[54] Kyzyma O A *et al* 2019 State of aggregation and toxicity of aqueous fullerene solutions *Appl. Surf. Sci.* **483** 69–75

[55] Dorel R and Echavarren A M 2019 From palladium to gold catalysis for the synthesis of crushed fullerenes and acenes *Acc. Chem. Res.* **52** 1812–23

[56] Lin H S, Jeon I, Chen Y, Yang X Y, Nakagawa T, Maruyama S, Manzhos S and Matsuo Y 2019 Highly selective and scalable fullerene-cation-mediated synthesis accessing cyclo[60] fullerenes with five-membered carbon ring and their application to perovskite solar cells *Chem. Mater.* **31** 8432–9

[57] Vidal S, Izquierdo M, Filippone S, Fernández I, Akin S, Seo J Y, Zakeeruddin S M, Grätzel M and Martín N 2019 Site-selective synthesis of β-[70] PCBM-like fullerenes: efficient application in perovskite solar cells *Chem. Eur. J.* **25** 3224–8

[58] Peng Z, Jiang Q, Peng P and Li F F 2021 NH_3-activated fullerene derivative hierarchical microstructures to porous Fe_3O_4/NC for oxygen reduction reaction and Zn–air battery *Eng. Sci.* **14** 27–38

[59] Yasuno T, Ohe T, Ikeda H, Takahashi K, Nakamura S and Mashino T 2019 Synthesis and antitumor activity of novel pyridinium fullerene derivatives *Int. J. Nanomed.* **14** 6325

[60] Tokimaru Y, Ito S and Nozaki K 2018 A hybrid of corannulene and azacorannulene: synthesis of a highly curved nitrogen-containing buckybowl *Angew. Chem.* **130** 9966–70

[61] Kinzyabaeva Z S and Sabirov D S 2020 Sonochemical synthesis of novel C_{60} fullerene 1,4-oxathiane derivative through the intermediate fullerene radical anion *Ultrason. Sonochem.* **67** 105169

[62] Sergeeva V *et al* 2019 Antioxidant properties of fullerene derivatives depend on their chemical structure: a study of two fullerene derivatives on HELFs *Oxid. Med. Cell. Longev.* **2019** 4398695

[63] Rad A S, Ardjmand M, Esfahani M R and Khodashenas B 2021 DFT calculations towards the geometry optimization, electronic structure, infrared spectroscopy and UV–vis analyses of Favipiravir adsorption on the first-row transition metals doped fullerenes; a new strategy for COVID-19 therapy *Spectrochim. Acta* A **247** 119082

[64] Afanasiev P 2018 Topotactic synthesis of size-tuned MoS_2 inorganic fullerenes that allows revealing particular catalytic properties of curved basal planes *Appl. Catal.* B **227** 44–53

[65] Barendt T A, Myers W K, Cornes S P, Lebedeva M A, Porfyrakis K, Marques I, Félix V and Beer P D 2019 The green box: an electronically versatile perylene diimide macrocyclic host for fullerenes *J. Am. Chem. Soc.* **142** 349–64

[66] Umeyama T and Imahori H 2019 Isomer effects of fullerene derivatives on organic photovoltaics and perovskite solar cells *Acc. Chem. Res.* **52** 2046–55

[67] Paul D, Bhattacharya B, Deb J and Sarkar U 2018 May. Optical properties of C28 fullerene cage: a DFT study *AIP Conf. Proc.* **1953** 030236

[68] Martinez-Agramunt V and Peris E 2019 Photocatalytic properties of a palladium metallosquare with encapsulated fullerenes via singlet oxygen generation *Inorg. Chem.* **58** 11836–42

[69] Barham J P, Tanaka S, Koyama E, Ohneda N, Okamoto T, Odajima H, Sugiyama J I and Norikane Y 2018 Selective, scalable synthesis of C_{60}-fullerene/indene monoadducts using a microwave flow applicator *J. Org. Chem.* **83** 4348–54

[70] Xu Y, Kaur R, Wang B, Minameyer M B, Gsänger S, Meyer B, Drewello T, Guldi D M and von Delius M 2018 Concave–convex Π–Π template approach enables the synthesis of [10]cycloparaphenylene–fullerene[2]rotaxanes *J. Am. Chem. Soc.* **140** 13413–20

[71] Donskyi I *et al* 2018 Interactions of fullerene-polyglycerol sulfates at viral and cellular interfaces *Small* **14** 1800189

[72] Bloodworth S *et al* 2018 Synthesis and properties of open fullerenes encapsulating ammonia and methane *Chem. Phys. Chem.* **19** 266–76

[73] Ding L, Jin B, Guo Z, Zhao Y, Chen J and Peng R 2019 Regioselective synthesis and crystallographic characterization of nontethered *cis*-1 and *cis*-2-bis-(benzofuro)[60] fullerene derivatives *Org. Lett.* **21** 9924–8

[74] Kiew S F, Kiew L V, Lee H B, Imae T and Chung L Y 2016 Assessing biocompatibility of graphene oxide-based nanocarriers: a review *J. Control. Release* **226** 217–28

[75] Elsaid K, Abdelkareem M A, Maghrabie H M, Sayed E T, Wilberforce T, Baroutaji A and Olabi A G 2021 Thermophysical properties of graphene-based nanofluids *Int. J. Thermofluids* **10** 100073

[76] Chu K, Wang X H, Li Y B, Huang D J, Geng Z R, Zhao X L, Liu H and Zhang H 2018 Thermal properties of graphene/metal composites with aligned graphene *Mater. Des.* **140** 85–94

[77] Novoselov K S, Colombo L, Gellert P R, Schwab M G and Kim K 2012 A roadmap for graphene *Nature* **490** 192–200

[78] Wallace P R 1947 The band theory of graphite *Phys. Rev.* **71** 622

[79] Geim A K and Novoselov K S 2010 The rise of graphene *Nanosci. Technol.* **6** 11–9

[80] Chen W, Li D, Tian L, Xiang W, Wang T, Hu W, Hu Y, Chen S, Chen J and Dai Z 2018 Synthesis of graphene quantum dots from natural polymer starch for cell imaging *Green Chem.* **20** 4438–42

[81] Martini L *et al* 2019 Structure-dependent electrical properties of graphene nanoribbon devices with graphene electrodes *Carbon* **146** 36–43

[82] Han S, Meng Q, Araby S, Liu T and Demiral M 2019 Mechanical and electrical properties of graphene and carbon nanotube reinforced epoxy adhesives: experimental and numerical analysis *Composites* A **120** 116–26

[83] Tiwari S K, Sahoo S, Wang N and Huczko A 2020 Graphene research and their outputs: status and prospect *J. Sci. Adv. Mater. Devices* **5** 10–29

[84] Liu F, Wang C, Sui X, Riaz M A, Xu M, Wei L and Chen Y 2019 Synthesis of graphene materials by electrochemical exfoliation: recent progress and future potential *Carbon Energy* **1** 173–99

[85] Zhang Q, Cheng X, Wang C, Rao A M and Lu B 2021 Sulfur-assisted large-scale synthesis of graphene microspheres for superior potassium-ion batteries *Energy Environ. Sci.* **14** 965–74

[86] Wang J, Guo L N, Lin W M, Chen J, Liu C L, Zhang S and Zhen T T 2019 Effect of the graphene content on the microstructures and properties of graphene/aluminum composites *New Carbon Mater.* **34** 275–85

[87] Wang J, Guo L N, Lin W M, Chen J, Zhang S, Zhen T T and Zhang Y Y 2019 The effects of graphene content on the corrosion resistance, and electrical, thermal and mechanical properties of graphene/copper composites *New Carbon Mater.* **34** 161–9

[88] Kim S *et al* 2020 Enhanced gas sensing properties of graphene transistor by reduced doping with hydrophobic polymer brush as a surface modification layer *ACS Appl. Mater. Interfaces* **12** 55493–500
[89] He D, Zhang P, Li S and Luo H 2018 A novel free-standing CVD graphene platform electrode modified with AuPt hybrid nanoparticles and L-cysteine for the selective determination of epinephrine *J. Electroanal. Chem.* **823** 678–87
[90] Huang M *et al* 2018 Highly oriented monolayer graphene grown on a Cu/Ni (111) alloy foil *ACS Nano* **12** 6117–27
[91] Afsahi S *et al* 2018 Novel graphene-based biosensor for early detection of Zika virus infection *Biosens. Bioelectron.* **100** 85–8
[92] Di Gaspare L, Scaparro A M, Fanfoni M, Fazi L, Sgarlata A, Notargiacomo A, Miseikis V, Coletti C and De Seta M 2018 Early stage of CVD graphene synthesis on Ge (001) substrate *Carbon* **134** 183–8
[93] Gao Y, Cao T, Cellini F, Berger C, De Heer W A, Tosatti E, Riedo E and Bongiorno A 2018 Ultrahard carbon film from epitaxial two-layer graphene *Nat. Nanotechnol.* **13** 133–8
[94] Hou D, Liu Q, Wang X, Quan Y, Qiao Z, Yu L and Ding S 2018 Facile synthesis of graphene via reduction of graphene oxide by artemisinin in ethanol *J. Materiomics* **4** 256–65
[95] Kalita H, Palaparthy V S, Baghini M S and Aslam M 2020 Electrochemical synthesis of graphene quantum dots from graphene oxide at room temperature and its soil moisture sensing properties *Carbon* **165** 9–17
[96] Pandey S, Karakoti M, Dhali S, Karki N, SanthiBhushan B, Tewari C, Rana S, Srivastava A, Melkani A B and Sahoo N G 2019 Bulk synthesis of graphene nanosheets from plastic waste: an invincible method of solid waste management for better tomorrow *Waste Manag.* **88** 48–55
[97] Puértolas J A, Castro M, Morris J A, Ríos R and Ansón-Casaos A 2019 Tribological and mechanical properties of graphene nanoplatelet/PEEK composites *Carbon* **141** 107–22
[98] Chu K, Wang F, Wang X H and Huang D J 2018 Anisotropic mechanical properties of graphene/copper composites with aligned graphene *Mater. Sci. Eng.* A **713** 269–77
[99] Vázquez-Moreno J M, Sánchez-Hidalgo R, Sanz-Horcajo E, Viña J, Verdejo R and López-Manchado M A 2019 Preparation and mechanical properties of graphene/carbon fiber-reinforced hierarchical polymer composites *J. Compos. Sci.* **3** 30
[100] Jadoon T, Carter-Fenk K, Siddique M B A, Herbert J M, Hussain R, Iqbal S, Iqbal J and Ayub K 2020 Silver clusters tune up electronic properties of graphene nanoflakes: a comprehensive theoretical study *J. Mol. Liq.* **297** 111902
[101] Lopez-Polin G, Gomez-Navarro C and Gomez-Herrero J 2022 The effect of rippling on the mechanical properties of graphene *Nano Mater. Sci.* **4** 18–26
[102] Du Y, Li D, Liu L and Gai G 2018 Recent achievements of self-healing graphene/polymer composites *Polymers* **10** 114
[103] Wu B, Zhang G, Yan M, Xiong T, He P, He L, Xu X and Mai L 2018 Graphene scroll-coated α-MnO_2 nanowires as high-performance cathode materials for aqueous Zn-ion battery *Small* **14** 1703850
[104] Qiu B, Zhao X, Hu G, Yue W, Ren J and Yuan X 2018 Optical properties of graphene/ MoS^2 heterostructure: first principles calculations *Nanomaterials* **8** 962
[105] Li C, Zhang X, Lv Z, Wang K, Sun X, Chen X and Ma Y 2021 Scalable combustion synthesis of graphene-welded activated carbon for high-performance supercapacitors *Chem. Eng. J.* **414** 128781

[106] La Notte L, Bianco G V, Palma A L, Di Carlo A, Bruno G and Reale A 2018 Sprayed organic photovoltaic cells and mini-modules based on chemical vapor deposited graphene as transparent conductive electrode *Carbon* **129** 878–83
[107] Bronner C *et al* 2018 Hierarchical on-surface synthesis of graphene nanoribbon heterojunctions *ACS Nano* **12** 2193–200
[108] Alam F, Choosri M, Gupta T K, Varadarajan K M, Choi D and Kumar S 2019 Electrical, mechanical and thermal properties of graphene nanoplatelets reinforced UHMWPE nanocomposites *Mater. Sci. Eng.* B **241** 82–91
[109] Li M, Gao H, Liang J, Gu S, You W, Shu D, Wang J and Sun B 2018 Microstructure evolution and properties of graphene nanoplatelets reinforced aluminum matrix composites *Mater. Charact.* **140** 172–8
[110] Bai S, Jiang L, Xu N, Jin M and Jiang S 2018 Enhancement of mechanical and electrical properties of graphene/cement composite due to improved dispersion of graphene by addition of silica fume *Constr. Build. Mater.* **164** 433–41
[111] Ding J, Rahman O, Peng W, Dou H and Yu H 2018 A novel hydroxyl epoxy phosphate monomer enhancing the anticorrosive performance of waterborne graphene/epoxy coatings *Appl. Surf. Sci.* **427** 981–91
[112] Chu K, Wang F, Li Y B, Wang X H, Huang D J and Geng Z R 2018 Interface and mechanical/thermal properties of graphene/copper composite with Mo_2C nanoparticles grown on graphene *Composites* A **109** 267–79
[113] Chernozatonskii L A, Sorokin P B, Kvashnin A G E and Kvashnin D G E 2009 Diamond-like C_2H nanolayer, diamane: simulation of the structure and properties *JETP Lett.* **90** 134–8
[114] Fyta M 2013 Nitrogen-vacancy centers and dopants in ultrathin diamond films: electronic structure *J. Phys. Chem.* C **117** 21376–81
[115] Muniz A R and Maroudas D 2012 Opening and tuning of band gap by the formation of diamond superlattices in twisted bilayer graphene *Phys. Rev.* B **86** 075404
[116] Chernozatonskii L A, Demin V A and Kvashnin D G 2020 Ultrawide-bandgap moiré diamanes based on bigraphenes with the twist angles $\Theta \sim 30°$ *Appl. Phys. Lett.* **117** 253104
[117] Qin G, Wu L and Gou H 2022 Diamane: design, synthesis, properties, and challenges *Funct. Diam.* **1** 83–92
[118] Ke F *et al* 2019 Synthesis and electronic structure characterization of diamane arXiv:1902.01570
[119] Bakharev P V *et al* 2020 Chemically induced transformation of chemical vapour deposition grown bilayer graphene into fluorinated single-layer diamond *Nat. Nanotechnol.* **15** 59–66
[120] Piazza F and Monthioux M 2021 Ultra-thin carbon films: the rise of sp^3-C-based 2D materials? *C J. Carbon Res.* **7** 30
[121] Chernozatonskii L A, Demin V A and Kvashnin D G 2021 Fully hydrogenated and fluorinated bigraphenes–diamanes: theoretical and experimental studies *C J. Carbon Res.* **7** 17
[122] Sahin H, Leenaerts O, Singh S K and Peeters F M 2015 GraphAne: from synthesis to applications arXiv:1502.05804
[123] Muniz A R and Maroudas D 2013 Superlattices of fluorinated interlayer-bonded domains in twisted bilayer graphene *J. Phys. Chem.* C **117** 7315–25
[124] Sorokin P B and Yakobson B I 2021 Two-dimensional diamond—diamane: current state and further prospects *Nano Lett.* **21** 5475–84

[125] Tiwari S K, Pandey R, Wang N, Kumar V, Sunday O J, Bystrzejewski M, Zhu Y and Mishra Y K 2022 Progress in diamanes and diamanoids nanosystems for emerging technologies *Adv. Sci.* **9** 2105770

[126] Chernozatonskii L A, Sorokin P B, Kuzubov A A, Sorokin B P, Kvashnin A G, Kvashnin D G, Avramov P V and Yakobson B I 2011 Influence of size effect on the electronic and elastic properties of diamond films with nanometer thickness *J. Phys. Chem.* C **115** 132–6

[127] Piazza F, Monthioux M, Puech P and Gerber I C 2020 Towards a better understanding of the structure of diamanoïds and diamanoïd/graphene hybrids *Carbon* **156** 234–41

[128] Momeni K *et al* 2020 Multiscale computational understanding and growth of 2D materials: a review *NPJ Comput. Mater.* **6** 22

[129] Paul S, Momeni K and Levitas V I 2020 Shear-induced diamondization of multilayer graphene structures: a computational study *Carbon* **167** 140–7

[130] Piazza F, Monthioux M, Puech P, Gerber I C and Gough K 2021 Progress on diamane and diamanoid thin film pressureless synthesis *C J. Carbon Res.* **7** 9

[131] Piazza F, Cruz K, Monthioux M, Puech P and Gerber I 2020 Raman evidence for the successful synthesis of diamane *Carbon* **169** 129–33

[132] Piazza F, Gough K, Monthioux M, Puech P, Gerber I, Wiens R, Paredes G and Ozoria C 2019 Low temperature, pressureless sp^2 to sp^3 transformation of ultrathin, crystalline carbon films *Carbon* **145** 10–22

[133] Son J *et al* 2020 Tailoring single-and double-sided fluorination of bilayer graphene via substrate interactions *Nano Lett.* **21** 891–8

[134] Chen X, Dubois M, Radescu Cioranescu S, Rawal A, Zhao C. Liquid-Phase Exfoliation of Poly(Dicarbon Monofluoride) (C2F)n (Cambridge: Cambridge Open Engage)

[135] Elias D C *et al* 2009 Control of graphene's properties by reversible hydrogenation: evidence for graphane *Science* **323** 610–3

[136] Antipina L Y and Sorokin P B 2015 Converting chemically functionalized few-layer graphene to diamond films: a computational study *J. Phys. Chem.* C **119** 2828–36

[137] Luo Z, Yu T, Kim K J, Ni Z, You Y, Lim S, Shen Z, Wang S and Lin J 2009 Thickness-dependent reversible hydrogenation of graphene layers *ACS Nano* **3** 1781–8

[138] Jones J D 2012 Modification of graphene properties: electron induced reversible hydrogenation, oxidative etching and layer-by-layer thinning *University of North Texas* Dissertation

[139] Jones J D, Morris C F, Verbeck G F and Perez J M 2013 Oxidative pit formation in pristine, hydrogenated and dehydrogenated graphene *Appl. Surf. Sci.* **264** 853–63

[140] Ahmad Y, Batisse N, Chen X and Dubois M 2021 Preparation and applications of fluorinated graphenes *C J. Carbon Res.* **7** 20

[141] Felten A, Eckmann A, Pireaux J J, Krupke R and Casiraghi C 2013 Controlled modification of mono-and bilayer graphene in O_2, H_2 and CF_4 plasmas *Nanotechnology* **24** 355705

[142] Mortazavi B, Shojaei F, Javvaji B, Azizi M, Zhan H, Rabczuk T and Zhuang X 2020 First-principles investigation of mechanical, electronic and optical properties of H-, F- and Cl-diamane *Appl. Surf. Sci.* **528** 147035

[143] Pakornchote T, Ektarawong A, Pinsook U and Bovornratanaraks T 2021 Modifying electronic and elastic properties of 2-dimensional [110] diamond by nitrogen substitution *C J. Carbon Res.* **7** 8

[144] Bianco A, Chen Y, Chen Y, Ghoshal D, Hurt R H, Kim Y A, Koratkar N, Meunier V and Terrones M 2018 A carbon science perspective in 2018: current achievements and future challenges *Carbon* **132** 785–801

[145] Ke F *et al* 2020 Synthesis of atomically thin hexagonal diamond with compression *Nano Lett.* **20** 5916–21

[146] Pakornchote T, Ektarawong A, Busayaporn W, Pinsook U and Bovornratanaraks T 2020 Roles of nitrogen substitution and surface reconstruction in stabilizing nonpassivated single-layer diamond *Phys. Rev.* B **102** 075418

[147] Li J, Yin H, Gao N, Zhang M, Mu J, Gao L and Li H 2019 First-principles calculations for Li, P dopants and vacancy defect in ultra-thin hydrogenated diamond nanofilms: structural, electronic and optical properties *Diam. Relat. Mater.* **99** 107526

[148] Zhang X *et al* 2016 Birch-type hydrogenation of few-layer graphenes: products and mechanistic implications *J. Am. Chem. Soc.* **138** 14980–6

[149] Machado A S, Maroudas D and Muniz A R 2013 Tunable mechanical properties of diamond superlattices generated by interlayer bonding in twisted bilayer graphene *Appl. Phys. Lett.* **103** 013113

[150] Raeisi M, Mortazavi B, Podryabinkin E V, Shojaei F, Zhuang X and Shapeev A V 2020 High thermal conductivity in semiconducting Janus and non-Janus diamanes *Carbon* **167** 51–61

[151] Li H, Cheng S, Li J and Song J 2015 A review on the low-dimensional and hybridized nanostructured diamond films *J. Nanomater.* **2015** 692562

[152] Chernozatonskii L A, Sorokin P B and Artukh A A 2014 Novel graphene-based nanostructures: physicochemical properties and applications *Russ. Chem. Rev.* **83** 251

[153] Muniz A R and Fonseca A F 2015 Carbon-based nanostructures derived from bilayer graphene with zero thermal expansion behavior *J. Phys. Chem.* C **119** 17458–65

[154] Chernozatonskii L A, Katin K P, Demin V A and Maslov M M 2021 Moiré diamanes based on the hydrogenated or fluorinated twisted bigraphene: the features of atomic and electronic structures, Raman and infrared spectra *Appl. Surf. Sci.* **537** 148011

[155] Odkhuu D, Shin D, Ruoff R S and Park N 2013 Conversion of multilayer graphene into continuous ultrathin sp^3-bonded carbon films on metal surfaces *Sci. Rep.* **3** 3276

[156] Gupta S, Yang J H and Yakobson B I 2018 Two-level quantum systems in two-dimensional materials for single photon emission *Nano Lett.* **19** 408–14

[157] Li J, Li H, Wang Z and Zou G 2013 Structure, magnetic, and electronic properties of hydrogenated two-dimensional diamond films *Appl. Phys. Lett.* **102** 073114

[158] Zhu L and Zhang T 2019 Suppressed thermal conductivity in fluorinated diamane: optical phonon dominant thermal transport *Appl. Phys. Lett.* **115** 151904

[159] Zhu L, Li W and Ding F 2019 Giant thermal conductivity in diamane and the influence of horizontal reflection symmetry on phonon scattering *Nanoscale* **11** 4248–57

[160] Hu Y, Li D, Yin Y, Li S, Ding G, Zhou H and Zhang G 2020 The important role of strain on phonon hydrodynamics in diamond-like bi-layer graphene *Nanotechnology* **31** 335711

[161] Zhang T and Zhu L 2020 Sensitively tuning the thermal conductivity of diamane via engineering the mass of functional groups *Nanotechnology* **31** 435409

[162] Sorokin P B and Chernozatonskii L A 2013 Graphene-based semiconductor nanostructures *Phys. Usp.* **56** 105

[163] Zheng Z, Zhan H, Nie Y, Xu X, Qi D and Gu Y 2020 Single layer diamond—a new ultrathin 2D carbon nanostructure for mechanical resonator *Carbon* **161** 809–15

[164] Fan L, Yao W and Zhang Z 2020 Regulation of failure mechanism of a bilayer Gr/h-BN staggered stacked heterostructure via interlayer sp^3 bonds, interface connection, and defects *Appl. Phys.* A **126** 752

[165] Balog R, Jørgensen B, Wells J, Lægsgaard E, Hofmann P, Besenbacher F and Hornekær L 2009 Atomic hydrogen adsorbate structures on graphene *J. Am. Chem. Soc.* **131** 8744–5

[166] Cheng T, Liu Z and Liu Z 2020 High elastic moduli, controllable bandgap and extraordinary carrier mobility in single-layer diamond *J. Mater. Chem.* C **8** 13819–26

[167] Huang M *et al* 2020 Large-area single-crystal AB-bilayer and ABA-trilayer graphene grown on a Cu/Ni(111) foil *Nat. Nanotechnol.* **15** 289–95

[168] Kvashnin A G, Avramov P V, Kvashnin D G, Chernozatonskii L A and Sorokin P B 2017 Features of electronic, mechanical, and electromechanical properties of fluorinated diamond films of nanometer thickness *J. Phys. Chem.* C **121** 28484–9

[169] Chernozatonskii L A, Mavrin B N and Sorokin P B 2012 Determination of ultrathin diamond films by Raman spectroscopy *Phys. Status Solidi* B **249** 1550–4

[170] Gao Y 2017 Force microscopy of two-dimensional materials *Doctoral Dissertation* Georgia Institute of Technology

[171] Martins L G P *et al* 2021 Hard, transparent, sp^3-containing 2D phase formed from few-layer graphene under compression *Carbon* **173** 744–57

[172] Subrahmanyam K S, Kumar P, Maitra U, Govindaraj A, Hembram K P S S, Waghmare U V and Rao C N R 2011 Chemical storage of hydrogen in few-layer graphene *Proc. Natl Acad. Sci. USA* **108** 2674–7

[173] Thümmel H T 1998 Theoretical study on X–H, –O, –OH, –NO, –ONO, and $–NO_2$ (X = CH_3, t-C_4H_9, $C_{13}H_{21}$) *J. Phys. Chem.* A **102** 2002–8

[174] Gall O Z, Meng C, Bhamra H, Mei H, John S W and Irazoqui P P 2019 A batteryless energy harvesting storage system for implantable medical devices demonstrated *in situ* *Circuits Syst. Signal Process.* **38** 1360–73

[175] Williams N X, Bullard G, Brooke N, Therien M J and Franklin A D 2021 Printable and recyclable carbon electronics using crystalline nanocellulose dielectrics *Nat. Electron.* **4** 261–8

[176] Qi L, Li H, Wu X, Zhang Z, Duan W and Yi M 2021 A hybrid piezoelectric–electromagnetic wave energy harvester based on capsule structure for self-powered applications in sea-crossing bridges *Renew. Energy* **178** 1223–35

[177] Jost K, Durkin D P, Haverhals L M, Brown E K, Langenstein M, De Long H C, Trulove P C, Gogotsi Y and Dion G 2015 Natural fiber welded electrode yarns for knittable textile supercapacitors *Adv. Energy Mater.* **5** 1401286

[178] Outeiral C, Strahm M, Shi J, Morris G M, Benjamin S C and Deane C M 2021 The prospects of quantum computing in computational molecular biology *Wiley Interdiscip. Rev. Comput. Mol. Sci.* **11** e1481

[179] Deka B K, Hazarika A, Lee S, Park Y B and Park H W 2020 Triboelectric-nanogenerator-integrated structural supercapacitor based on highly active P-doped branched Cu–Mn selenide nanowires for efficient energy harvesting and storage *Nano Energy* **73** 104754

[180] Hao D, Zhang T, Guo L, Feng Y, Zhang Z and Yuan Y 2021 A high-efficiency, portable solar energy-harvesting system based on a foldable-wings mechanism for self-powered applications in railways *Energy Technol.* **9** 2000794

[181] Pei P, Wang K and Ma Z 2014 Technologies for extending zinc–air battery's cyclelife: a review *Appl. Energy* **128** 315–24

[182] Collins M N, Ren G, Young K, Pina S, Reis R L and Oliveira J M 2021 Scaffold fabrication technologies and structure/function properties in bone tissue engineering *Adv. Funct. Mater.* **31** 2010609

[183] Thomas N, Lähdesmäki I and Parviz B A 2012 A contact lens with an integrated lactate sensor *Sensors Actuators* B **162** 128–34

[184] Duschl A 2016 Nanomedicine *Immune Rebalancing* (New York: Academic) pp 251–74

IOP Publishing

Nanocarbon Allotropes Beyond Graphene
Synthesis, properties and applications
Arpan Kumar Nayak and Santosh K Tiwari

Chapter 2

Synthesis and application of graphene nanoribbons

Benalia Kouini, Hossem Belhamdi, Amina Hachaichi and Asma Nour El Houda Sid

Graphene, a two-dimensional carbon nanomaterial, has been of great interest to industry and academia in recent decades. The extensive research that has been conducted on graphene has led to the discovery of graphene nanoribbons (GNRs), graphene-based structures of nanoscale dimensions, which possess excellent metallic or semiconducting electronic properties that depend upon the dimensions, chirality, and geometry.

The exceptional properties of GNRs are in strong contrast to those of graphene, which mainly exhibits zero band gap characteristics. Owing to their one-dimensional nature, GNRs differ extensively from 2D graphene sheets. This review chapter discusses the current development of GNRs and carbon nanotubes (CNTs) to be used in electrical applications. The development that has been reported concerning the manufacture of GNRs and CNTs with predefined dimensions using the approaches of bottom-up chemical synthesis are examined thoroughly. The properties of GNRs and CNTs along with the modeling and analytical works on GNR- and CNT-based interconnects and field-effect transistors are also discussed. At the end of this chapter, the applications of GNRs in the fields of biomedicine, energy, electronic devices, dielectric devices, logic devices, optoelectric device, etc, are discussed.

2.1 Introduction

Technological advances linked to carbon based materials have received a lot of attention in the twenty-first century. Specifically, the fields of materials science and engineering are benefitting from this wonder material by exploring the materials to realize various properties and applications. In this respect, graphene, fullerene, carbon nanotubes (CNTs), graphene nanoribbons (GNRs), diamane, diamanoids, etc, have been discovered as novel carbon nanomaterials present in the nanometer

doi:10.1088/978-0-7503-5177-5ch2

regime which demonstrate a vital role in shaping the technological advances of materials science and nanotechnology in the modern world. These nanomaterials have been exciting scientists to research more and more in the field to achieve a excellent features. Even though CNTs were discovered more than two decades ago, in 1991, by Japanese physicist Sumio Iijima, graphene and graphene-based nanoribbons (GNRs), however, were only discovered more recently. The world of materials science is indebted to Andre Geim and Konstantin Novoselov for discovering graphene in 2004 while they were at the University of Manchester. These remarkable discoveries, CNTs and GNRs, have demonstrated excellent properties, growth technologies, and importance to the modern world such that astonishing applications can be realized. In this chapter, we discuss CNTs, graphene, and GNRs, their properties, growth regimes, and the modeling and analysis of GNR and CNT internconnects to be used in transistors and practical circuits. At the end of the chapter, we discuss some of the important applications of GNRs.

2.2 Graphene

The two-dimensional (2D) material known as 2D graphite or graphene is fabricated from three-dimensional (3D) graphite and constitutes a single layer. An allotrope of carbonaceous materials, graphene consists of sp^2-bonded planar sheets that are one atom thick and are tightly packed in a crystal lattice representing a honeycomb-like structure [1]. The name graphene arose from the word *graphite* and the suffix *-ene* [2]. The most straightforward way to explain graphene is that it is atomic-scale chicken netting comprised of carbon atoms and their bonds (figure 2.1).

Graphene is made up of carbon–carbon bonds which have a length of 0.142 nm [3]. Similarly, graphite is produced by stacking graphene sheets with an interplanar distance of 0.335 nm [4]. Some carbon allotropes, such as CNTs, graphite, fullerenes, charcoal, etc, all contain graphene as their fundamental structural component but the orientation can vary. Discussing the history of two-dimensional carbon structures, in 1947 P R Wallace conducted extensive research on the band theory of 2D graphite layers, or graphene [5]. However, the 2D crystals were thought to be thermodynamically unstable and the creation of these structures not possible, until 2004. The experimental discovery of graphene in 2004 defied this above mentioned conventional wisdom, and Andre Geim and Konstantin Novoselov were awarded the 2010 Nobel Prize in Physics 'for their ground-breaking discoveries with the two-dimensional substance graphene' [6].

2.3 Graphene nanoribbons

Graphene nanoribbons (GNRs) (also called nanographene ribbons) are graphene strips with an ultrathin width which is less than 50 nm. Fujita *et al* provided a study on the electronic states of graphene ribbons as a theoretical model [7].

The orientation and structure of GNRs are determined by observing the shapes of the carbon edges of the sheets. The edges are either zigzag or armchair, indicating the structure to be either zigzag GNR or armchair GNR (figure 2.2). The basic characteristic of zigzag GNRs is their metallic nature, which is helpful in

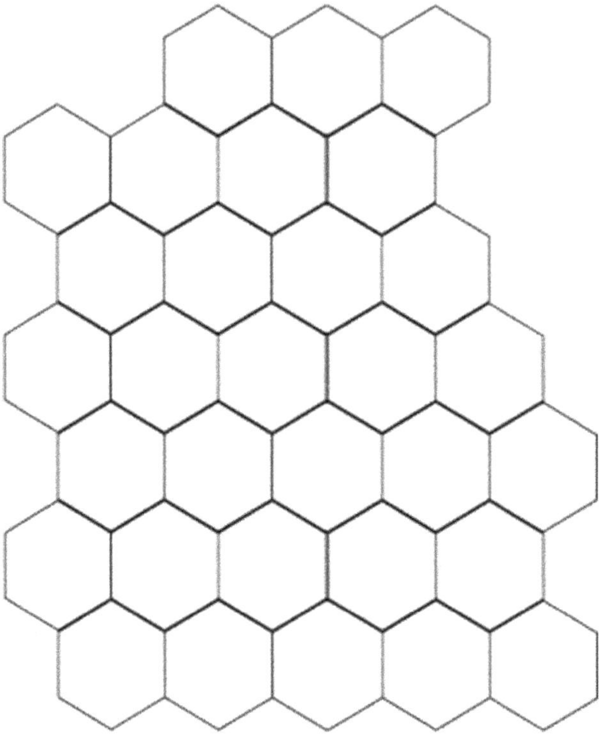

Figure 2.1. Graphene sheet. (Reproduced with permission from [101]. Copyright 2015 Taylor and Francis.)

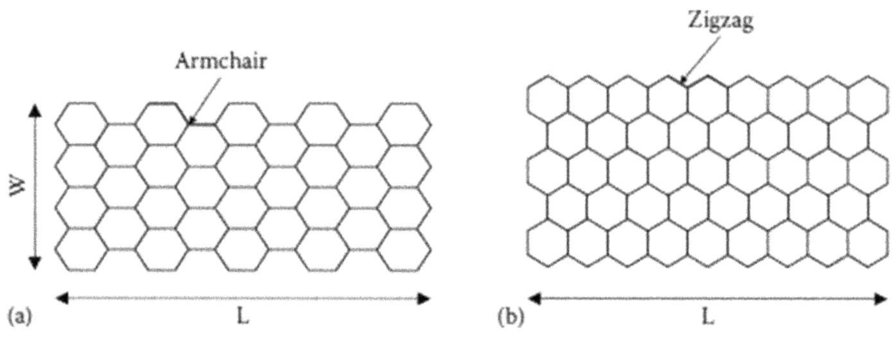

Figure 2.2. GNRs with (a) armchair and (b) zigzag chirality. (Reproduced with permission from [101]. Copyright 2015 Taylor and Francis.)

applications as interconnects. However, armchair GNRs are either metallic or semiconducting in nature, which mainly depends upon the geometry (chirality).

2.4 Carbon nanotubes (CNTs)

The accidental discovery of carbon nanotubes (CNTs) in 1991 by Sumio Iijima laid the foundation for carbon nanomaterials. CNTs are those carbon nanomaterials

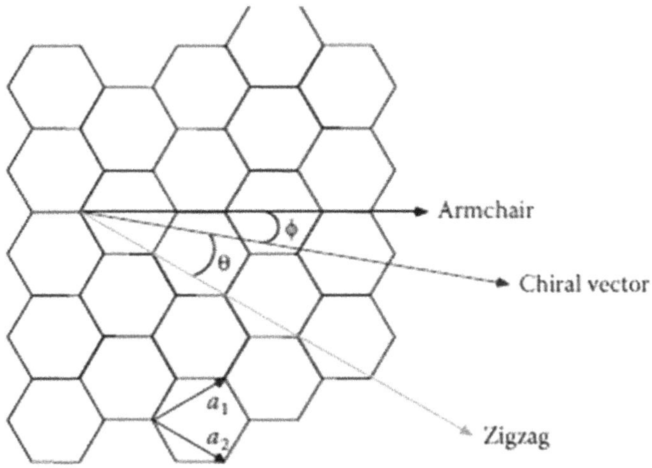

Figure 2.3. Schematic representation of a graphene sheet. (Reproduced with permission from [101]. Copyright 2015 Taylor and Francis.)

that are basically rolled up graphene sheets. These structures have been shown to demonstrate both metallic and semiconducting characteristics which depend mainly on their chirality or structure. In the applications of CNTs, those that are semiconducting in nature have been used in nanoelectronic devices, while metallic CNTs have been observed to be used in nanointerconnects. The structure of CNTs can be classified as single-walled or multi-walled. Single-walled CNTs (SWCNTs) are those nanotubes that are fabricated by rolling a graphene sheet to form a tubular structure, whereas multi-walled CNTs (MWCNTs) are fabricated by concentrically rolling up graphene sheets with multiple layers or they are concentrical rolled up SWCNTs in tubular structures. The diameter of SWCNTs has been observed to range from 0.7 to 5 nm, while the diameter of MWCNTs ranges from a few to tens of nanometers. Figure 2.3 depicts the schematic structure of a graphene sheet.

2.5 Properties of CNTs

The most vital properties of GNRs and CNTs are shown in table 2.1, which mainly consist of the electrical, thermal, and mechanical properties. 2D graphite is the fundamental source of both CNTs and GNRs. An allotrope of carbon, graphite is composed of hexagonal carbon atom lattices that are stacked in three dimensions. The term '2D graphite' or 'graphene' refers to a single layer of graphite. Each carbon atom in graphite has four valence electrons. In the plane, three of these electrons have the capability to interact strongly with nearby atoms to create a robust bond. The created covalent bonds are referred to as σ bonds with 2s, $2p_x$, and $2p_y$ orbitals. The $2p_z$ orbital, which is perpendicular to the plane, is formed by the fourth electron. The conductivity is unaffected and is not contributed to by electrons while in formation of σ bonds. The fourth electron from the $2p_z$ orbital is the only electron that participates in conduction. While in case of graphene, three σ bonds and a single

Table 2.1. Properties of CNTs and GNRs.

Property	Tungsten	Copper	SWCNTs	MWCNTs	GNRs
Maximum current density, J_{max} (A cm^{-2})	10^8	10^7	$>10^9$	$>10^9$	$>10^8$
Mean free path (nm) at 300 K	33	40	$>10^3$	2.5×10^4	1×10^3
Melting point (K)	3695	1357		3800	
Density (g cm^{-3})	19.25	8.94	1.3–1.4	1.75–2.1	2.09–2.33
Tensile strength (GPa)	1.51	0.22	22.2 +2.2	11–63	
Thermal conductivity ($\times 10^3$ W m^{-1} K^{-1})	0.173	0.385	1.75–5.8	3	3–5
Temperature coefficient of resistance ($\times 10^3$ K^{-1})	4.5	4	<1.1	−1.37	−1.47

$2p_z$ orbital are present in the structure. These bonds are perpendicular to the plane which is referred to as a π covalent bond.

2.6 Synthesis methods of graphene nanoribbons (GNRs) and carbon nanotubes (CNTs)

CNTs and GNRs are two recently discovered carbon materials that have provided technology with extensive new properties and applications. Graphene consists of carbon atoms which are organized in a hexagonal array while CNTs are rolled up graphene sheets shaped in the form of a tube or a cylinder. Both these materials have demonstrated exceptional properties and great potential in formulating the future in the field of large-scale integration (VLSI) circuits. However, the core bottleneck in the application of these structures in VLSI circuits is related to the appropriate growth technology and compatibility with the current silicon-based complimentary metal–oxide–semiconductor (CMOS) incorporated circuit technology. However, researchers have been working continuously to find the solution to these problems. With regard to this, some mitigation materials and special techniques for the synthesis of these materials are discussed later in this chapter.

2.6.1 Works related to GNR and CNT technologies

CNTs have various synthesis methods that are being researched continuously for modification, efficient operation, and suitable application. The early synthesis methods for CNTs consisted of laser ablation techniques and arc-discharge techniques. Both these procedures have been observed to fabricate SWCNTs and MWCNTs). The laser ablation technique has been observed to produce SWCNTs with very high purity (90%). However, scaling up or industrial production is not amenable. In the case of the arc-discharge method, the process can fabricate large quantities of SWCNTs but with the cost of limited purity. In the last few years,

CNTs have been synthesized effectively and efficiently using the technique of chemical vapor deposition (CVD). The fabrication of electronic devices, sensors, field emitters, and other applications where controlled growth over masked areas is necessary for other alternative processing are among the uses of CVD that have been mentioned. The ability of plasma-enhanced CVD (PECVD) to create nanotubes with a vertical alignment has been taken into consideration. The current prominent synthesis method of CNTs are the CVD method and arc-discharge method [1]. The arc-discharge method has the ability to synthesize high-quality CNTs consisting of nanotubes with diameters of 2–200 nm and a length of 1–100 μm [2]. However, disadvantages persists in interconnect applications when CNTs grown using the arc-discharge technique are used. The arc-discharge method has been observed to produce CNTs in soot form which consists of a mixture of various types of carbon materials. This soot of CNTs plus other materials needs proper purification, sonication, and substrate dispersion to obtain pure CNT structures. In addition, the synthesized nanotubes are fabricated through a random and slow process and are located on a flat substrate, horizontally. This limits the electron transport of nanotubes, through a Schottky barrier that is created at the metal–CNT junction, to side-contacted electrons. For its employment in VLSI, a more reliable and consistent architecture and structure for nanotube synthesis must be employed.

CNTs with exceptional electrical properties can be produced using the CVD technique and then the formed structures can be integrated into on-chip connection systems. With this approach, Delzeit *et al* [3] have demonstrated the synthesis of MWCNT structures on various substrates by employing the technique of thermal CVD using catalysts of multilayered metals. In another study conducted by Huang *et al* [4] the group demonstrated the synthesis of highly oriented MWCNTs using substrates of polished polycrystalline and single-crystal nickel. The technique of hot filament CVD exhausting a plasma-enhanced beam was used, which had temperatures below 666 °C. The group fabricated CNTs with diameters ranging from 10 to 500 nm and lengths ranging from 0.1 to 50 μm. However, these structure depend mostly on the growth conditions. In another experiment, well-aligned, uniform films of CNTs were produced using the microwave PECVD technique by Bower *et al* [5]. They demonstrated how nanotubes could be grown on curved surfaces with their alignment always being perpendicular to the surface of the local substrate. They were able to create multi-walled nanotubes that were 12 μm long and roughly 30 nm in diameter. The nanotube density was observed to be 4.4×10^9 nanotubes/cm^2. In another experiment, Yao *et al* [6] measured the properties of intrinsic high-field transport for metallic SWCNTs. They concluded that individual nanotubes have the capability to transport currents along with a density exceeding 109 A cm^{-2}. In an inductively coupled plasma reactor, Delzeit *et al* [7] produced high-density plasma from a methane–hydrogen combination so that the synthesis of MWCNTs on silicon substrates using multilayered Al/Fe catalysts could be performed. They were able to produce vertically aligned nanotubes where the alignment was superior to the frequently observed samples that are thermally produced. In another experiment using the technique of thermal CVD of an ethylenediamine precursor, Zhang *et al* [8] demonstrated a synthesis method of well-aligned CNTs with diameters ranging from

100 to 150 nm on silicon wafers that had nickel deposited on them. Vertically aligned structures with high density around the large areas on the surface of the silicon were observed for these CNTs. They discovered that the thickness of the nickel layer had an impact on the size and density of the nanotubes. By adjusting the CVD time, the nanotube array's length was managed. Similarly, Delzeit *et al* [9] utilized the technique of ion beam sputtering so that sequential deposition of multilayered metals can be performed on various substrates. This technique is applied to SWCNTs that are produced by CVD to tune its density. From this process, the diameter of the structure was achieved to be between 0.9 and 2.7 nm in which the average tubes had a diameter of 1.3 nm. Cui *et al* [10] proposed a molecular memory device fabricated sing semiconducting SWCNTs with a channel length of 150 nm. Sweeping gate voltages ranging from 3 V allowed them to store data, and the resulting storage stability was observed to be more than 12 days at normal temperature. In another experiment, Li *et al* [11] developed a bottom-up method to include MWCNTs into multilevel interconnects while building silicon integrated circuits. MWCNTs are vertically produced from designed catalyst spots using PECVD.

Discussing some other works related to carbon nanotubes, Li *et al* [12] studied the electric transport characteristics of a single vertical MWCNT *in situ* at room temperature using a scanning electron microscope chamber. Their observations demonstrate a multi-channel quasi-ballistic conductivity that occurs in the large-diameter MWCNTs. The involvement of numerous walls in electrical transportation and the huge diameter of the MWCNTs both contribute to this explanation. In another analysis, Chen *et al* [13] have provided an innovative method which employs microwave plasma chemical vapor deposition to generate free-standing, highly oriented and organized CNTs across two substrates. Two CNT-terminal contacts and a CNT bundle were combined to test the overall resistance and it was discovered to be around 14.7 kΩ. Studying the frequency response using CNTs, Gomez-Rojas *et al* [14] demonstrated a thorough analysis of the SWCNT's radio frequency (RF) response. After analyzing the S-parameters, they discovered an AC response over the frequency range from 30 kHz to 6 GHz. In the case of the electrical response, Rice *et al* [15] analyzed the electrical response shown by individual MWCNTs, which provided a value of 24 GHz. For kinetic inductance, Plombon *et al* [16] analyzed the kinetic inductance of a bundle of SWCNTs. They noted the dependence of the frequency impedance on the number of tubes, implying that the increment in the number of tubes will result in high-frequency impedance. Similarly, Close and Wong [17, 18] of Stanford University showed the electrical characterization and construction of interconnects based on MWCNTs. They investigated four types of materials for metal contacts (aluminum (Al), gold (Au), palladium (Pd), and titanium (Ti)) and discovered that Au and Pd are the finest contact metals. This work is mentioned in [19] as well. With the use of interconnects, Wu *et al* [20] reported a thorough electrical characterization associated with contact resistance in carbon based nanostructures.

Analyzing geometry based structures, Harutyunyan *et al* [21] through their analysis have shown the chiral-selective development of CNTs. Through this process they were able to tune the metallic fraction of the developed nanotubes by altering the ambient inert gas that is utilized during the thermal annealing process of the catalyst with the combination of reductive and oxidative species. This procedure helped the authors to formulate the metallic fraction of 0.91 in CNTs. In another analysis, Patil *et al* [22] demonstrated the growth of SWCNTs at a wafer size. CNTs from quartz wafers were moved toward silicon wafers and then the transferred CNTs were utilized to manufacture CNT field-effect transistors (CNTFETs). Analyzing the performances of CNTs in applications, Franklin *et al* and Chen *et al* [23] demonstrated that nanotube transistors operate as expected with excellent performance when their channel length is changed from 3 μm to 15 nm. Nanotube transistors did not exhibit the short-channel effects which are common in conventional metal–oxide–semiconductor (MOS) devices. They further showed the functionality of a nanotube transistor with channel and contact lengths of 20 nm, an on-current of 10 μA, an on/off current ratio of 1×10^5, and a peak transconductance of 20 μS. Li *et al* [24] provided a method to determine the bulk resistivity and contact resistance of vertically produced carbon nanofibers (CNFs). In another analysis, Chen *et al* [25] provided the first experimental findings regarding the monolithic integration of interconnects of graphene with CMOS technology, which has the reputation of industry standard. The results of the findings compared the functionality of MWCNT interconnects and high-speed on-chip graphene. Chai *et al* [26] demonstrated how a nickel layer catalyzed with an interfacial layer containing graphite has the capability to greatly improve electric contact to the CNT. Similarly, Ward *et al* [27] demonstrated a method to reduce the CNT interconnect sheet resistance. They showed that interconnects made with SWCNTs have lower resistance than those made with MWCNTs. They achieved a sheet resistance of 50 Ω sq^{-1} using a CNT functionalization and alignment technique. It is also projected that the sheet resistance can further be reduced to 10 Ω sq^{-1} using their techniques. Thus, analytical works of the synthesis and properties of CNTs have influenced the technological applications in the case of interconnects.

2.6.2 Modeling and analysis works on GNR- and CNT-based interconnects

In order to realize excellent applications of interconnects in nanoelectronics, modeling and analytical research on efficient and effective use, along with the investigation of the performance of carbon nanomaterial based interconnects needs proper discussion. In this light, Burke *et al* [28, 29] first proposed the electrical equivalent circuit model for CNTs using the Luttinger liquid theory. Salahuddin *et al* [30] presented a model for 3D conductors to 1D CNTs to develop a transmission-line model. Pop *et al* [31, 32] presented a temperature-dependent model for CNTs considering acoustic and optical phonon scattering. Srivastava and Banerjee *et al* [33] examined the viability of CNT bundles to be used as interconnects for VLSI circuits. They created a model of electrical equivalent for interconnects based on CNT bundles. Analyses of local interconnects, intermediate interconnects, and global interconnects were performed which showed that CNT bundles outperformed copper interconnects at global and intermediate lengths.

Raychowdhury and Roy *et al* [34] demonstrated and analyzed the performances of metallic SWCNT interconnects with the circuit-compatible resistance, inductance, and capacitance (RLC) model for metallic SWCNT interconnects. Along with this, a bias-dependent model of CNT resistance was also analyzed in their work. Naeemi and Meindl *et al* [35] also presented various models for MWCNTs. The authors developed a model to provide the number of conducting channels in MWCNTs based on the diameter of MWCNTs and temperature.

Naeemi and Meindl *et al* [36] analyzed the performance of interconnects at local, intermediate, and global lengths in which the interconnects were fabricated using SWCNTs. In another analysis, Naeemi and Meindl *et al* [37] further created a model which consists of MWCNT and SWCNT interconnects to analyze the resistance of the temperature coefficient. Furthermore, Naeemi and Meindl *et al* [38] studied the performance of power and signal interconnects based on SWCNT and MWCNT materials. Massoud and Nieuwoudt *et al* [39] presented a comprehensive modeling of CNT-based interconnects. Nieuwoudt and Massoud *et al* [40] modeled the magnetic inductance of the CNT bundle accurately using the method of partial element equivalent circuit (PEEC). Nieuwoudt and Massoud *et al* [41] predicted the resistance of the CNT bundle and created a diameter-dependent models of ohmic and contact resistance. Nieuwoudt and Massoud *et al* [42] further looked at how changes in the manufacturing process might affect upcoming SWCNT bundle-based interconnects. Ten different causes of process variability were modeled. Nieuwoudt and Massoud *et al* [43] studied the impact of kinetic and magnetic inductance on voltage overshoot and delay. Their research showed that the kinetic inductance needed in the future interconnect systems consisting of SWCNT bundles will not be sufficient. In another analysis, Nieuwoudt and Massoud *et al* [44] presented well known models and design techniques for CNT-based interconnects. Here, the authors accurately approximated the MWCNT bundle's coupling capacitances and electrostatic ground. Additionally, they discovered that SWCNT bundles are less susceptible to changes to their characteristics than large diameter MWCNT bundles. Haruehanroengra and Wang *et al* [45] analyzed the conductivity of combined SWCNT and MWCNT bundles. Their research demonstrates that by choosing the right parameters, such as metallic tube ratio, tube density, and bundle width, mixed CNT bundles may achieve conductivity improvements of two to five times over copper. Rossi *et al* [46] performed crosstalk analysis in CNT bus architecture. Their work showed that crosstalk-induced delay can be reduced up to 59% and the crosstalk-induced voltage peak can be reduced up to 81% using their proposed CNT bus architecture. Koo *et al* [47] compared the performance of optical interconnects and CNTs. They showed that at lower bandwidth density and switching activity lower than 20%, CNTs are the most energy efficient provided their mean free path (MFP) and packing density are improved. Li *et al* [48] studied the performance of MWCNT interconnects by using a compact model. They also compared the performance of interconnects based on MWCNTs to those of SWCNT-based and conventional copper-based interconnects. Through this comparison, they discovered that MWCNT interconnects have less delay than copper interconnects at the global and intermediate lengths.

Srivastava *et al* [49], for the first time, analyzed the use of interconnects based on SWCNTs in integrated circuits that are in the nanoscale region. They performed analysis of the CNT-based vias which demonstrated that the CNT bundles that are densely packed experienced a reduction in power by four times and eight times at the 22 nm and 14 nm technology nodes, respectively, when compared with copper interconnects. Chen *et al* [50] reported findings regarding the electrothermal characterization of a metallic interconnect array based on SWCNTs. They examined the effects of self-heating towards a signal integrity interconnect array based on SWCNTs. Pu *et al* [51] conducted a crosstalk study on interconnect systems based on bundles of single- and double-walled carbon nanotubes (SWCNTs and DWCNTs). According to their study, the SWCNT bundle counterpart will not be as well suited to the future generation of interconnect technology when compared to the DWCNT bundle-based interconnect. Li *et al* [52] reviewed the research that had been conducted on GNR and CNT interconnects. They demonstrated thermal and electrical model interconnects based on CNTs and GNRs. They also demonstrated that copper interconnects cannot compete with the performance of SWCNT, DWCNT, and MWCNT based interconnects. Furthermore, it was demonstrated that CNTs may be used in 3D integrated circuits as through-silicon vias (TSVs). Li and Banerjee *et al* [53] examined the effects that are experienced on the CNT interconnects at high frequencies. They created a method of frequency-linking impedance removal procedure for interconnects made of MWCNTs and SWCNTs. The authors further demonstrated that planar spiral inductors made of CNTs may achieve a Q factor that is more than three times greater than their counterparts made of copper without the need for any magnetic components or Q-factor development methodologies. Fathi and Forouzandeh *et al* [54], using CNT-based interconnects, implemented stability investigation and analysis based on the modeling of the transmission line and the Nyquist stability criteria. Similarly, Nasiri *et al* [55] completed the analysis of stability for GNR interconnects. For the investigation of signal propagation through bundles of SWCNTs and metallic and quasi-metallic SWCNTs, Sarto *et al* [56] suggested a hybrid transmission line–quantum mechanical model. The model is created using the time and frequency domains of an electron waveguide while accounting for the dampening effect brought on by electron scattering. Sarto and Tamburrano *et al* [57] developed an equivalent single-conductor (ESC) model that is used for MWCNT interconnects. Similarly, Sarto and Tamburrano *et al* [58] completed a comparative analysis of interconnects based on MWCNTs and multilayer GNRs (MLGNRs). The authors also analyzed the performance of radio frequency (RF) up to 100 GHz using these interconnects. D'Amore *et al* [59] carried out a fast transient analysis of interconnects based on MWCNT and SWCNT bundles. Pasricha *et al* [60] performed a system-level evaluation of the performance of CNT global interconnects. According to their experimental findings, global MWCNT buses can boost performance of the fabricated device while SWCNTs are less appropriate for global interconnect buses. Performance improvements over copper global connection buses are also seen when SWCNT and mixed bundles are used. Naeemi and Meindl *et al* [61] proposed a circuit model that is based on a compact for GNR interconnects. Xu *et al* [62] used the Landauer formula of linear response and a tight-binding model to develop a conductance model for GNRs. Additionally, they conducted delay analyses on GNR interconnects and contrasted the outcomes with those obtained from other interconnect materials, including CNT, copper, and others. Nasiri *et al* [55] investigated the stability of GNR interconnects. In their work, they evaluated the stability of MLGNR interconnects by using the parameter of Nyquist stability. Their study demonstrated that MLGNR interconnects stabilize with increasing length and breadth. Lee *et al* [63] examined the performance of graphene interconnects that were monolithically implemented on a prototype CMOS chip of 0.35 μm. They created graphene wires that were narrow and had a length up to 1 mm by further processing the huge graphene sheets using conventional lithography techniques. On these graphene wires, they achieved end-to-end data transfer with a bit-error rate (BER) under 2×10^{-10}. Sarkar *et al* [64, 65], through their analysis, provided a thorough and accurate approach for determining the high-frequency impedance of graphene structures that are important for low-loss on-chip inductors and on-chip interconnects. Yu *et al* [66] examined the critical reliability-limiting elements in a copper/bilayer graphene (BLG) hybrid interconnect system. The findings reveal that BLG has a remarkable current carrying capacity that is 100 times greater than that of copper. Yu *et al* [67] showed that stacked multilayer graphene (s-MLG) has the capability to be a novel material system for carbon based interconnects with enhanced performance. They demonstrated that when more layers are stacked, the wire sheet resistance of the s-MLG decreases and exhibits superior conductivity compared to the exfoliated MLG sample that consists of the same number of layers. Lee *et al* [68] used large-area MLG sheets fabricated by utilizing CVD and used these to create graphene wires. Multilayer graphene sheets typically have a thickness of 10–20 nm and a sheet resistance of 500–1000 Ω sq^{-1}. Cui *et al* [69] examined the properties of signal transmission in the case of MLGNR interconnects that use 14 nm and 22 nm technology nodes. In their research, a transient response is established together with an ESC model. Rakheja and Naeemi *et al* [70] differentiated the energy consumption and efficiency of graphene spin interconnects that are used in a nonlocal spin-torque (NLST) circuit from those of a CMOS circuit developed at the end of the silicon technological road plan [71]. Thus, the many modeling and analysis works performed by researchers and scientists have made the application of these material a less difficult task, as we can know the complications, characteristics, and properties of the modeled interconnects to be used in applications.

2.6.3 Works related to CNT- and GNR-based field-effect transistors

A transistor is an electrical device that uses semiconducting materials to regulate the flow of electrical signals by using the principles of both conduction and insulation of electrical voltage or current. Similarly, field-effect transistors (FETs) are those transistors that regulate the flow of electrical signals by applying voltage to a gate, which then changes the conductance of the source and drain. Silicon is used extensively in FETs as a semiconductive material but now CNTs and GNRs are also joining this race and to our surprise, they are leading this race. This is due to their excellent properties and their versatile nature. Analyzing FETs based on CNTs, Javey *et al* [72] fabricated a Schottky barrier CNTFET (SB-CNTFET) using nanotube metal junctions. The authors used semiconducting SWCNTs that displayed conductance similar to room temperature near the limit of the ballistic transport of $4e^2/h$ along with a high capacity of carrying current (~25 mA per tube). Similarly, Durkop *et al* [73] fabricated transistors using semiconducting CNTs with a channel length exceeding 300 μm. They developed nanotubes in a tube furnace at 900 °C. In another analysis, Hoenlein *et al* [74] discussed the applicability of CNTs in FETs. To support the applicability, Zhou *et al* [75] studied semiconducting SWCNTs in the diffusive transport regime. Similarly, moving toward another device fabrication, Guo *et al* [76] presented the physics for device fabrication using CNTFETs. They classified CNTFETs into two broad classes: (i) CNT MOSFETs, which are analogous to the conventional silicon MOSFET, and (ii) CNT MSDFETs, that are used as metal sources or drain FETs. In the case of ballistic transport, they self-consistently helped to solve the Schrödinger and Poisson equations by applying the non-equilibrium Green function (NEGF) equation.

Owing to their near ballistic electron transportation, semiconducting CNTs are taken into consideration to be used as a channel region for high-speed transistors. Three types of CNTFET structures have been proposed: band-to-band tunneling CNTFETs (BTBT-CNTFETs), MOSFET-like CNTFETs, and SB-CNTFETs. The CNTFETs' structure is comparable to that of the MOSFETs. Future nanoelectronic devices and circuits will find the CNTFET an interesting device, due to its low OFF current and operation of ballistic electron transport. It has been demonstrated in another work conducted by Raychowdhury *et al* [77] that MOSFET-like CNTFETs are superior to SB-CNTFETs. The projected MOSFET-like CNTFET [78–80] has been observed to have a probable scaled down channel length of 10 nm. Similarly, Deng *et al* and Wong *et al* [79, 80] suggested an equal circuit of the MOSFET-like CNTFET that is compatible with the simulation program with integrated circuit emphasis (SPICE) simulation tool. Sinha *et al* [81] have developed a compact modeling tool which uses non-iterative physics for the construction of CNT interconnects and transistors. The group demonstrated that, with the allowance for contact material variation, an SB-CNTFET with a range of diameter of 11.5 nm may operate with a circuit that is more than 8 times faster than a 22 nm CMOS. In another analysis, Lin *et al* [82] created static random access memory (SRAM) cells by utilizing CNTFETs that are similar to MOSFETs. The dual-chirality characteristic of the device was utilized in the design. It is observed that the CNTFET-based SRAM cell outperforms its CMOS counterpart in terms of temperature fluctuations, power supply voltage, and resistance to process. Similarly, Zhang *et al* [83] presented a design technique to overcome two major sources of CNTFET imperfections: metallic CNTs and CNT density variations. Thus, FET device fabrication using CNTs and GNRs is currently receiving a lot of attention.

2.6.4 Practical circuits

Practical circuits are those circuits that have the capability to force electrons to move against the opposing resistance using energy and, as a result, voltage manifestation is observed between two circuit points that have the resistance between each other. The first integrated stand-alone circuit, which combined individual CNT interconnect wires and silicon transistors on the same chip, was the subject of discussion in analysis conducted by Close *et al* [84]. The group reported that the operating frequency of the device is greater than 1 GHz. They designed an interconnect of MWCNTs placed on top of a chip of CMOS. The chip was produced by utilizing a 0.25 μm silicon CMOS that consists of around 11 000 transistors. The group interconnected various stages of traditional ring oscillators using MWCNTs as electrical cables. Thus, the fabrication of practical circuits was performed.

2.7 Modeling of GNR and CNT interconnects

Integrated circuits have two major components: semiconducting devices, such as diodes, or transistors, and interconnects. Interconnects are the films that are made up of thin metal sheets, used as a physical joining medium between various electrical nodes on a semiconductor chip. Interconnects help to transmit signals from one point to another without being distorted. Depending on the interconnects' length, they are classified into three types: (i) global interconnect (length > 100 μm), (ii) semiglobal or intermediate interconnect (10 μm > length > 100 μm), and (iii) local interconnect (length < 10 μm). In terms of technology, interconnects can be classified into three types. They are:

- Metallic interconnects.
- Optical interconnects.
- Superconducting interconnects.

Previously, aluminum served as the connection material in terms of metallic interconnect technology. However, copper eventually took the position of aluminum because aluminum demonstrated lower resistance than copper.

The number of constituents that are integrated into a chip also increases dramatically as very large-scale integration (VLSI) technology develops. The devices and the interconnects are miniaturized to a large extent. In state-of-the-art VLSI chips, the devices are placed so compactly that there is not enough space for the interconnects. Also due to the huge number of connections required in a chip, the interconnections are not possible in a single layer. Several layers of interconnects are required to complete all the required connections between the devices. In the multilayer interconnection structure, alternate layers of interconnects are drawn in a perpendicular direction. Low-k dielectric materials are used to isolate several metal layers. The vertical interconnection structure is known as a via. Vias helps to establish connections between alternate metal layers.

The interconnect's cross-section typically grows from the bottom to the upper metal layers. The bottom metal layers are used to create the local interconnects. The middle metal layers are used to create the intermediate interconnects, whilst the top metal layers are used to create the global interconnects. The vertical dimension of the wires exceeds the lateral dimension due to their unique cross-section. This allows for the placement of more interconnects in a given space.

Surface scattering and grain boundary scattering become more noticeable when the interconnect size is shrunk, and the bulk resistivity of the material increases considerably. For example, the bulk resistivity of copper, which is 1.7 μΩ-cm, is increased by 3–4 times in the sub-nanometer regime. Another significant issue is the vulnerability towards electromigration. The reduction of cross-sectional area is observed when the dimension of the interconnect is scaled down. At the same time, due to the huge integration of the devices, the possibility of simultaneous switching also increases. Therefore, the current density of the interconnects increases significantly. When the current density increases beyond a threshold, the electron knocks some of the host metal atoms to migrate from their original location. This phenomenon is known as electromigration. Electromigration causes the disruption of thin films of metal which will further form hillocks or create voids, ultimately leading to the creation of short circuits.

The interconnect delay currently contributes to 50% of the total path delay. The interconnects also cause problems in signal integrity and device reliability in addition to timing issues. A precise circuit equivalent model needs to be created for complicated interconnects in order to quantify the effects of the interconnects on circuit performance. The equivalent model consists of resistance, inductance, and capacitance (RLC). Resistance is mainly determined by the geometry of the interconnects only and does not depend on the distribution of the interconnects in its surroundings. In contrast, capacitance and inductance are strongly affected by the geometry and the distribution of neighboring conductors.

After the discovery of CNTs and graphene, there has been significant effort put toward their use as VLSI interconnects in the nanometer regime. Exceptional electrical, mechanical, and thermal characteristics are demonstrated by CNTs and graphene. The mean free path (MFP) of CNTs (λCNTs) is several times larger than that of copper (λCu $\approx$ 40 nm at room temperature). Thus, modeling of interconnects helps to visualize the performance of the device.

2.7.1 Modeling of graphene nanoribbon interconnects

In graphene, the carbon atoms are arranged in a honeycomb structure, as shown in figure 2.4. There are differences in the basic structure of graphene from which GNRs and CNTs are generated. The boundary conditions are what cause the differences. The wave function in CNTs is periodic along its circumference. However, in GNRs, the border of the two edges does not have the wave function.

The graphene sheet has an armchair or zigzag edge orientation depending upon the arrangement of the carbon atoms. In contrast to armchair GNRs, which have either metallic nature or semiconducting nature depending on the number of rings of carbon across the width, zigzag GNRs are always metallic. Zigzag GNRs are recommended for connection applications because of their metallic quality.

For the interconnects, a multilayer GNR structure has been suggested due to the extraordinary resistance of monolayer GNRs [62, 85]. Figure 2.5 illustrates a multilayer GNR structure modified for VLSI interconnects.

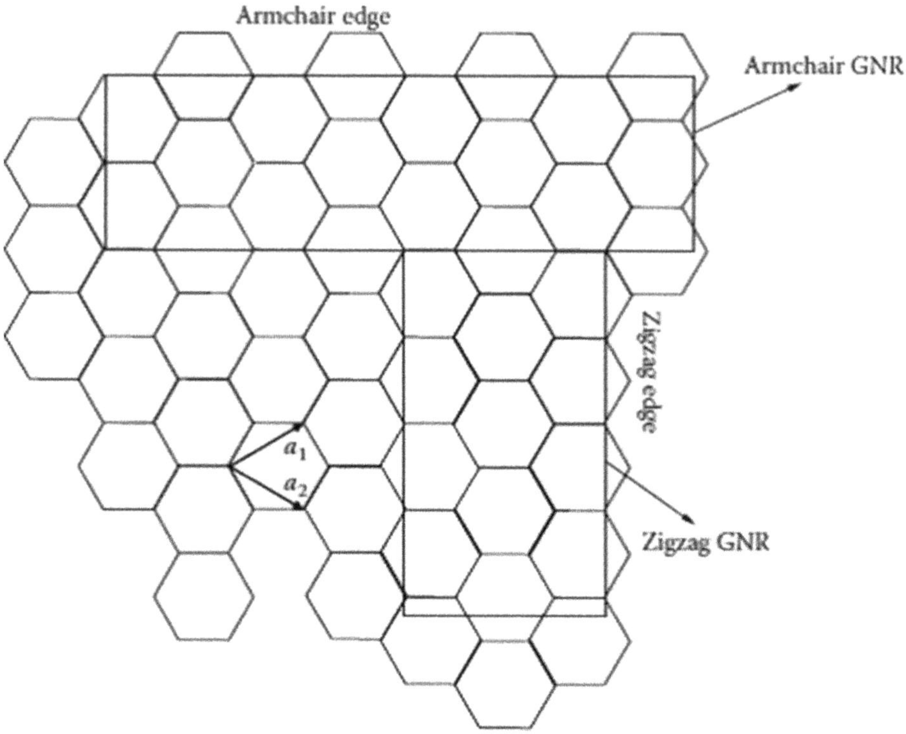

Figure 2.4. Lattice structure of a graphene sheet. (Reproduced with permission from [101]. Copyright 2015 Taylor and Francis.)

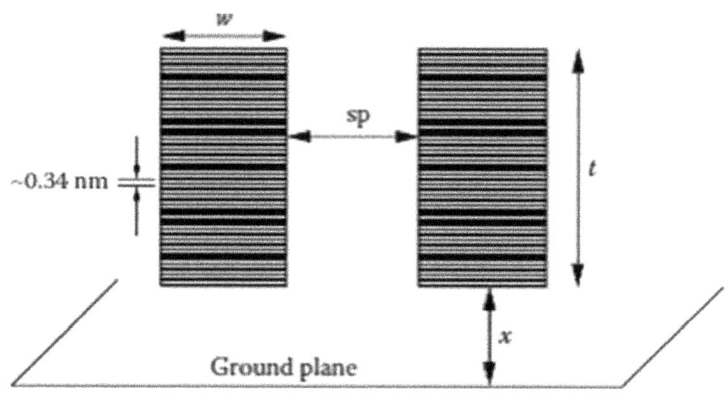

Figure 2.5. Multilayer GNR structure. (Reproduced with permission from [101]. Copyright 2015 Taylor and Francis.)

The interconnect dimensions in figure 2.5 are as follows: thickness (t), width (w), height (x) above ground plane, and spacing (sp) between interconnects. The van der Waals gap, which is the distance between each graphene layer, is δ (= 0.34 nm) [62]. According to the specifications provided by the International Technology Roadmap

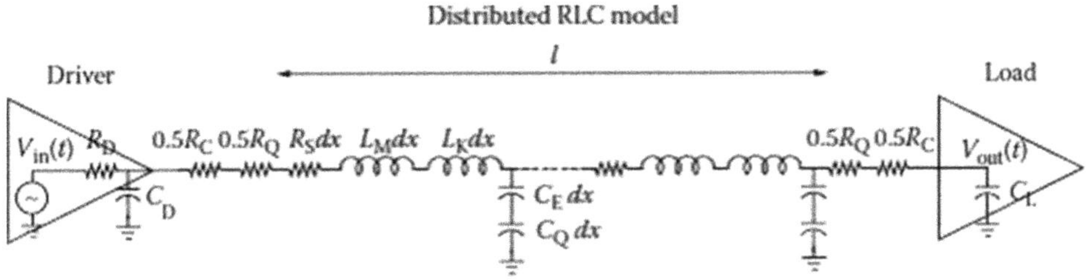

Figure 2.6. Schematic of the RLC interconnect model of GNR interconnects. (Reproduced with permission from [101]. Copyright 2015 Taylor and Francis.)

for Semiconductors (ITRS), the interconnect is thought to have the measurement of the 16 nm technology node [86].

Figure 2.6 depicts the schematic model of an interconnect [87] along with the driver and loads present at both interconnect ends. Also, modeling of an interconnect using a distributed RLC network is shown where RC = resistance due to imperfect contacts, RQ = quantum resistance, and RS (= RQ/λ) = scattering resistance p.u.l., where λ = MFP of electrons in the structure of GNRs. The quantum resistance is defined extensively in [62].

2.7.2 Applications of structurally uniform GNRs

2.7.2.1 Application of GNRs in biomedicine and toxicity

The application of GNRs in the biomedical field has been proven effective by many researchers. Olga *et al* discussed the application of GNRs in the field of biomedicine while keeping in mind the safety aspects as well. Several areas of application such as in chemical sensors, mechanical sensors, acoustic sensors, photosensors, DNA sequencing devices, tissue engineering, and drug and gene delivery vehicles, etc, are benefitted by GNR-based devices and materials. Enhanced compatibility was observed to be present in GNRs while using them in human cells but genotoxicity and cytotoxicity are some of the issues that have been experienced by researchers. Additionally, the effects are more severe for nanoribbons than for graphene oxide nanoplates. The mechanical destruction of the cell membrane, stimulation or production of reactive oxidative stress (ROS), proliferation inhibition and autophagy, apoptosis induction, fragmentation of DNA, and chromosomal aberration are some of the possible mechanisms that have been uncovered with the toxicity. Mitigation measures of this toxicity have also been researched extensively. In terms of other properties, biodegradability under environmental conditions was observed while using GNRs. Thus, it can be implied with the above discussion that GNRs have the capability to be used in therapeutic agents and high-precision nanodevices which provides extensive biomedical applications, but the safety measures and mitigation measures regarding the toxicity need further research. Research on ecological toxicity, migration, accumulation, and destruction inside ecosystems needs be done with proper emphasis [85].

2.7.2.2 Application of GNRs as potential on-chip interconnect materials

Modeling and analysis of interconnect materials have been discussed. The application part of using GNRs as materials for interconnects to be used on a chip should also be analyzed. Arnab *et al* undertook a comprehensive study in order to examine the possible use of graphene nanoribbons (GNRs) as one of the future on-chip interconnects. The electrical characteristics of the graphene nanoribbon are significantly influenced by its shape. The GNR zigzag edge is still determined to be semiconductive in nature whereas the GNR armchair edge exhibits semiconductive as well as metallic behavior depending on the amount of dimers. The GNR interconnect has been investigated thoroughly following the investigation of the structural and electrical properties, synthesis, and fabrication procedures. The modeling of GNRs from graphene sheets is a challenging operation since the conductivity of GNRs largely depends on the roughness and shape of its edges. The number of layers in the GNR affects its conductivity as well. High resistivity is displayed by single layers, and graphitic nature with low conductivity is displayed by several layers. The controlled production of GNRs is therefore of key interest for the progress of technology to replace traditional metal interconnects with GNRs, as only few layered GNRs are observed to be useful for interconnect applications. In order to link the structural characteristics of GNR interconnects with dispersed RLC networks to their electrical characteristics, electrical circuit modeling is investigated. The model demonstrates a high correlation between the GNRs' conductivity and width, particularly below a range of 20 nm. Due to scattering of charged impurity and phonon scattering, the mean free path (MFP) of suspended graphene on a silicon dioxide (SiO_2) substrate decreases drastically at about 100 nm [86]. Thus, the application of GNR in interconnect materials has helped to increase the performance of devices.

2.7.2.3 Application of GNRs as composite electrolyte membranes in direct methanol fuel cells

Fuel cells also have benefitted from GNRs as they can be incorporated as electrolytic membranes in the cells. In a simple one-step procedure using ultrasonication in chlorosulfonic acid, Avanish Shukla *et al* described the concept for the synthesis of graphene nanoribbon–graphene quantum dot (GNR–GQD) hybrids along with grafting of GQDs on GNR sheets. Further, the 4-benzenediazonium sulfonate precursor is prepared using the diazotization procedure, which sulfonates both GNRs and GQDs to create nanohybrids of sulfonated graphene nanoribbons–sulfonated graphene quantum dots (sGNR–sGQD). In order to create a nanocomposite membrane, nanohybrids are dispersed in sulfonated poly(ether ether ketone) (sPEEK), which has a synergistic and structure-dependent impact. With a 40% increase in peak power density and greater durability (up to 100 h), the sPEEK/sGNR–sGQD (1.5 wt%) nanocomposite membrane outperforms pure sPEEK and Nafion 117 in straight methanol fuel cells (DMFCs). This is due to the membrane's superior physical and chemical properties, such as proton conductivity, increased water uptake, decreased methanol crossing, and ion exchange capacity so that its potential can be realized when applied in a membrane [87].

2.7.2.4 Application of GNRs as semiconductors

GNRs have been used in electronic applications through the simple process of changing tool applications. In general, the produced GNRs in solution have functional groups on the side to increase the solubility of materials, but the solution also has involvement in π-conjugation backbones. Alkyl side chains have been used frequently as functional groups in construction. Electrical conductivity and solubility must be traded off because insulating side chains prevent intermolecular electron hopping. For solvent-soluble GNRs to be used in equipment and devices for electronics, these side chains must be removed. Mai *et al* and Feng *et al* have synthesized poly(ethylene oxide) PEO-functionalized GNRs at a liquid–solid interface. With the production of thin-film-GNRs, the development of thin-film-based field-effect transistors (FETs) on SiO_2/Si substrates were conducted [88–92]. Thus, semiconductor materials have been fabricated using GNRs.

2.7.2.5 Application of GNRs in the dielectrics field

Insulators also have a vital role in electronic devices. Discussing the application of GNR-based devices, Itami *et al*, Miyauchi *et al*, and Ito *et al* synthesized a cove-type GNR 109 with a width of ~0.71 nm. They used peripheral alkyl chains that allowed the GNRs to be soluble in several types of organic solvents. A study predicted that the GNRs of cove-type consist of partly flat bands at $2\pi/3 < |k| < \pi$. Also, two separated valleys in momentum space can be found that were analogous when compared with the zigzag-type GNRs. However, in the case of cove-type GNRs 109, the structures are thought to have unclear flat bands due to their ultra-narrow structure, mentioned in the example of the $N \leqslant 4$ zigzag-type GNRs [93–95].

2.7.2.6 Application of GNRs in electroluminescence devices

GNRs ($N = 7$) with armchair orientation have been proven effective for the fabrication of optical devices in which the electroluminescence liberated from them has been observed to be red. Similarly, Prezzi *et al* and Schull *et al* used the armchair orientated GNRs ($N = 7$) present at the tip of a Au (111) substrate and Scanning tunneling microscopy as well. Emission thus originates with the help of intra-GNR excitonic transitions that are concerned with the electronic states present at the terminal level of the GNRs. Yano *et al* have further discussed the applications of structurally uniform GNRs [96, 97].

2.7.2.7 Application of GNRs in electron transport

The theory of electron transport in graphene and GNRs has been analyzed, with a particular emphasis on the help that atomic pseudopotentials (self-consistent and empiric) provided in determining the band structure and other crucial transport factors such as electron–phonon matrix components and line-edge roughness scattering. The summary takes into account the potential application of graphene in devices with enormous scale integration technology [98] and includes the properties of electron–phonon scattering in the case of suspended graphene sheets, impurity and remote-phonon scattering in the case of supported and gated graphene, and electron–phonon and line-edge roughness scattering in the case of armchair-edge nanoribbons.

2.7.2.8 Application of GNRs in electrical property utilization

Due to their intriguing electronic characteristics, GNRs are currently being actively researched for potential electronic applications. Direct band gap creation in perfect ribbons with controlled gap size was provided through theoretical calculations and analysis. Additionally, zigzag edged GNRs were anticipated to have exceptional magnetic characteristics, providing a new material system for the application of the next generation of spintronic devices. More experimental research led to the development of many approaches for obtaining GNR structures in a fathomable nanometer regime, including top-down lithography, unzipping of carbon nanotubes, chemical sono-exfoliation, etc. In fact, it has been demonstrated that using nanowire masks, a GNR in the sub-10 nm range may be produced in a controlled manner. In general, the most recent synthesis method of GNR structures points to a feasible route of large-scale block copolymer lithography for the manufacture of semi-conducting graphene film. The permanent scalability of the block copolymer lithography technique, that is used to create the nano mesh devices, should make it possible to rationally design and produce graphene-based devices and circuits using conventional semiconductor manufacturing.

On the other hand, compared to what was anticipated in theoretical analysis of the ideal GNRs, the electrical transport investigations of experimentally constructed GNRs showed more complex transport behavior. The abnormalities (charge impurity and edge roughness) that caused Anderson localization and the Coulomb blockade effect were blamed for this disparity. However, there is no question that the parameters of disorder can provide a significant impact on the performance of the current GNR devices, and that the device performance can be increased with better tuning of the graphene nanostructures [99].

2.7.2.9 Application of GNRs for logic application devices

GNRs have been used in logic based devices. Juan *et al* presented the development of logic applications in GNR devices. A simulated study conducted on new carbon structures for nano-based devices along with an innovative synthesis method proved the usefulness of GNRs in electrical applications for digital circuit design. The amount and quality of innovative research works on graphene-based structures have inspired several researchers to fully work and progress in this field of material science. Several newer outcomes and articles have provided hopeful predictions for these graphene-based devices so that the incorporation of graphene into industry to realize its mind-blowing applications can be a reality. This will help scientists and researchers to gain access to materials and devices which have superior and unique properties to formulate exceptional performances. The advantages of new lucrative methods are sufficiently competitive to allow partial or complete transformations, but the current industrial processes will only allow the substitution of current materials for graphene [100].

2.8 Conclusions

Carbon based nanomaterials have helped to modernize the field of material science and engineering to a greater degree. The use of fullerene, carbon nanotubes, graphene, graphene nanoribbons, etc, has helped to achieve what was thought to be impossible in previous decades. These materials have fulfilled all the necessary demands that have been put forward by science and technology in the field of materials science. In this review chapter, we have discussed the current situation of carbon nanotubes and graphene nanoribbons. Current works on the synthesis and property modifications of carbon nanotubes and graphene nanoribbons are discussed thoroughly. Also, the modeling and analysis works related to carbon nanotubes and graphene nanoribbon based interconnects and field-effect transistors were studied extensively. The studies showed the possibility of using graphene nanoribbons and carbon nanotubes in applications of interconnects, practical circuits, and field-effect transistors. At the end of the chapter, applications that are linked with graphene nanoribbons are discussed. Thus, graphene nanoribbons and carbon nanotubes are considered as foundational pillars to be used in electronics and biomedical applications.

References

[1] Ngo Q, Petranovic D, Krishnan S, Cassell A M, Ye Q, Li J, Meyyappan M and Yang C Y 2004 Electron transport through metal-multiwall carbon nanotube interfaces *IEEE Trans. Nanotechnol.* **3** 311–7

[2] Iijima S 1991 Helical microtubules of graphitic carbon *Nature* **354** 56–8

[3] Delzeit L, Nguyen C V, Chen B, Stevens R, Cassell A, Han J and Meyyappan M 2002 Multiwalled carbon nanotubes by chemical vapor deposition using multilayered metal catalysts *J. Phys. Chem.* B **106** 5629–35

[4] Huang Z P, Xu J W, Ren Z F, Wang J H, Siegal M P and Provencio P N 1998 Growth of highly oriented carbon nanotubes by plasma-enhanced hot filament chemical vapor deposition *Appl. Phys. Lett.* **73** 3845–7

[5] Bower C, Zhu W, Jin S and Zhou O 2000 Plasma-induced alignment of carbon nanotubes *Appl. Phys. Lett.* **77** 830–2

[6] Yao Z, Kane C L and Dekker C 2000 High-field electrical transport in single-wall carbon nanotubes *Phys. Rev. Lett.* **84** 2941–4

[7] Delzeit L, McAninch I, Cruden B A, Hash D, Chen B, Han J and Meyyappan M 2002 Growth of multiwall carbon nanotubes in an inductively coupled plasma reactor *J. Appl. Phys.* **91** 6027–33

[8] Zhang W D, Wen Y, Li J, Xu G Q and Gan L M 2002 Synthesis of vertically aligned carbon nanotubes films on silicon wafers by pyrolysis of ethylenediamine *Thin Solid Films* **422** 120–5

[9] Delzeit L, Chen B, Cassell A, Stevens R, Nguyen C and Meyyappan M 2001 Multilayered metal catalysts for controlling the density of single-walled carbon nanotube growth *Chem. Phys. Lett.* **348** 368–74

[10] Cui J B, Sordan R, Burghard M and Kern K 2002 Carbon nanotube memory devices of high charge storage stability *Appl. Phys. Lett.* **81** 3260–2

[11] Li J, Ye Q, Cassell A, Ng H T, Stevens R, Han J and Meyyappan M 2003 Bottom-up approach for carbon nanotube interconnects *Appl. Phys. Lett.* **82** 2491–3

[12] Li H J, Lu W G, Li J J, Bai X D and Gu C Z 2005 Multichannel ballistic transport in multiwall carbon nanotubes *Phys. Rev. Lett.* **95** 086601–4

[13] Chen Z, Cao G, Lin Z, Koehler I and Bachmann P K 2006 A self-assembled synthesis of carbon nanotubes for interconnects *Nanotechnology* **17** 1062–6
[14] Gomez-Rojas L, Bhattacharyya S, Mendoza E, Cox D C, Rosolen J M and Silva S R 2007 RF response of single-walled carbon nanotubes *Nano Lett.* **7** 2672–5
[15] Rice P, Wallis T M, Russek S E and Kabos P 2007 Broadband electrical characterization of multiwalled carbon nanotubes and contacts *Nano Lett.* **7** 1086–90
[16] Plombon J J, O'Brien K P, Gstrein F, Dubin V M and Jiao Y 2007 High-frequency electrical properties of individual and bundled carbon nanotubes *Appl. Phys. Lett.* **90** 063106
[17] Close G F and Wong H-S P 2008 Assembly and electrical characterization of multiwall carbon nanotube interconnects *IEEE Trans. Nanotechnol.* **7** 596–600
[18] Close G F and Wong H-S P 2009 Measurement of subnanosecond delay through multiwall carbon-nanotube local interconnects in a CMOS integrated circuit *IEEE Trans. Electron Devices* **56** 43–9
[19] Patel-Predd P 2008 Carbon-nanotube wiring gets real *IEEE Spectr.* **45** 14
[20] Wu W, Krishnan S, Yamada T, Sun X, Wilhite P, Wu R, Li K and Yang C Y 2009 Contact resistance in carbon nanostructure via interconnects *Appl. Phys. Lett.* **94** 163111-1–3
[21] Harutyunyan A R, Chen G, Paronyan T M, Pigos E M, Kuznetsov O A, Hewaparakrama K, Kim S M, Za-kharov D, Stach E A and Sumanasekera G U 2009 Preferential growth of single-walled carbon nanotubes with metallic conductivity *Science* **326** 116–20
[22] Patil N, Lin A, Myers E R, Ryu K, Badmaev A, Zhou C, Wong H-S P and Mitra S 2009 Wafer-scale growth and transfer of aligned single-walled carbon nanotubes *IEEE Trans. Nanotechnol.* **8** 498–504
[23] Franklin A D and Chen Z 2010 Length scaling of carbon nanotube transistors *Nat. Nanotechnol.* **5** 858–62
[24] Li K, Wu R, Wilhite P, Khera V, Krishnan S, Sun X and Yang C Y 2010 Extraction of contact resistance in carbon nanofiber via interconnects with varying lengths *Appl. Phys. Lett.* **97** 253109-1–3
[25] Chen X, Akinwande D, Lee K-J, Close G F, Yasuda S, Paul B C, Fujita S, Kong J and Wong H-S P 2010 Fully integrated graphene and carbon nanotube interconnects for gigahertz high-speed CMOS electronics *IEEE Trans. Electron Devices* **57** 3137–43
[26] Chai Y, Hazeghi A, Takei K, Chen H-Y, Chan P C H, Javey A and Wong H-S P 2012 Low-resistance electrical contact to carbon nanotubes with graphitic interfacial layer *IEEE Trans. Electron Devices* **59** 12–9
[27] Ward J W, Nichols J, Stachowiak T B, Ngo Q and Egerton E J 2012 Reduction of CNT interconnect resistance for the replacement of Cu for future technology nodes *IEEE Trans. Nanotechnol.* **11** 56–62
[28] Burke P J 2002 Luttinger liquid theory as a model of the gigahertz electrical properties of carbon nanotubes *IEEE Trans. Nanotechnol.* **1** 119–44
[29] Burke P J 2003 An RF circuit model for carbon nanotubes *IEEE Trans. Nanotechnol.* **2** 55–8
[30] Salahuddin S, Lundstrom M and Datta S 2005 Transport effects on signal propagation in quantum wires *IEEE Trans. Electron Devices* **52** 1734–42
[31] Pop E, Mann D, Reifenberg J, Goodson K and Dai H 2005 Electro-thermal transport in metallic single-wall carbon nanotubes for interconnect applications *Proc. of the IEEE Int. Electron Devices Meeting (Washington, DC, December 2005)* pp 256–9

[32] Pop E, Mann D A, Goodson K E and Dai H 2007 Electrical and thermal transport in metallic single-wall carbon nanotubes on insulating substrates *J. Appl. Phys.* **101** 093710-1–10

[33] Srivastava N and Banerjee K 2005 Performance analysis of carbon nanotube interconnects for VLSI applications *Proc. of the IEEE/ACM Int. Computer Aided Design Conf. (San Jose, CA, November)* pp 383–90

[34] Raychowdhury A and Roy K 2006 Modeling of metallic carbon-nanotube interconnects for circuit simulations and a comparison with Cu interconnects for scaled technologies *IEEE Trans. Comput. Aided Des. Integr. Circuits Syst.* **25** 58–65

[35] Naeemi A and Meindl J D 2006 Compact physical models for multiwall carbon-nanotube interconnects *IEEE Electron Device Lett.* **27** 338–40

[36] Naeemi A and Meindl J D 2007 Design and performance modeling for single-walled carbon nanotubes as local, semiglobal, and global interconnects in gigascale integrated systems *IEEE Trans. Electron Devices* **54** 26–37

[37] Naeemi A and Meindl J D 2007 Physical modeling of temperature coefficient of resistance for single- and multi-wall carbon nanotube interconnects *IEEE Electron Device Lett.* **28** 135–8

[38] Naeemi A and Meindl J D 2008 Performance modeling for single- and multiwall carbon nanotubes as signal and power interconnects in gigascale systems *IEEE Trans. Electron Devices* **55** 2574–82

[39] Massoud Y and Nieuwoudt A 2006 Modeling and design challenges and solutions for carbon nanotube-based interconnect in future high performance integrated circuits *ACM J. Emerg. Technol. Comput. Syst.* **2** 155–96

[40] Nieuwoudt A and Massoud Y 2006 Evaluating the impact of resistance in carbon nanotube bundles for VLSI interconnect using diameter-dependent modeling techniques *IEEE Trans. Electron Devices* **53** 2460–6

[41] Nieuwoudt A and Massoud Y 2006 Understanding the impact of inductance in carbon nanotube bundles for VLSI interconnect using scalable modeling techniques *IEEE Trans. Nanotechnol.* **5** 758–65

[42] Nieuwoudt A and Massoud Y 2007 On the impact of process variations for carbon nanotube bundles for VLSI interconnect *IEEE Trans. Electron Devices* **54** 446–55

[43] Nieuwoudt A and Massoud Y 2007 Performance implications of inductive effects for carbon-nanotube bundle interconnect *IEEE Electron Device Lett.* **28** 305–7

[44] Nieuwoudt A and Massoud Y 2008 On the optimal design, performance, and reliability of future carbon nanotube-based interconnect solutions *IEEE Trans. Electron Devices* **55** 2097–110

[45] Haruehanroengra S and Wang W 2007 Analyzing conductance of mixed carbon-nanotube bundles for interconnect applications *IEEE Electron Device Lett.* **28** 756–9

[46] Rossi D, Cazeaux J M, Metra C and Lombardi F 2007 Modeling crosstalk effects in CNT bus architectures *IEEE Trans. Nanotechnol.* **6** 133–45

[47] Koo K-H, Cho H, Kapur P and Saraswat K C 2007 Performance comparisons between carbon nanotubes, optical, and Cu for future high-performance on-chip interconnect applications *IEEE Trans. Electron Devices* **54** 3206–15

[48] Li H, Yin W-Y, Banerjee K and Mao J-F 2008 Circuit modeling and performance analysis of multi-walled carbon nanotube interconnects *IEEE Trans. Electron Devices* **55** 1328–37

[49] Srivastava N, Li H, Kreupl F and Banerjee K 2009 On the applicability of single-walled carbon nanotubes as VLSI interconnects *IEEE Trans. Nanotechnol.* **8** 542–59

[50] Chen W C, Yin W-Y, Jia L and Liu Q H 2009 Electrothermal characterization of single-walled carbon nanotube (SWCNT) interconnect arrays *IEEE Trans. Nanotechnol.* **8** 718–28

[51] Pu S-N, Yin W-Y, Mao J-F and Liu Q H 2009 Crosstalk prediction of single- and double-walled carbon-nanotube (SWCNT/DWCNT) bundle interconnects *IEEE Trans. Electron Devices* **56** 560–8

[52] Li H, Xu C, Srivastava N and Banerjee K 2009 Carbon nanomaterials for next-generation interconnects and passives: physics, status, and prospects *IEEE Trans. Electron Devices* **56** 1799–821

[53] Li H and Banerjee K 2009 High-frequency analysis of carbon nanotube interconnects and implications for on-chip inductor design *IEEE Trans. Electron Devices* **56** 2202–14

[54] Fathi D and Forouzandeh B 2009 A novel approach for stability analysis in carbon nanotube interconnects *IEEE Electron Device Lett.* **30** 475–7

[55] Nasiri S H, Moravvej-Farshi M K and Faez R 2010 Stability analysis in graphene nanoribbon interconnects *IEEE Electron Device Lett.* **31** 1458–60

[56] Sarto M S, Tamburrano A and D'Amore M 2009 New electron-waveguide-based modeling for carbon nanotube interconnects *IEEE Trans. Nanotechnol.* **8** 214–25

[57] Sarto M S and Tamburrano A 2010 Single-conductor transmission-line model of multiwall carbon nanotubes *IEEE Trans. Nanotechnol.* **9** 82–92

[58] Sarto M S and Tamburrano A 2010 Comparative analysis of TL models for multilayer graphene nanoribbon and multiwall carbon nanotube interconnects *IEEE Int. Symp. on Electromagnetic Compatibility (EMC) (Fort Lauderdale, FL, July)* pp 212–7

[59] Sarto M S, D'Amore M and Tamburrano A 2010 Fast transient analysis of next-generation interconnects based on carbon nanotubes *IEEE Trans. Electromagn. Compat.* **52** 496–503

[60] Kurdahi F J, Pasricha S and Dutt N 2010 Evaluating carbon nanotube global interconnects for chip multiprocessor applications *IEEE Trans. Very Large Scale Integr. VLSI Syst.* **18** 1376–80

[61] Naeemi A and Meindl J D 2009 Compact physics-based circuit models for graphene nanoribbon interconnects *IEEE Trans. Electron Devices* **56** 1822–33

[62] Xu C, Li H and Banerjee K 2009 Modeling, analysis, and design of graphene nano-ribbon interconnects *IEEE Trans. Electron Devices* **56** 1567–78

[63] Lee K-J, Qazi M, Kong J and Chandrakasan A P 2010 Low-swing signaling on monolithically integrated global graphene interconnects *IEEE Trans. Electron Devices* **57** 3418–25

[64] Sarkar D, Xu C, Li H and Banerjee K 2011 High-frequency behavior of graphene-based interconnects. Part I: impedance modeling *IEEE Trans. Electron Devices* **58** 843–52

[65] Sarkar D, Xu C, Li H and Banerjee K 2011 High-frequency behavior of graphene-based interconnects. Part II: impedance analysis and implications for inductor design *IEEE Trans. Electron Devices* **58** 853–9

[66] Yu T, Lee E-K, Briggs B, Nagabhirava B and Yu B 2011 Bilayer graphene/copper hybrid on-chip interconnect: a reliability study *IEEE Trans. Nanotechnol.* **10** 710–4

[67] Yu T, Liang C-W, Kim C, Song E-S and Yu B 2011 Three-dimensional stacked multilayer graphene interconnects *IEEE Electron Device Lett.* **32** 1110–2

[68] Lee K-J, Chandrakasan A P and Kong J 2011 Breakdown current density of CVD-grown multilayer graphene interconnects *IEEE Electron Device Lett.* **32** 557–9

[69] Yin W-Y, Cui J-P, Zhao W-S and Hu J 2012 Signal transmission analysis of multilayer graphene nano-ribbon (MLGNR) interconnects *IEEE Trans. Electromagn. Compat.* **54** 126–32
[70] Rakheja S and Naeemi A 2012 Graphene nanoribbon spin interconnects for nonlocal spin-torque circuits: comparison of performance and energy per bit with CMOS interconnects *IEEE Trans. Electron Devices* **59** 51–9
[71] Wilson L 2006 *International Technology Roadmap for Semiconductors (ITRS) Reports* https://www.semiconductors.org/wp-content/uploads/2018/06/0_2015-ITRS-2.0-Executive-Report.pdf
[72] Javey A, Guo J, Wang Q, Lundstrom M and Dai H 2003 Ballistic carbon nanotube field-effect transistors *Nature* **424** 654–7
[73] Durkop T, Getty S A, Cobas E and Fuhrer M S 2004 Extraordinary mobility in semiconducting carbon nanotubes *Nano Lett.* **4** 35–9
[74] Hoenlein W, Kreupl F, Duesberg G S, Graham A P, Liebau M, Seidel R V and Unger E 2004 Carbon nanotube applications in microelectronics *IEEE Trans. Compon. Packag. Technol.* **27** 629–34
[75] Zhou X, Park J-Y, Huang S, Liu J and McEuen P L 2005 Band structure, phonon scattering, and the performance limit of single-walled carbon nanotube transistors *Phys. Rev. Lett.* **95** 146805-1–4
[76] Guo J, Koswatta S O, Neophytou N and Lundstrom M 2006 Carbon nanotube field-effect transistors *Int. J. High Speed Electron. Syst.* **16** 897–912
[77] Raychowdhury A, Keshavarzi A, Kurtin J, De V and Roy K 2006 Carbon nanotube field-effect transistors for high-performance digital circuits—DC analysis and modeling toward optimum transistor structure *IEEE Trans. Electron Devices* **53** 2711–7
[78] Wong H-S P 2014 *Stanford University CNFET Model* http://nano.stanford.edu/
[79] Deng J and Wong H-S P 2007 A compact SPICE model for carbon-nanotube field-effect transistors including nonidealities and its application. Part I: model of the intrinsic channel region *IEEE Trans. Electron Devices* **54** 3186–94
[80] Deng J and Wong H-S P 2007 A compact SPICE model for carbon-nanotube field-effect transistors including nonidealities and its application. Part II: full device model and circuit performance benchmarking *IEEE Trans. Electron Devices* **54** 3195–205
[81] Sinha S, Balijepalli A and Cao Y 2009 Compact model of carbon nanotube transistor and interconnect *IEEE Trans. Electron Devices* **56** 2232–42
[82] Lin S, Kim Y and Lombardi F 2010 Design of a CNTFET-based SRAM cell by dual-chirality selection *IEEE Trans. Nanotechnol.* **9** 30–7
[83] Zhang J, Patil N, Lin A, Wong H-S P and Mitra S 2010 Carbon nanotube circuits: living with imperfections and variations *Proc. of the Design Automation and Test in Europe Conf. and Exhibition (Dresden, 8–12 March)* pp 1159–64
[84] Close G F, Yasuda S, Paul B, Fujita S and Wong H-S P 2008 A 1 GHz integrated circuit with carbon nanotube interconnects and silicon transistors *Nano Lett.* **8** 706–9
[85] Zakharova O V, Mastalygina E E, Golokhvast K S and Gusev A A 2021 Graphene nanoribbons: prospects of application in biomedicine and toxicity *Nanomaterials* **11** 2425
[86] Arnab H and Basu S 2018 Graphene nanoribbon as potential on-chip interconnect material —a review *C J. Carbon Res.* **4** 49

[87] Shukla A, Dhanasekaran P, Nagaraju N, Bhat S D and Pillai V K 2019 A facile synthesis of graphene nanoribbon-quantum dot hybrids and their application for composite electrolyte membrane in direct methanol fuel cells *Electrochim. Acta.* **297** 267–80

[88] Narita A *et al* 2014 Synthesis of structurally well defined and liquid-phase-processable graphene nanoribbons *Nat. Chem.* **6** 126–33

[89] Mitoma N, Yano Y, Ito H, Miyauchi Y and Itami K 2019 Graphene nanoribbon dielectric passivation layers for graphene electronics *ACS Appl. Nano Mater.* **2** 4825–31

[90] Yano Y, Mitoma N, Matsushima K, Wang F, Matsui K, Takakura A, Miyauchi Y, Ito H and Itami K 2019 Living annulative π-extension polymerization for graphene nanoribbon synthesis *Nature* **571** 387–92

[91] Abbas A N, Liu G, Narita A, Orosco M, Feng X, Müllen K and Zhou C 2014 Deposition, characterization, and thin-film-based chemical sensing of ultra-long chemically synthesized graphene nanoribbons *J. Am. Chem. Soc.* **136** 7555–8

[92] Huang Y *et al* 2016 Poly(ethylene oxide) functionalized graphene nanoribbons with excellent solution processability *J. Am. Chem. Soc.* **138** 10136–9

[93] Nakada K, Fujita M, Dresselhaus G and Dresselhaus M S 1996 Edge state in graphene ribbons: nanometer size effect and edge shape dependence *Phys. Rev.* B **54** 17954–61

[94] Wakabayashi K, Takane Y, Yamamoto M and Sigrist M 2009 Effect on electronic transport properties of graphene nanoribbons and presence of perfectly conducting channel *Carbon* **47** 124–37

[95] Yamabe T, Tanaka K, Ohzeki K and Yata S 1982 Electronic structure of polyacenacene. A one-dimensional graphite *Solid State Commun.* **44** 823–5

[96] Chong M C, Afshar-Imani N, Scheurer F, Cardoso C, Ferretti A, Prezzi D and Schull G 2018 Bright electroluminescence from single graphene nanoribbon junctions *Nano Lett.* **18** 175–81

[97] Yuuta Y, Mitoma N, Ito H and Itami K 2020 A quest for structurally uniform graphene nanoribbons: synthesis, properties, and applications *J. Org. Chem.* **85** 4–33

[98] Fischetti M V, Kim J, Narayanan S, Ong Z-Y, Sachs C, Ferry D K and Aboud S J 2013 Pseudopotential-based studies of electron transport in graphene and graphene nanoribbons *J. Phys.: Condens. Matter* **25** 473202

[99] Bai J and Huang Y 2010 Fabrication and electrical properties of graphene nanoribbons *Mater. Sci. Eng.* R **70** 341–53

[100] Marmolejo-Tejada J M and Velasco-Medina J 2016 Review on graphene nanoribbon devices for logic applications *Microelectron. J.* **48** 18–38

[101] Das D and Rahaman H 2015 *Carbon Nanotube and Graphene Nanoribbon Interconnects* (Boca Raton, FL: CRC Press/Taylor and Francis)

IOP Publishing

Nanocarbon Allotropes Beyond Graphene
Synthesis, properties and applications
Arpan Kumar Nayak and Santosh K Tiwari

Chapter 3

Synthesis and application of hetero-atom doped graphene nanoribbons

Umer Mehmood and Rabia Nazar

Graphene nanoribbons (GNRs) are graphene sheets that are only a few nanometers thick. GNRs have additional advantages over graphene sheets due to their quasi-one-dimensional nature. For example, when used in conductive films and polymer composites, the high aspect ratio of GNRs lowers the percolation threshold and allows them to be spun into fiber from their liquid crystalline alignment. GNRs have a wide range of structural and physical properties depending on the method of synthesis. GNRs can be made using one of three methods: graphene cutting with lithography, unzipping of carbon nanotubes (CNTs), and bottom-up synthesis. Owing to their distinctive structure and characteristics, GNRs are used extensively in numerous applications, including biomedicine, nanoelectronics, and fuel cells. In this chapter, we will cover the fundamentals of GNRs, including their structure, properties, methods of synthesis, and applications.

3.1 Introduction

One of the most abundant elements present in the world is carbon. It qualifies as the most essential element of all living things [1]. A special feature of this element is its tendency to form various kinds of strong bonds with a large array of other elements. The hybridization of carbon in sp, sp^2, and sp^3 forms allows it to exist in different allotropic forms [2]. The three conventional allotropic forms studied and explored are diamond, coal, and graphite. However, by the end of the twentieth-century different procedures were adopted to synthesize special kinds of allotropes, e.g. graphene, carbon nanotubes (CNTs), fullerenes, etc. With the advent of nanotechnology, a specialized class of different types of nanostructured carbon has been developed, i.e. nanoribbons, nanowires, and quantum dots [3]. The special characteristics of carbon nanomaterials have paved the way for their use in electronics [5, 6], energy storage, medical [7], and engineering applications [8].

doi:10.1088/978-0-7503-5177-5ch3

Graphene was discovered in 2004, which took nanotechnology to the next level [4]. Geim and Novoselov were the first scientists who synthesized single-layered graphene [12]. A tremendous amount of research has been carried out utilizing graphene. It has a two-dimensional layered structure and can be used in high-performance devices for different energy-related applications [9]. It has emerged as the most outstanding material in this century due to its remarkable mechanical, optical, thermal, and electrical properties [10]. Focus has shifted to improving the morphological/structural features of graphene to exploit its performance in energy-related applications. In order to achieve this modification, several approaches are in use and under vigorous research globally, i.e. doping with different heteroatoms, the use of metal nanoparticles or metal oxides, and making its composite with different polymers [11].

With this new structural feature, a new horizon of nanotechnology applications has opened. In particular, the high mobility of the charge carrier of graphene is one of the most important features of electronic devices. However, the 2D structure of graphene lacks an electronic bandgap which hinders its use as the active layer in field-effect transistors (FETs). This is because these devices cannot be switched off as there is no bandgap [13]. This issue can be remedied by decreasing the dimensions of graphene, specifically through the production of quasi-zero-dimensional graphene quantum dots (GQDs) and quasi-one-dimensional graphene nanoribbons (GNRs) [15].

The quantum confinement of this nanoscale graphene is the source of the potential for this specialized class of material to be used in next-generation semiconductor devices with finite bandgaps. The doping of nanographene can help in the efficient tuning of their electronic and intrinsic properties. The doping leads to good electrochemical properties as it facilitates the chemical reactions at their interfaces [16]. Different atoms such as nitrogen and sulfur are being doped on graphene, GNRs, graphene oxides, reduced graphene oxides, and GQDs. A considerable amount of research is under way to explore their uses in mostly energy-related applications.

3.1.1 Graphene nanoribbons

Graphene nanoribbons are one of the available specialized classes of different nanographene materials. As the name suggests, they have arbitrary long lengths. They are thin slices of graphene sheets with widths in the range of nanometers to tens of nanometers. They are considered one-dimensional nanomaterials due to their high aspect ratio. This is a comparatively new class of nanographene and, based on its structure, it can have metallic or semiconducting characteristics.

GNRs are categorized into two types based on their end pattern:

- Armchair graphene nanoribbons (AGNRs).
- Zig zag graphene nanoribbons (ZGNRs).

Another type, with an alternative end pattern having zig zag and armchair edges, is designated as chiral graphene nanoribbons (CGNRs). The chiral angle is not directly proportional to the zig zag segment's length of the nanoribbons. The orientation of these three types is shown in figure 3.1. The properties of these GNRs are different due to a difference in their electronic edge states. In the case of ZNGRs,

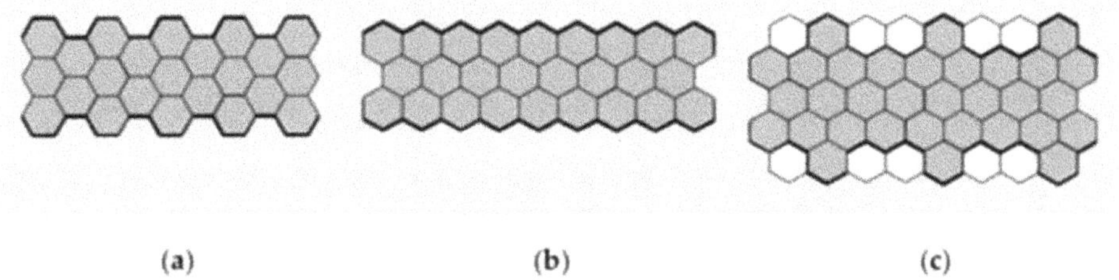

Figure 3.1. Different types of GNRs: (a) armchair, (b) zig zag, and (c) cove. (Reproduced from [14]. CC BY 4.0.)

a non-bonding π state exists, which acts as an electronic conduction network. Due to this channel of electronic conduction, ZGNRs exhibit magnetic behavior. However, no such state is present in AGNRs. Based on these electronic edge differences, a metallic behavior is shown by ZGNRs. However, depending on the width of the AGNRs, they can exist in two states, i.e. metallic or semiconducting [17]. It has been observed that the electronic behavior of GNRs is highly dependent on any kind of structural modification. The structural defects present in planar ZGNRs make them a strong candidate for use in bio-diagnostics, nanoelectronics, biosensors, and spintronics. The width of AGNRs and ZGNRs plays a significant role in dictating their end-use properties. Their widths can be represented by N_a and N_z, respectively. This number shows the number of carbon atoms in the width of the ribbon we choose. The nanocomponents can be defined as N_a-AGNR and N_z-ZGNR. In the case of AGNRs, the electronic structure is dependent on the width and can be classified into three sub-families [18].

These three families of AGNRs behave as semiconducting if they are narrower than 10 nm, and the bandgap increases with the decrease in their widths. However, the bandgaps of AGNRs in each subfamily having the same n values follow the trend $3n + 2 \ll 3n < 3n + 1$. With an increase in the bandgap and effective mass of AGNRs, the charge carrier mobility is observed to be low [19]. It is very important to achieve an optimum value of bandgap and charge carrier mobility of AGNRs, which can be carried out by varying their widths.

On the other hand, ZGNRs have localized electronic edge states and smaller bandgaps. A theoretical prediction states that the edge states of these types of nanoribbons can be spin-polarized. This feature makes them potential materials for spintronic applications. As an illustration, figure 3.2 [20] can be referred to for counting the number of chains for 9-AGNR and 6-ZNGR structures.

3.2 Synthesis mechanism of GNRs

Two methods are known for the synthesis of graphene nanoribbon: top-down and bottom-up. The top-down method involves the breaking down of large carbon structures into nanoscopic forms. Creating a composite structure by collecting small building blocks is known as the bottom-up approach (shown in figure 3.3) [21]. The bottom-up method is known to be more advantageous, as it has more control over

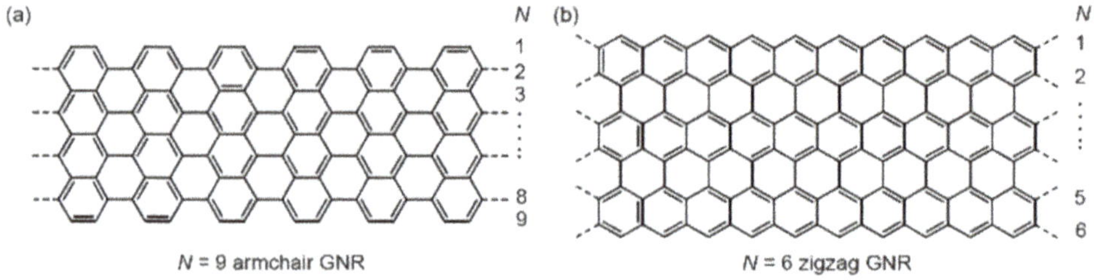

Figure 3.2. The finite width honeycomb structure of GNRs. (a) The lattice of a 9-AGNR. (b) The lattice of a 6-ZGNR. The dashed lines constituting a box show the unit cell. Hydrogen atoms are represented by the open circles at the edges. (Reproduced from [20]. CC BY 4.0.)

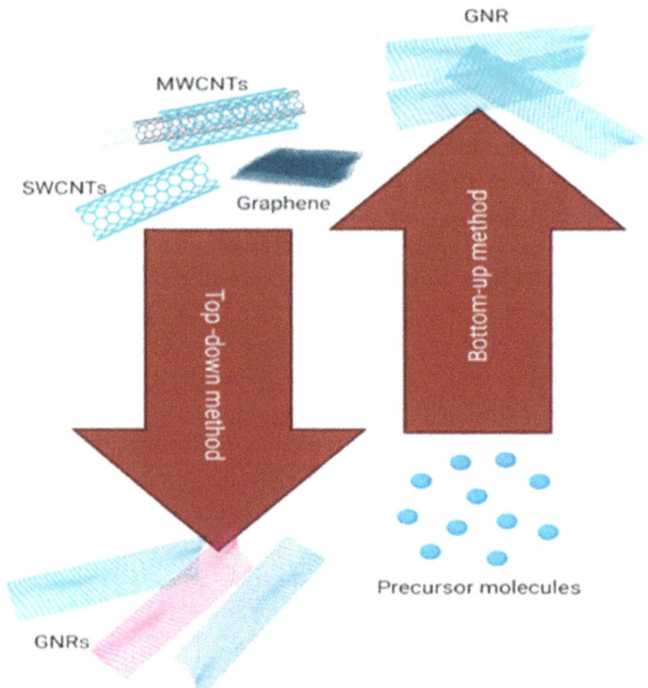

Figure 3.3. Synthesis of GNRs using top-down and bottom-up techniques. (Reproduced with permission from [21]. Copyright 2020 Elsevier.)

the product obtained at the end. The graphene nanoribbons synthesized with this technique have better structural properties, and fewer or no edge abnormalities. The polydispersity index of these ribbons is also very low. This technique is more attractive as it results in outstanding electronic, magnetic, and optical properties in GNRs obtained. This technique does not give a high yield. The benefit of the top-down approach is the ability to achieve high yield and the production of these nanoribbons in micrometer length. This method includes the breaking up of larger parent particles, so a high production rate is possible [22]. However these are complicated procedures and are not easy to control. The graphene nanoribbons obtained are also not of constant width and edges [23]. Hence, the bottom-up approach to produce GNRs is easier, and the product obtained from this process has well-defined edges [24].

3.2.1 Top-down synthesis

Top-down synthesis entails lithographic patterning, unzipping of carbon nanotubes, sonochemical cutting, and metal-catalyzed cutting and etching of graphene.

3.2.1.1 Lithographic patterning

This method involves the etching of graphene flakes, graphene on a metal surface by CVD, or epitaxial graphene on a silicon carbide substrate. GNRs can be synthesized by the lithography patterning technique. In order to do so, a graphene sheet is patterned with a mask. By applying oxygen plasma, the surface of the graphene can be etched away except for the surface covered by the mask [25, 26]. The nanoribbons in this method are almost 10 nm in width [27]. A schematic of the process is shown in figure 3.4(a). A nanowire mask is patterned on the surface, and the image of the

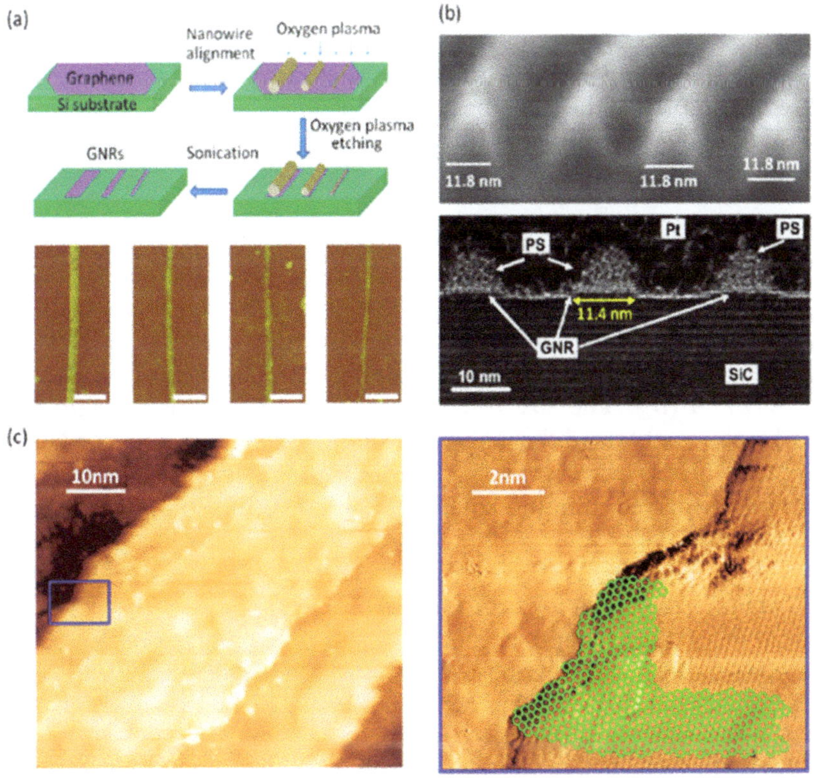

Figure 3.4. Lithographic patterning. (a) Oxygen plasma etched GNRs. Reprinted with permission from [28]. Copyright (2009) American Chemical Society. (b) Microscopic images (SEM and TEM) of GNRs obtained through lithography. (c) GNRs patterned on SiC substrate, the zoomed-in area studied with the help of STM. (b) and (c) reprinted with permission from [29]. Copyright (2012) American Chemical Society.

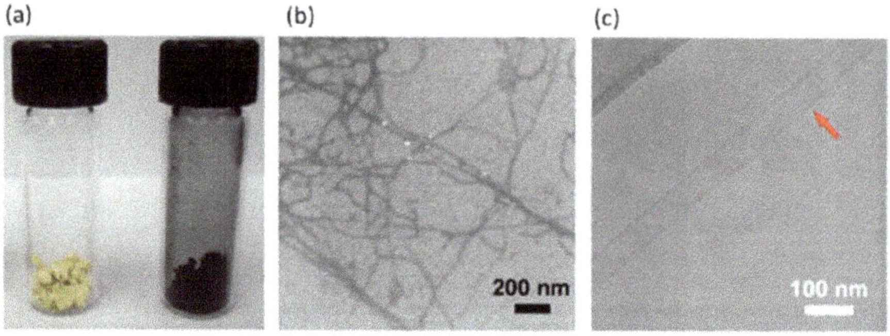

Figure 3.5. Chemical methods to synthesize GNRs. (a) GNR powder as a black powder. (Reproduced with permission from [31]. Copyright 2014 Springer Nature.) (b) SEM image of the dispersed GNRs in a solution. (c) TEM image of a single GNR.

resulting nanoribbons is captured by a scanning tunneling microscope (STM) after the removal of the mask. The images taken by scanning electron microscope (SEM) and transmission electron microscope (TEM) give a clear image of nanoribbons obtained of approximately 10 nm in width. The quality of the edge is important to know the further details of the obtained product. For this purpose, STM is a better option. Figure 3.5(c) reveals a mix of armchair and zig zag patterns as highlighted in green. The instability and lack of control of the electron beam result in rough edges. This is a problem with this technique. Well-defined and smooth edges are very difficult to obtain by employing this technique. Recent progress has reported relatively higher mobility, i.e. 3500 $cm^2\ V^{-1}\ s^{-1}$.

3.2.1.2 Chemical methods

Better control of the size distribution of graphene nanoribbons is obtained with chemical reactions. Appropriate solvents are needed in this method in which the graphite-based precursors can undergo a chemical reaction. GNRs obtained through this technique are in a powder or a liquid form (as shown in figures 3.5(a) and (b)). The size distribution of the ribbon is around $\geqslant$ 1 nm. A network of ribbons obtained through this technique is shown with the help of TEM (figure 3.5(c)). THz spectroscopy is used to find the mobility of charge carriers of the ribbons dispersed in liquid solution. The range of mobility is reported to be 150–15 000 $cm^2\ V^{-1}\ s^{-1}$ [30].

3.2.1.3 Graphene cutting with catalytic particles

The dissociation of carbon bonds is another way of producing nanoribbons. Utilizing a catalytic process, the graphene nanostructure is excised from a graphene flake. This is done by depositing catalytic metal particles on the surface of a graphene flake. Metallic particles such as iron or nickel are used to etch a graphene flake in a hydrogen environment. The interaction of metals with graphene results in the by-product formation of gases such as methane. The essential condition of obtaining nanoribbons is by cutting the paths of particles parallel to each other. This method produces graphene nanoribbons with smooth and well-defined edges, almost 10 nm wide [33].

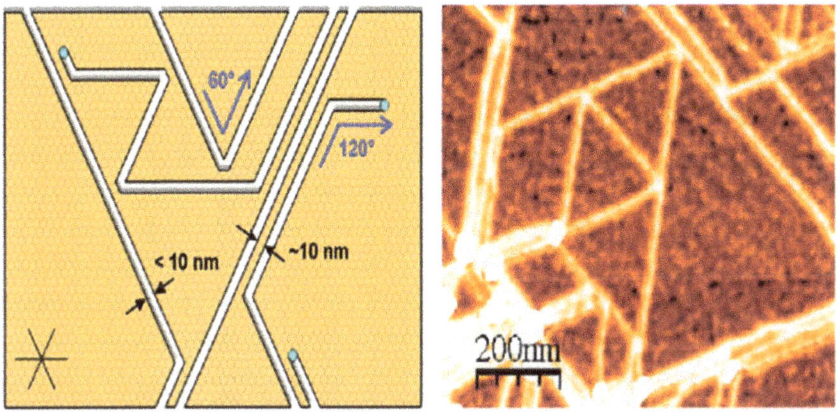

Figure 3.6. GNR synthesis by cutting with catalytic particles. (Reproduced with permission from [33]. Copyright 2009 American Chemical Society.)

In figure 3.6(a) the direction of cutting the paths is shown. The AFM image (figure 3.6(b)) represents the graphene surface when exposed to metal particles. The problem with this technique is related to the cutting direction. This is not known beforehand, so many graphene nanostructure shapes are obtained. Thus further experimental work is needed to control the shapes of nanoribbons achieved by this process.

3.2.1.4 Unzipping of carbon nanotubes

Unzipping can be accomplished using a variety of processes, including oxidation cutting, plasma etching, laser-induced unzipping, STM tip unzipping, electrical unwrapping, and intercalation exfoliation [34, 35]. SWCNT and MWCNT unzipping for GNR synthesis is by far the most frequent top-down synthesis method for generating graphene nanoribbons, but other top-down synthesis methods exist.

The most important characteristic of this technique is the use of CNTs rather than a graphite flake as a precursor. The starting materials are mostly MWCNTs having diameters in the range of 40–80 nm. The nanotubes are treated with a strong acid, e.g. concentrated sulfuric acid, and potassium permanganate ($KMnO_4$) is used as an oxidizing agent. The process is carried out at room temperature and followed by heating at around 55 °C–70 °C. The schematic of the process is given in figure 3.7. The resulting nanoribbon obtained with the unwrapping of CNTs is estimated to be 4 mm long, 100–500 nm wide, and have thicknesses of 1–30 graphene layers. The oxidation of C=C (double bonds of carbon) results in the unzipping of these CNTs. The products obtained are soluble both in water and in polar organic solvents. The GNRs obtained have great potential to be used in electronic devices.

The problem with this synthesis technique is the introduction of oxygen-containing groups during the oxidation process. This increases the structural defects in the resulting nanoribbons. Another drawback is related to the higher width of the nanoribbons produced. Chemical reduction methods are used to recover the electrical properties of graphene nanoribbons [36].

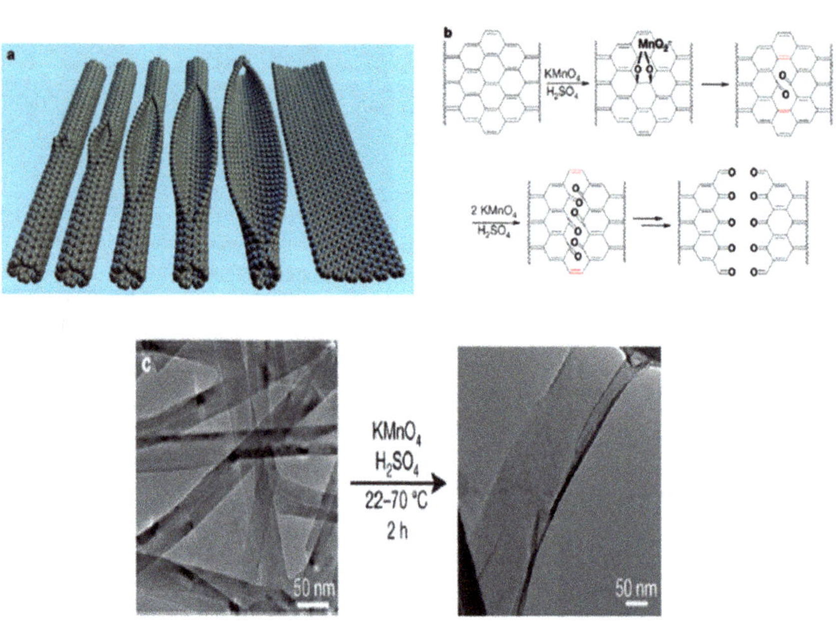

Figure 3.7. (a) Unzipping of MWCNTs. (b) The mechanism proposed for unzipping. (c) TEM images of before and after unzipping. (Reproduced with permission from [36]. Copyright 2009 Macmillan Publishers Limited.)

3.2.2 Bottom-up approaches

So far, three kinds of bottom-up synthesis have been reported:

i. Conversion of molecular precursors inside nanotubes.
ii. Chemical vapor deposition.
iii. Building block coupling.

The first two methods are discussed briefly here.

3.2.2.1 Conversion of molecular precursors

A reaction of monomers is carried out at the surface of catalytic metals to synthesize graphene nanoribbons. Different monomers can be used to produce GNRs with this technique. The first step is the reaction of a monomer, e.g. DBBA (10,10′-dibromo-9,9′-bianthryl) or its derivatives at the surface of metals at around 200 °C to produce polymer chains. The next step is the annealing at a higher temperature (400 °C) to dehydrogenate the polymer chains which results in the fabrication of graphene nanoribbons (shown in figure 3.8) [37]. The electronic properties of prepared GNRs can be estimated with photoemission experiments. Figure 3.8(b) shows the electronic states of less dispersed armchair graphene nanoribbons, having 1 eV binding energy. It shows the localized nature of these nanoribbons. STM can be employed to obtain additional property data, as shown in figure 3.8(c) [38] the edges can be observed to change from the armchair to zig zag pattern. Although monomer precursors are

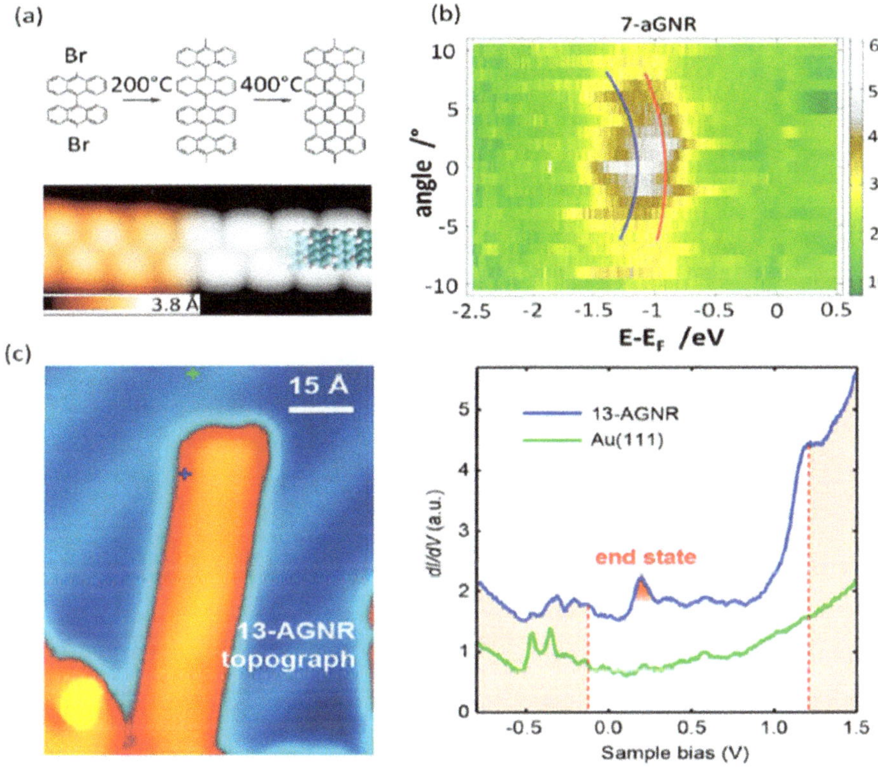

Figure 3.8. Synthesis of GNRs from molecular precursors. (a) Polymeric chain assembly. STM image of armchair GNRs. (Reproduced with permission from [39]. Copyright 2010 Springer Nature.) (b) Electronic chain of armchair GNRs. (Reproduced with permission from [40]. Copyright 2012 the American Physical Society.) (c) STM image of an armchair nanoribbon and dI/dV spectra on the right side. (Reproduced with permission from [41]. Copyright 2013 American Chemical Society.)

grown just on the catalytic metals' surfaces, the GNRs can be synthesized with controlled specifications.

3.2.2.2 Chemical vapor deposition

This is very similar to the molecular precursor-based conversion method. The only difference is that it is a single step method. A metallic surface is used as a catalyst at which the decomposition of gases such as methane is carried out at 700 °C–1000 °C temperature. The graphene is produced with the dissociation of carbon hydrogen bonds. The size of nanoribbons depends on the template on which the reaction occurs [42, 43]. Schematics of the growth of nanoribbons are shown in figure 3.9. On the SiO_2/Si substrate, a nickel nanobar is initially evaporated. Then ethylene is exposed to the substrate at high temperature, which initializes the growth of nanoribbons on the nickel nanobar. Here also the size and shape of the nanoribbons depends on the template used to produce them. An image of graphene nanoribbons seeded on nickel nanobars of different sizes is shown in figure 3.9.

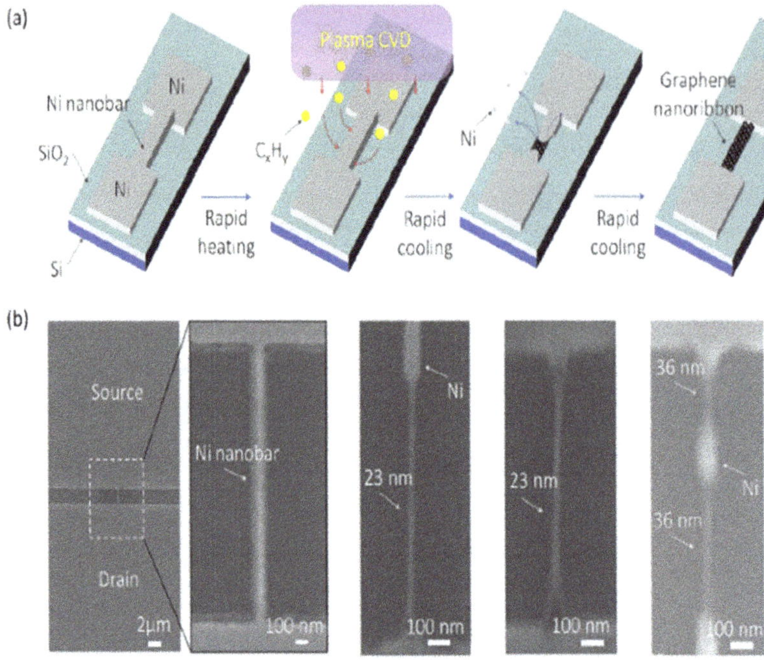

Figure 3.9. Chemical vapor deposition. (a) Schematics of the CVD process to synthesize GNRs. The seeding of graphene starts with this reaction, ultimately transforming into graphene nanoribbons. (b) SEM images of GNRs synthesized with this process. (Reproduced with permission from [44]. Copyright 2012 Springer Nature.)

3.3 Applications of graphene nanoribbons

GNRs are narrow, enlongated, single-layer graphene strips. GNRs are extraordinary nanomaterials of the graphene family due to their exceptional physical, chemical, electrical, mechanical, thermal, and optical capabilities. They have an exceptionally large surface area. Due to their distinctive structure and features, GNRs are utilized in a wide variety of applications, for example:

i. Nanoelectronics.
ii. Biomedicine.
iii. Catalysts.

3.3.1 Applications of GNRs in nanoelectronics

GNRs are considered one of the most promising models for future nanoelectronics. GNRs are good candidates for quantum nanoelectronics applications because they have high mobility and the ability to carry current, a sizeable bandgap, and a variety of electronic properties [45, 46]. GNRs synthesized from the bottom up are a far more promising platform for making devices and doing research because they can have their electrical properties changed by manipulating their atomically specific structures [47]. GNRs are often used in field-effect transistors (FETs), Schottky diodes, P–N junctions, liquid crystals, and transparent conductive electrodes.

Schottky barriers (SKBs) at the electrodes control most of the flow of electrons. This is likely because the GNRs have a narrow width and a large bandgap. Still, devices with n-type characteristics were found to have a high on/off ratio of 3.6×10^3 when measured in a vacuum. Low-bandgap 9- and 13-AGNRs were studied to eliminate SKBs and increase device performance [48]. Under UHV and sub-monolayer coverage, both GNRs were grown on Au(111)/mica surfaces to produce isolated GNRs for device fabrication. GNR-based devices with metal electrodes of 20 nm and a gate dielectric of 50 nm SiO_2 had ambipolar transport behavior. Instead of thermionic emission through the SKB, the devices were observed to use tunneling transport. This was determined by how little the current–voltage characteristics changed with temperature [47]. Also, the normalized conductance was expected to reach 20 mA m^{-1} in devices with closely spaced GNRs and bandgaps of 1 eV, where the SKB might be completely stopped [48]. FET applications could benefit greatly from UHV-fabricated GNRs, according to these results. Adjustments to the GNR bandgap and device set-up will be necessary to further optimize GNR FET performance. An *in situ* electronic measurement method utilizing an STM tip and an Au (111) substrate was invented as a result of the difficulties in fabricating GNR-based devices [49].

In the last few decades, there has been an urgent need to find ways to store energy that are reliable, efficient, and good for the environment [50, 51]. Micro-supercapacitors (MSCs) have a great deal of potential to supplement or even replace microbatteries due to their high rates of charging/discharging cycles, lifetimes, high capacitances, as well as high power densities. The most favorable aspects of graphene, including its large specific surface area as well as its superior mechanical and electrical properties, are integrated in the recently developed graphene-based MSCs [52]. Additionally, the normalized open-edge capacitance of graphene-based electrodes is higher than that of basal plane electrodes [53]. In this way, GNRs, which are strips of graphene that are one nanometer wide, could be useful for MSC applications. Recent studies have shown that chemical vapor deposition (CVD) can be used to make GNRs quickly and on the surface [22]. The inherent mobilities of CVD-grown $N = 9$ armchair GNRs were 350 $cm^2\ V^{-1}\ s^{-1}$, resulting in strong photoconductivity [54]. This shows that GNRs could be used as new materials for electrochemical storage applications that need a high charge mobility. Bottom-up produced GNR films were used for the first time as MSC electrode materials by Zhaoyang *et al* [55]. They found that the microdevice possessed an extremely high power density in addition to an outstanding volumetric capacitance. According to pump probe terahertz spectroscopy experiments, the electrochemical performance of MSCs may relate to the charge carrier mobility within the variously applied GNRs. Another group of experts aimed to prepare and process single-layer 1D GNRs [56]. The results demonstrate that the capacitance of 75 mF cm^{-2} at 100 kHz is substantially higher than the highest limit of the capacitance of the carbon-based electrodes, which is around 0.035 mF cm^{-2}.

The most intriguing feature of GNRs that distinguishes them from other graphene family members is their higher area normalization edge-plane structure. GNRs also feature a wide surface area, residual oxygen functions, approachable catalytic domains, high conductivity, and long-term stability. As a result, nanostructures based on GNRs have lately gained favor in electrochemical and biosensing technologies [57]. Due to their well-defined edge structure and high homogeneity, GNRs must be synthesized from the bottom up in order to be utilized in sensor applications. Using a variety of chemical synthesis approaches, structurally perfect GNRs with varying lengths and edge topologies have been generated recently. However, topics such as GNR deposition, characterization, electronic characteristics, and applications have not been investigated in detail. Deposition, characterization, and device manufacturing of ultra-long chemically produced GNRs have been optimized by Ahmad and associates [58]. They also showed that the GNR film devices were very sensitive to NO_2 gas down to parts per billion (ppb) levels. The findings make it possible to use chemically made GNRs in electronics and sensing in the years ahead. Govindasamy *et al* [23] reported silver coated GNRs made using a simple wet chemical process, with good organophosphorus methyl parathion sensing characteristics. Using Ag@GNRs/SPCE, a very sensitive methyl parathion sensor was constructed, and its practical application in food samples was demonstrated. Due to the synergistic impacts of Ag and the superior physicochemical features of GNRs, the composite is ideally suited for pesticide sensing. Several nanostructures predicated on GNRs have been developed, and their potential in electrochemical sensing technologies has been established [59–65]. Graphene quantum dots (GQDs) aided by GNRs were created using a one-step synchronous reduction procedure, and the resulting nanomaterial demonstrated exceptionally high electrocatalytic activity for oxygen reduction [62]. TEM images of GNRs/GQDs show uniform dispersion and hierarchical structure (figure 3.10).

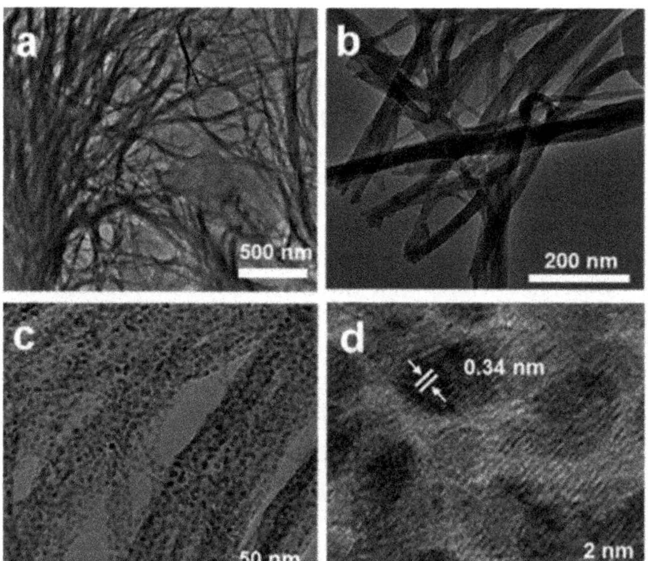

Figure 3.10. (a)–(c) TEM and (d) HRTEM images of a GNR-based GQD hybrid. (Reproduced with permission from [62]. Copyright 2015 Springer Nature.)

GQDs/GNRs hybrids produced better current density and smaller overpotential than platinum, as well as superior selectivity and robustness in alkaline media [66].

The sensor shows superb performance and demonstrates low detection limit, high sensitivity, fast response, and broad linear range for nimesulide detection.

3.3.2 Biomedical applications

GNRs that have been oxygenated are established as graphene oxide nanoribbons (GONRs). GONRs' oxygen-containing capabilities are responsible for their functionalization with diverse biomolecules, increasing their utility in biological applications. Possible uses include drug delivery, antitumor therapy, detection, preventing infections, scanning, genomics, photodynamic therapy, spinal cord injury care, bone healing, and more [21].

GNRs are a novel medication delivery agent that can be used in cancer treatment. Aromatic nanoribbons can be loaded with anticancer medicines for their usefulness. Covalently or non-covalently functionalized, its surface can deliver a large amount of the medicine precisely and correctly to the appropriate location. The surface can be modified with a polymer or ligand to deliver drugs directly to tumors [67]. The physicochemical attributes of GNRs could be used to improve drug delivery and therapy that responds to a stimulus. Graphene oxide was used to make a high-efficiency drug carrier for doxorubicin, which is used to treat cancer. Traditional nanocarriers cannot carry more than 100% doxorubicin by weight, but GO can carry 235% [68].

Multiple cell lines were subjected to a variety of bioassays to determine the cytotoxicity of GNRs. GNR that has been functionalized with 1,2-distearoyl-sn-glycero-3-phosphoethanolamine polyethylene glycol (DSPE-PEG) displays specific cytotoxicity against tumor cells. In comparison to its affinity for other types of cells, DSPE-PEG-GONR is particularly drawn to those cells that have human papillomavirus (HPV) genomes or exhibit EGFR receptors. The modified receptors and improved absorption are a result of the integrated genome. The micropinocytic absorption of DSPE-PEG-GONR is primarily responsible for the enhanced uptake of this molecule in EGFR-activated cells [21]. Doxorubicin was 100% more effective when it was sent to cells with high EGFR expression using the DSPE-PEG-GONR drug delivery system or the HPV genome. Glioblastoma multiforme cells may exhibit higher absorption and cytotoxicity if EGFR is overexpressed. MCF-7 and glial progenitor cells lack EGFR expression and display little to no cytotoxicity [67]. Therefore, nanomaterials may function as a method for delivering medications to precise locations without ligands. If you connect a ligand to this drug carrier, you can make it easier to target the medicine [67].

Cancer, hemophilia, cystic fibrosis, and sickle cell anemia are just a few of the diseases that can be cured via gene therapy. DNA and RNA are extremely fragile polymers with a half-life in serum of only 10 and 1 min, respectively. This means that genes need a steady carrier to get where they need to go. Solidity, superior loading capacity, cell entry via endocytosis, and delivery of genes to the nucleus are all requirements for the carrier. RNA and DNA delivery efficiency has recently been demonstrated using GNRs [69]. GNRs can transport genes. It is easier to mass-produce and more biocompatible than alternative pathogenic vectors. The GNR enables anisotropic assembly of ssDNA, a characteristic not shared by any other GO nanomaterial. GONRs may load a large number of genetic materials of varied sizes without modifying positively charged groups or non-essential vectors. High transfection efficiency is achieved in dividing or nondividing cells. GONRs at possible dosages (20–60 g ml^{-1}) have reduced cytotoxicity compared to PEI and fungene-6 [69]. PEI-g-GNR is an effective nucleic acid carrier for micro-RNA recognition. This vector can keep nucleases and other proteins from breaking down molecular beacon probes that have modified nucleic acid. Due to a GNR's huge surface area and high charge density, it transfects cells better than PEI or PEI-g-MWCNTs.

Selective imaging and photothermotherapy are enabled by ligand and cyanine-loaded reduced GONRs that have been PEG-functionalized. Active targeting as well as fluorescence imaging can both be accomplished using ligand and cyanine dye 3. The NIR absorption of reduced GONR suspended in PEG at a concentration of 100 g ml^{-1} was found to be 2.4 times higher than that of GONR and reduced rGO-PEG, respectively. The best concentration for use in living organisms is 1 g ml^{-1}, because higher concentrations are harmful to cells and genes [70]. Chemo-photothermal therapy combines chemotherapy with photothermal therapy to treat cancer. There are many advantages to GNR chemo-photothermal therapy over conventional chemo-photothermal therapy. PEG-GONRs loaded with doxorubicin were found to have an IC50 value 6.7-fold lower than conventional chemotherapy when used to treat glioma cells (U87) (figure 3.11) [71].

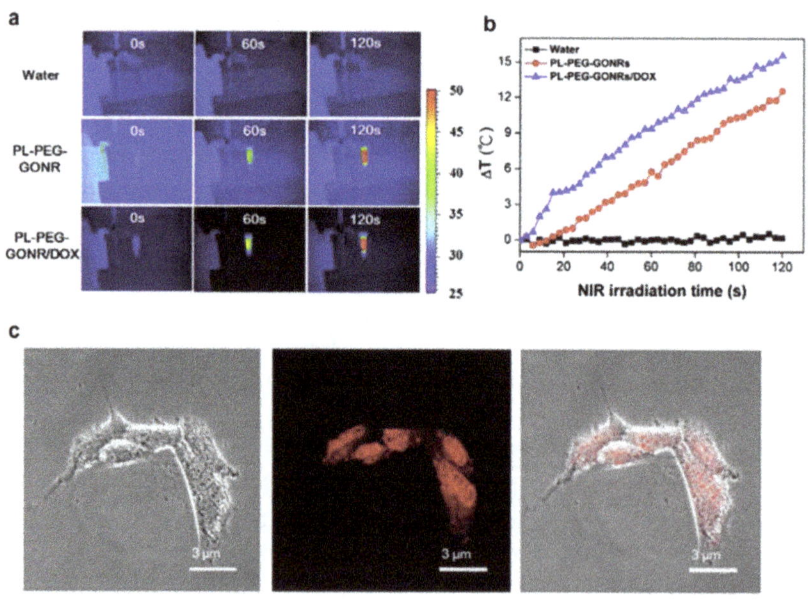

Figure 3.11. (a) NIR thermal images of PL-PEG-GONR in aqueous, free, and drug-loaded states. (b) Graph illustrating 120 s NIR irradiation during the increase of temperature. (c) Confocal microscopy photos of U87 cells. (Reproduced from [72]. CC BY 4.0.)

The electron network decreases in GNRs, causing photoluminescence. The nanoribbons' visible and IR properties enable cellular imaging. Nanoribbons' remarkable light transmittance, charge mobility, and photoluminescence make it possible to perform magnetic resonance imaging (MRI) and biological imaging [73]. Researchers recently created an azide-functionalized cove-shaped GNR. Click reactions can take place between edge azide groups, functional groups, and terminal alkynes when copper is used as the catalyst. Fluorescent dyes that are attached to clickable GNRs allow fluorescence and super resolution microscopy (SRM) be used to take pictures of GNRs that are spread out in small amounts [74].

3.3.3 Catalytic applications

The technology of the fuel cell is one of the most effective sustainable energy options for mitigating environmental degradation, depletion of natural resources, and global warming because it generates electricity using just water and heat [75]. In order to turn the chemical energy of fuels directly into electrical energy, fuel cells need a very good electrocatalyst that speeds up the slow ORR. Two-electron and four-electron reduction of O_2 to H_2O_2 are the two main ORR processes in acidic and alkaline electrolytes [76]. The ORR reaction catalyzed by platinum (Pt), a commonly used commercial catalyst, is generally performed via the $4e^-$ route with high initiation potential and high current density. Nevertheless, the advancement and wider implementation of fuel cells are impeded by the massive cost, lack of resources, and CO reactivity of Pt-based catalysts. A current subject among fuel cell researchers is the development of less expensive and more efficient electrocatalysts that do not need precious metals, such as those made from nonprecious metals [77]. In fuel cells and metal–air batteries, GNRs have drawn great attention because of their unusual structural properties, such as plentiful activation sites and edge effects [78].

Liu's group produced GNRs from CNTs via longitudinal unzipping [79]. The N-GNRs in figure 3.12 were made using a three-step procedure that included a reduction reaction, *in situ* polymerization, and heat treatment. The TEM image shows GNRs bonded after opening pristine CNTs. PANI nanorods are evenly disseminated across GNR sheets after polymerization. As shown in figures 3.12(d)–(g), the N-GNRs have a much higher current density than the GNRs. This suggests that the doping of the n element makes the ORR work better. The n value for N8.3-GNRs is approximately 3.91, indicating that the $4e^-$ ORR process predominates [79, 80].

GNRs can also be seen as a very favorable support for ORR catalysts. GNRs were initially utilized by Wang *et al* as a brand-new support material for a Pt nanoparticle based catalyst for the electro-oxidation of methanol [65]. After oxidizing and cutting MWCNTs, extremely dispersive GONRs were produced, on which PtCl6 $2e^-$ can be placed uniformly. Pt/GNR hybrids have a greater electrochemically active surface area, stronger electrochemical stability, and better CO tolerance than Pt/MWCNT and commercial Pt/XC72R catalysts. GNR is therefore a suitable two-dimensional alternative support material for electrocatalysts in direct methanol fuel cells.

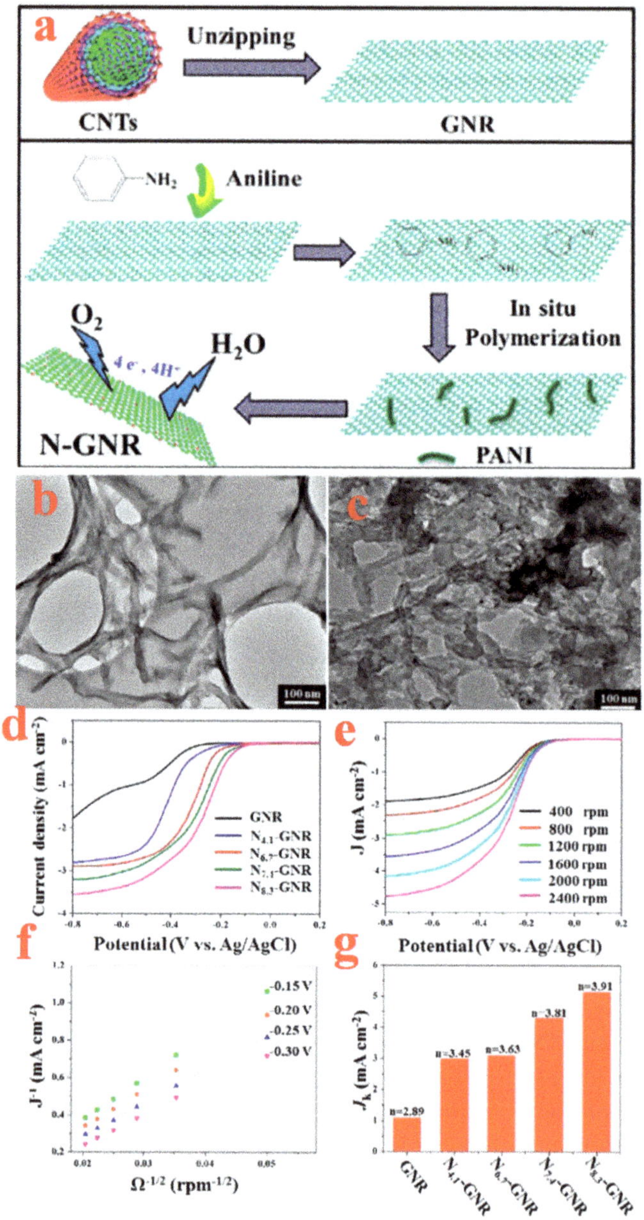

Figure 3.12. (a) Diagrammatic depictions of N-GNR synthesis. TEM micrographs of (b) sheets of pristine GNRs and (c) composites of GNR/polyaniline. (d) GNR and N-GNR LSV curves in O_2-saturated 0.1 M KOH solution. (e) Effect of rotating speed on LSV curves for $N_{8.3}$-GNRs. (f) LSVs derived K-L plots of GNRs and N-GNRs. (g) Current density of GNR and N-GNR at 0.7 V. Reproduced from [78] (CC BY 4.0.) and [79] (courtesy of ACS Publications).

3.4 Challenges and outlook

GNRs have high mobility and the ability to carry a lot of current. They also have a large bandgap and a wide range of electronic properties, which makes them a good choice for quantum electronic applications and medical uses. Over the last five years, progress has been made toward atomically precise bottom-up synthesis of GNRs and heterojunctions, which provide an ideal platform for functional molecular devices, as well as successful production of semiconducting GNR arrays on insulating substrates, which could be used in large-scale digital circuits. Nonetheless, the most challenging aspect of GNR synthesis is the scalable manufacture of narrow GNRs with high accuracy and efficiency, on diverse substrates, and with flexible placement and/or alignment. No synthesis technique currently fits all these conditions. A hybrid strategy that combines atomically precise production and the scalability of templated growth could take the field to a new level. Edge structure management of common GNRs remains difficult. Due to a limited library of precursors, to date only GNRs with specific chirality structures have been synthesized. However, it is still not viable to synthesize GNRs in artificially created structures. Using artificial intelligence to design a library of antecedents could provide feasible answers in the long term [45, 81]. For devices to form reliably, GNR length and yield need to keep getting better. Surface diffusivity and kinetic factors on substrates limit the length of bottom-up assemble GNRs to tens of nanometers or less. Increasing surface diffusion or generating lengthy, smooth steps can boost the end-fusion of GNRs [82]. It is possible to boost GNR yield and organization by eliminating transfer phases if the technology can be used on substrates made of crystalline insulators. To reach these targets, we need to make scalable substrate templates with surfaces that are accurate to the atomic level [45].

Characterization of GNRs is yet another difficult task. For large-scale GNR characterization, high-throughput metrology must carry the edge-specific GNR fingerprint. The combination of scanning tunneling microscopy and scanning transmission electron microscopy (STM/STS), and atomic force microscopy have been three of the most used techniques for characterizing the atomic structures of GNRs up to this point. Although these instruments are precise, their throughput is minimal. The chirality of CNTs can be determined using a combination of Raman spectroscopy and optical absorption [83]. GNRs may have a similar database, but it would be far more complicated due to the open edges.

Similarly, GNRs' exceptional electrical and mechanical capabilities make them a viable material for biomedical applications. Even though GNRs have become increasingly commonplace, careful consideration must be given to the possible harmfulness of these compounds, from both purposeful and unintentional exposure. Medical GNRs are used to make ultra-small devices including molecular sensors, light and thermal detectors,and sequencers. They are also utilized to transport drugs or genes to cells, and to construct tissue engineering structures [14]. The reproducibility and reliability of the GNRs in biomedical research are the most important issues to solve. Also, the biggest problems with using GNRs to deliver drugs and genes are making materials that can hold more drugs, have predictable *in vitro* and *in vivo* release profiles, and have lower toxicity profiles [84]. It will be difficult to produce GNRs using environmentally acceptable methods to limit the biohazards of carbon residues. Transporting GNRs requires retaining nanoscale features in macroscale devices. For GNRs to be used on a larger scale and sold in stores, there must be a lot of toxicology research, standardization of manufacturing processes, and regulatory provisions.

There is little dispute about the fascinating qualities and possible uses of these GNRs, however, due to the presence of several roadblocks, neither of these aspects has been completely investigated. Therefore, in conclusion, an appropriate interdisciplinary strategy will provide further understanding of the biological, electronic, and catalytic properties of these GNRs for their nanoelectronics, biomedical, and fuel cell applications, and can also assist in improving the current applications in a variety of different ways.

Acknowledgments

The authors would like to thank the Higher Education Commission of Pakistan (HEC) for their support in sponsoring this work through the CPEC-Collaborative Research Grant (Project No. CPEC-8). In addition, the authors would like to thank the PPE Department at UET Lahore for providing laboratory facilities.

References

[1] Palmer B S 2013 A review on the spontaneous formation of the building blocks of life and the generation of a set of hypotheses governing universal abiogenesis *Int. J. Astrobiol.* **12** 39–44

[2] Banhart F 2020 Elemental carbon in the sp^1 hybridization *Chem. Texts* **6** 3

[3] Dass D 2018 Structural analysis, electronic properties, and band gaps of a graphene nanoribbon: a new 2D materials *Superlattices Microstruct.* **115** 88–107

[4] Yang X, Yu X and Liu X 2018 Obtaining a sustainable competitive advantage from patent information: a patent analysis of the graphene industry *Sustainability* **10** 4800

[5] Sahoo M, Wang J C, Nishina Y, Liu Z, Bow J S and Lai C S 2020 Robust sandwiched fluorinated graphene for highly reliable flexible electronics *Appl. Surf. Sci.* **499** 143839

[6] Ballesio A, Parmeggiani M, Verna A, Frascella F, Cocuzza M, Pirri C F and Marasso S L 2019 A novel hot embossing graphene transfer process for flexible electronics *Microelectron. Eng.* **209** 16–9

[7] Garg K and Negi S 2020 Exploring the charge configuration of an armchair single walled carbon nanotube for drug delivery *Mater. Today Proc.* **28** 185–7

[8] Wang Y, Yang S, Chang H, Li W, Chen X, Hou R, Yan J, Yi X, Wang J and Wei T 2020 Flexible graphene-assisted van der Waals epitaxy growth of crack-free AlN epilayer on SiC by lattice engineering *Appl. Surf. Sci.* **520** 146358

[9] Wei H *et al* 2020 Advanced porous hierarchical activated carbon derived from agricultural wastes toward high performance supercapacitors *J. Alloys Compd.* **820** 153111

[10] Kumar R *et al* 2019 A review on synthesis of graphene, h-BN and MoS_2 for energy storage applications: recent progress and perspectives *Nano Res.* **12** 2655–94

[11] Hou C *et al* 2019 One-pot synthesized molybdenum dioxide–molybdenum carbide heterostructures coupled with 3D holey carbon nanosheets for highly efficient and ultrastable cycling lithium-ion storage *J. Mater. Chem.* A **7** 13460–72

[12] Novoselov K S, Geim A K, Morozov S V, Jiang D, Zhang Y, Dubonos S V, Grigorieva I V and Firsov A A 2004 Electric field in atomically thin carbon films *Science* **306** 666–9
[13] Schwierz F 2010 Graphene transistors *Nat. Nanotechnol.* **5** 487–96
[14] Zakharova O V, Mastalygina E E, Golokhvast K S and Gusev A A 2021 Graphene nanoribbons: prospects of application in biomedicine and toxicity *Nanomaterials* **11** 2425
[15] Chen L, Hernandez Y, Feng X and Müllen K 2012 From nanographene and graphene nanoribbons to graphene sheets: chemical synthesis *Angew. Chem. Int. Ed.* **51** 7640–54
[16] Sun L, Zhou H, Li L, Yao Y, Qu H, Zhang C, Liu S and Zhou Y 2017 Double soft-template synthesis of nitrogen/sulfur-codoped hierarchically porous carbon materials derived from protic ionic liquid for supercapacitor *ACS Appl. Mater. Interfaces* **9** 26088–95
[17] Dutta S and Pati S K 2010 Novel properties of graphene nanoribbons: a review *J. Mater. Chem.* **20** 8207–23
[18] Yang L, Park C H, Son Y W, Cohen M L and Louie S G 2007 Quasiparticle energies and band gaps in graphene nanoribbons *Phys. Rev. Lett.* **99** 186801
[19] Wang J, Zhao R, Yang M, Liu Z and Liu Z 2013 Inverse relationship between carrier mobility and bandgap in graphene *J. Chem. Phys.* **138** 084701
[20] Wang X Y, Yao X and Müllen K 2019 Polycyclic aromatic hydrocarbons in the graphene era *Sci. China Chem.* **62** 1099–144
[21] Johnson A P, Gangadharappa H V and Pramod K 2020 Graphene nanoribbons: a promising nanomaterial for biomedical applications *J. Control. Release* **325** 141–62
[22] Chen Z *et al* 2016 Synthesis of graphene nanoribbons by ambient-pressure chemical vapor deposition and device integration *J. Am. Chem. Soc.* **138** 15488–96
[23] Niu W, Liu J, Mai Y, Müllen K and Feng X 2019 Synthetic engineering of graphene nanoribbons with excellent liquid-phase processability *Trends Chem.* **1** 549–58
[24] Barzegar H R, Pham T, Talyzin A V and Zettl A 2016 Synthesis of graphene nanoribbons inside boron nitride nanotubes *Phys. Status Solidi* **253** 2377–9
[25] Gardener J A and Golovchenko J A 2012 Ice-assisted electron beam lithography of graphene *Nanotechnology* **23** 185302
[26] Sinitskii A and Tour J M 2012 Patterning graphene nanoribbons using copper oxide nanowires *Appl. Phys. Lett.* **100** 103106
[27] Hwang W S *et al* 2015 Graphene nanoribbon field-effect transistors on wafer-scale epitaxial graphene on SiC substrates a *APL Mater.* **3** 011101
[28] Bai J, Duan X and Huang Y 2009 Rational fabrication of graphene nanoribbons using a nanowire etch mask *Nano Lett.* **9** 2083–7
[29] Naguib M, Mashtalir O, Carle J, Presser V, Lu J, Hultman L, Gogotsi Y and Barsoum M W 2012 Two-dimensional transition metal carbides *ACS Nano* **6** 1322–31
[30] Narita A *et al* 2014 Synthesis of structurally well-defined and liquid-phase-processable graphene nanoribbons *Nat. Chem.* **6** 126–32
[31] Vo T H, Shekhirev M, Kunkel D A, Morton M D, Berglund E, Kong L, Wilson P M, Dowben P A, Enders A and Sinitskii A 2014 Large-scale solution synthesis of narrow graphene nanoribbons *Nat. Commun.* **5** 3189
[32] Wu Z S, Ren W, Gao L, Liu B, Zhao J and Cheng H M 2010 Efficient synthesis of graphene nanoribbons sonochemically cut from graphene sheets *Nano Res.* **3** 16–22
[33] Campos L C, Manfrinato V R, Sanchez-Yamagishi J D, Kong J and Jarillo-Herrero P 2009 Anisotropic etching and nanoribbon formation in single-layer graphene *Nano Lett.* **9** 2600–4

[34] Jiao L, Zhang L, Ding L, Liu J and Dai H 2010 Aligned graphene nanoribbons and crossbars from unzipped carbon nanotubes *Nano Res.* **3** 387–94
[35] Ozden S, Autreto P A S, Tiwary C S, Khatiwada S, Machado L, Galvao D S, Vajtai R, Barrera E V and Ajayan M P 2014 Unzipping carbon nanotubes at high impact *Nano Lett.* **14** 4131–7
[36] Kosynkin D V, Higginbotham A L, Sinitskii A, Lomeda J R, Dimiev A, Price B K and Tour J M 2009 Longitudinal unzipping of carbon nanotubes to form graphene nanoribbons *Nature* **458** 872–6
[37] Cançado L G *et al* 2004 Anisotropy of the Raman spectra of nanographite ribbons *Phys. Rev. Lett.* **93** 047403
[38] Loh K P, Bao Q, Eda G and Chhowalla M 2010 Graphene oxide as a chemically tunable platform for optical applications *Nat. Chem.* **2** 1015–24
[39] Cai J *et al* 2010 Atomically precise bottom-up fabrication of graphene nanoribbons *Nature* **466** 470–3
[40] Linden S *et al* 2012 Electronic structure of spatially aligned graphene nanoribbons on Au (788) *Phys. Rev. Lett.* **108** 216801
[41] Chen Y C, De Oteyza D G, Pedramrazi Z, Chen C, Fischer F R and Crommie M F 2013 Tuning the band gap of graphene nanoribbons synthesized from molecular precursors *ACS Nano* **7** 6123–8
[42] Sokolov A N *et al* 2013 Direct growth of aligned graphitic nanoribbons from a DNA template by chemical vapour deposition *Nat. Commun.* **4** 2402
[43] Norimatsu W and Kusunoki M 2010 Formation process of graphene on SiC (0 0 0 1) *Physica* E **42** 691–4
[44] Kato T and Hatakeyama R 2012 Site- and alignment-controlled growth of graphene nanoribbons from nickel nanobars *Nat. Nanotechnol.* **7** 651–6
[45] Wang H, Wang H S, Ma C, Chen L, Jiang C, Chen C, Xie X, Li A P and Wang X 2021 Graphene nanoribbons for quantum electronics *Nat. Rev. Phys.* **3** 791–802
[46] Kim W Y and Kim K S 2008 Prediction of very large values of magnetoresistance in a graphene nanoribbon device *Nat. Nanotechnol.* **3** 408–12
[47] Chen Z, Narita A and Müllen K 2020 Graphene nanoribbons: on-surface synthesis and integration into electronic devices *Adv. Mater.* **32** 2001893
[48] Llinas J P *et al* 2017 Short-channel field-effect transistors with 9-atom and 13-atom wide graphene nanoribbons *Nat. Commun.* **8** 633
[49] Koch M, Ample F, Joachim C and Grill L 2012 Voltage-dependent conductance of a single graphene nanoribbon *Nat. Nanotechnol.* **7** 713–7
[50] Bonaccorso F, Colombo L, Yu G, Stoller M, Tozzini V, Ferrari A C, Ruoff R S and Pellegrini V 2015 2D materials. Graphene, related two-dimensional crystals, and hybrid systems for energy conversion and storage *Science* **347** 6217
[51] Beidaghi M and Gogotsi Y 2014 Capacitive energy storage in micro-scale devices: recent advances in design and fabrication of micro-supercapacitors *Energy Environ. Sci.* **7** 867–84
[52] Wu Z S *et al* 2017 Bottom-up fabrication of sulfur-doped graphene films derived from sulfur-annulated nanographene for ultrahigh volumetric capacitance micro-supercapacitors *J. Am. Chem. Soc.* **139** 4506–12
[53] Xiang Z, Dai Q, Chen J F and Dai L 2016 Edge functionalization of graphene and two-dimensional covalent organic polymers for energy conversion and storage *Adv. Mater.* **28** 6253–61

[54] Chen Z *et al* 2017 Chemical vapor deposition synthesis and terahertz photoconductivity of low-band-gap $N = 9$ armchair graphene nanoribbons *J. Am. Chem. Soc.* **139** 3635–8
[55] Liu Z, Chen Z, Wang C, Wang H I, Wuttke M, Wang X Y, Bonn M, Chi L, Narita A and Müllen K 2020 Bottom-up, on-surface-synthesized armchair graphene nanoribbons for ultra-high-power micro-supercapacitors *J. Am. Chem. Soc.* **142** 17881–6
[56] Qorbani M, Esfandiar A, Mehdipour H, Chaigneau M, Irajizad A and Moshfegh A Z 2019 Shedding light on pseudocapacitive active edges of single-layer graphene nanoribbons as high-capacitance supercapacitors *ACS Appl. Energy Mater.* **2** 3665–75
[57] Ananda Murthy H C, Gebremedhn Kelele K, Ravikumar C R, Nagaswarupa H P, Tadesse A and Desalegn T 2021 Graphene-supported nanomaterials as electrochemical sensors: a mini review *Results Chem.* **3** 100131
[58] Abbas A N, Liu G, Narita A, Orosco M, Feng X, Müllen K and Zhou C 2014 Deposition, characterization, and thin-film-based chemical sensing of ultra-long chemically synthesized graphene nanoribbons *J. Am. Chem. Soc.* **136** 7555–8
[59] Li L, Raji A R O, Fei H, Yang Y, Samuel E L G and Tour J M 2013 Nanocomposite of polyaniline nanorods grown on graphene nanoribbons for highly capacitive pseudocapacitors *ACS Appl. Mater. Interfaces* **5** 6622–7
[60] Zhang R, Sun C L, Lu Y J and Chen W 2015 Graphene nanoribbon-supported PtPd concave nanocubes for electrochemical detection of TNT with high sensitivity and selectivity *Anal. Chem.* **87** 12262–9
[61] Li N, Ma H, Cao W, Wu D, Yan T, Du B and Wei Q 2015 Highly sensitive electrochemical immunosensor for the detection of alpha fetoprotein based on PdNi nanoparticles and N-doped graphene nanoribbons *Biosens. Bioelectron.* **74** 786–91
[62] Jin H, Huang H, He Y, Feng X, Wang S, Dai L and Wang J 2015 Graphene quantum dots supported by graphene nanoribbons with ultrahigh electrocatalytic performance for oxygen reduction *J. Am. Chem. Soc.* **137** 7588–91
[63] Davis D J, Raji A R O, Lambert T N, Vigil J A, Li L, Nan K and Tour J M 2014 Silver–graphene nanoribbon composite catalyst for the oxygen reduction reaction in alkaline electrolyte *Electroanalysis* **26** 164–70
[64] Liu Y, Wei L, Hu Y, Huang X, Wang J, Li J, Hu X and Zhuang N 2016 Influence of Pd-doping concentration on activity and CO anti-poisoning ability of PtPd/GNRs alloy catalyst for ethanol oxidation and density functional theory analysis *J. Alloys Compd.* **656** 452–7
[65] Wang C, Li H, Zhao J, Zhu Y, Yuan W Z and Zhang Y 2013 Graphene nanoribbons as a novel support material for high performance fuel cell electrocatalysts *Int. J. Hydrog. Energy* **30** 13230–7
[66] Govindasamy M, Mani V, Chen S M, Maiyalagan T, Selvaraj S, Chen T W, Lee S Y and Chang W H 2017 Highly sensitive determination of non-steroidal anti-inflammatory drug nimesulide using electrochemically reduced graphene oxide nanoribbons *RSC Adv.* **7** 33043–51
[67] Chowdhury S M, Surhland C, Sanchez Z, Chaudhary P, Suresh Kumar M A, Lee S, Peña L A, Waring M, Sitharaman B and Naidu M 2015 Graphene nanoribbons as a drug delivery agent for lucanthone mediated therapy of glioblastoma multiforme *Nanomedicine* **11** 109–18
[68] Yang X, Zhang X, Liu Z, Ma Y, Huang Y and Chen Y 2008 High-efficiency loading and controlled release of doxorubicin hydrochloride on graphene oxide *J. Phys. Chem.* C **112** 17554–8
[69] Mousavi S M, Soroshnia S, Hashemi S A, Babapoor A, Ghasemi Y, Savardashtaki A and Amani A M 2019 Graphene nano-ribbon based high potential and efficiency for DNA, cancer therapy and drug delivery applications *Drug Metab. Rev.* **51** 91–104

[70] Akhavan O, Ghaderi E and Emamy H 2012 Nontoxic concentrations of PEGylated graphene nanoribbons for selective cancer cell imaging and photothermal therapy *J. Mater. Chem.* **22** 20626–33
[71] Krasteva N, Staneva D, Vasileva B, Miloshev G and Georgieva M 2021 Bioactivity of PEGylated graphene oxide nanoparticles combined with near-infrared laser irradiation studied in colorectal carcinoma cells *Nanomaterials* **11** 3061
[72] Burdanova M G, Kharlamova M V, Kramberger C and Nikitin M P 2021 Applications of pristine and functionalized carbon nanotubes, graphene, and graphene nanoribbons in biomedicine *Nanomaterials (Basel)* **11** 3020
[73] Goenka S, Sant V and Sant S 2014 Graphene-based nanomaterials for drug delivery and tissue engineering *J. Control. Release* **173** 75–88
[74] Joshi D, Hauser M, Veber G, Berl A, Xu K and Fischer F R 2018 Super-resolution imaging of clickable graphene nanoribbons decorated with fluorescent dyes *J. Am. Chem. Soc.* **140** 9574–80
[75] Yang Z, Ren J, Zhang Z, Chen X, Guan G, Qiu L, Zhang Y and Peng H 2015 Recent advancement of nanostructured carbon for energy applications *Chem. Rev.* **115** 5159–223
[76] Nie Y, Li L and Wei Z 2015 Recent advancements in Pt and Pt-free catalysts for oxygen reduction reaction *Chem. Soc. Rev.* **44** 2168–201
[77] Dai L, Xue Y, Qu L, Choi H J and Baek J B 2015 Metal-free catalysts for oxygen reduction reaction *Chem. Rev.* **115** 4823–92
[78] Tong X, Wei Q, Zhan X, Zhang G and Sun S 2016 The new graphene family materials: synthesis and applications in oxygen reduction reaction *Catalysis* **7** 1
[79] Liu M, Song Y, He S, Tjiu W W, Pan J, Xia Y Y and Liu T 2014 Nitrogen-doped graphene nanoribbons as efficient metal-free electrocatalysts for oxygen reduction *ACS Appl. Mater. Interfaces* **6** 4214–22
[80] Wang X, Li X, Zhang L, Yoon Y, Weber P K, Wang H, Guo J and Dai H 2009 N-doping of graphene through electrothermal reactions with ammonia *Science* **324** 768–71
[81] Segler M H S, Preuss M and Waller M P 2018 Planning chemical syntheses with deep neural networks and symbolic AI *Nature* **555** 604–10
[82] Moreno C, Paradinas M, Vilas-Varela M, Panighel M, Ceballos G, Peña D and Mugarza A 2018 On-surface synthesis of superlattice arrays of ultra-long graphene nanoribbons *Chem. Commun.* **54** 9402–5
[83] Anderson N, Hartschuh A and Novotny L 2007 Chirality changes in carbon nanotubes studied with near-field Raman spectroscopy *Nano Lett.* **7** 577–82
[84] Banerjee A N 2018 Graphene and its derivatives as biomedical materials: future prospects and challenges *Interface Focus* **8** 20170056

Chapter 4

Synthesis and application of graphene nanowires

Siddheswar Rudra and Arpan Kumar Nayak

A lot of research has been done on graphene since its initial exfoliation in 2004 because of its very desirable features in many areas of materials engineering. Recently, graphene-based composites have drawn more attention for electrochemical energy storage because they combine the advantages of graphene and other electrochemical materials to produce excellent electrochemical performances. Unique features, including the bipolar transistor effect, ballistic transport of charges, and massive quantum oscillations, are produced by its interactions with other substances and with light as well as by its naturally two-dimensional nature. This chapter will explain approaches to further enhance graphene-based composites in order to create the next generation of electrochemical energy storage devices, with a focus on the coupling of graphene with other electrochemical materials to maximize their performance.

4.1 Introduction

In the current state of affairs, dwindling of fossil fuel resources pose a serious problem for the entire world. Due to the scarcity of non-renewable fossil fuel resources such as coal and oil, both the expanding population and the sustainable growth of contemporary society have been hampered severely. The most promising approach for society to address the aforementioned issues is the highly efficient use of green and renewable resources. With its numerous uses, including carbon dioxide (CO_2) extraction, hydrogen storage, dye-sensitized solar cells, water purification, and energy storage, biomass is one of the planet's most abundant renewable resources [1–6]. Supercapacitors share Li-ion batteries' high power density, quick charge–discharge rate, and extended cycle life [7]. Although in the past, activated carbon was used as an electrode material for superconductors, biomass-derived activated carbons (ACs) are now preferred because of their intriguing characteristics, including their high thermal and chemical stabilities, ability to generate high specific capacitance due to their large surface areas, and superior electrical

doi:10.1088/978-0-7503-5177-5ch4

conductivity, which is crucial for power density [8, 9]. In another aspect, carbon-based material electrical double-layer capacitors (EDLCs) have gained attention for electrical energy storage followed by the electrostatic accretion of charges. The most commonly used carbon materials are graphene [10], graphene oxide [11], and carbon aerogels [12]. In addition to this, the uses of supercapacitors (SCs) are also noticeable as they have fast charge or discharge rates, including higher power densities and super-long cycle lives [13]. Carbon-based materials from ACs [14] and carbon nanotubes [15] receive attention for their various applications, including energy applications. In energy-related applications, they are mostly used in electrodes because of their interesting properties, such as high surface area, excellent electrical conductivity, and low cost [13]. These properties are ideal for EDLC electrode materials as they are efficiently designed, porous, and most importantly carbon-based materials with a high surface area and the presence of a sufficient number of electroactive sites [16, 17]. For these types of materials porosity plays an important role compared to other materials. The reason is that when these types of porous materials are present they can be adjusted easily within a narrow range of length scale compared to any other conventional porous materials. A common example of these kinds of porous materials are hierarchical porous materials that are made up of a variety of pores such as macropores (>50 nm), mesopores (2–50 nm), and micropores (<2 nm). As a result, structurally they can be coordinated with various kinds of morphologies with various different length scales [18]. When the size is comparatively large, such as a hollow macro-chamber, they are applicable for ion transportation at a higher rate. As a result, they exhibit an important role in ion reservoirs. In addition to this, for ion diffusion at the inner-pore surface, the low-resistance pathways gain interest by interconnecting the mesopores. A recent research area in the field of energy storage gave much importance to constructing novel porous architectures, specifically the hierarchically porous materials [19]. Generally, two types of techniques are available for developing porous-based carbon composites, they are templating techniques including hard-template and soft-template procedures. In the hard-template synthesis procedure, the hard templates are synthesized by impregnation following carbonization and template removal [20], whereas the soft-template synthesizing techniques proceed through synthesizing the first soft templates as a substrate for self-assembly through high-temperature carbonization and etching [21]. Although both techniques are capable of synthesizing hierarchical materials, compared to other methods these have some complex steps that take a long time and are costly, thus reducing profitability [22]. As bio-renewable sources are receiving much attention for both their eco-friendly and simple nature, scientists are interested in developing an easy, environmentally friendly and template-free procedure for bio-renewable sources that have readily available carbon sources [13]. In addition to this hierarchical porous material, many research reports have highlighted introducing heteroatoms, as heteroatoms have the ability to improve the crystallinity of the respective materials and contribute a certain pseudocapacitance [23]. Currently, some important procedures, such as hydrothermal synthesis, steam explosions, and microwave heating, are being introduced for the production of carbon materials with excellent electrochemical properties [24]. These type of simpler methods are important for making precursors expand before carbonization, without excessive activation steps [25]. Therefore, researchers are working to change the aforementioned procedures so that they can make a specific material because it is better than the others and will also be better for the advancement of sophisticated energy materials. In this field, calcium alginate gel is sometimes considered as the carbon precursor. Some researchers are able to synthesize three-dimensional (3D) porous honeycomb carbon (PHC) through a one-step carbonization process by applying different heating rates. As a result, in this above process of carbonization the volume is expanded. [26] (Figure 4.1). These synthesized materials have excellent electrochemical properties including superior electrochemical performance. Compared to other traditional ACs, this synthesized PHC has a higher rate performance and better cycle stability. This research provided new directions to establishing environmental improvement, biomass conversion, and renewable energy utilization [26]. Designing and assembling flexible, lightweight, wearable sensors has attracted interest due to the present demand for point-of-care testing devices, portable equipment for *in vivo* clinical diagnostics, and on-site environmental monitoring. Due to their short response time, great sensitivity, excellent selectivity, simplicity of use, and good reproducibility, electrochemical sensors are an intriguing choice.

One of the critical challenges in producing elastic electrochemical sensors is the lack of free-standing paper-like electrode materials. When a person takes out a pencil and begins writing on a piece of paper, they often have no idea that these same materials have such scientific significance. The scientific community, however, showed interest in the materials used for writing on a piece of paper, specifically graphite. Andre Geim and Konstantin Novoselov were awarded the Nobel Prize in Physics on 5 October 2010 for 'groundbreaking experiments relating to the

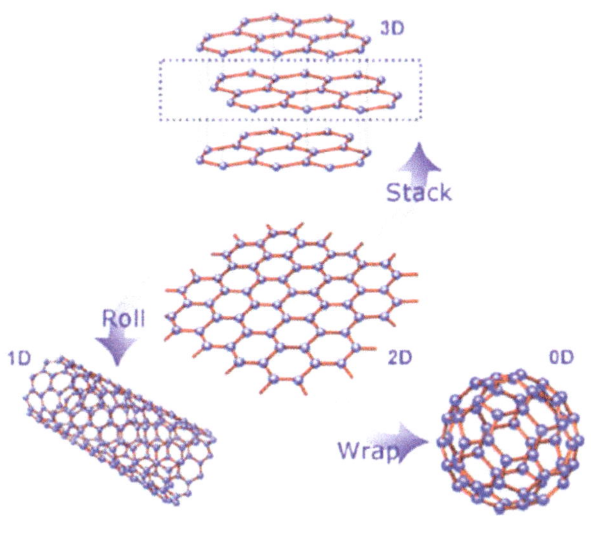

Figure 4.1. Graphene, as a primary building block, is an allotrope of carbon with variable dimensional structures. (Reproduced with permission from [27]. Copyright 2016 Taylor and Francis.)

two-dimensional substance graphene'. Graphene is a single atomic layer of graphite [28]. Graphene, a novel class of two-dimensional carbon allotropes, has sparked tremendous interest over the last two decades in various fields because of its unique physical and chemical features, including exceptional electronic conductivity, superior mechanical qualities, and enormous theoretical specific surface area [28–37]. Graphene is a single-layer carbon sheet with a hexagonal packed lattice structure resembling a honeycomb with long-range π-conjugation and has been demonstrated to have unique properties such as a large theoretical specific surface area (~2620 m^2 g^{-1}), excellent mechanical strength and flexibility, good optical transparency (~97.7%), a high Young's modulus (~1 TPa), quantum hall effect (QHE), excellent thermal conductivity (~3000–5000 W m^{-1} K^{-1}), and high carrier mobility at room temperature (~10 000 cm^2 V^{-1} s^{-1}), etc [29, 38–41]. Graphene, as a wonderful 2D building block, can be assembled into a variety of functional macroscopic structures, such as graphene fibers, films, paper, and sponges [42, 43]. One of graphene's most surprising qualities is that its charge carriers behave like massless relativistic particles, sometimes known as Dirac fermions, and may move around with little scattering under ambient conditions. This one-of-a-kind activity has resulted in a slew of unusual graphene occurrences [40]. According to research, graphene is a zero-gap 2D semimetal with a limited overlap between the conductance and valance bands and little scattering for charge carrier motion in natural environments [40]. When the gate voltage is applied, it also exhibits a high ambipolar electric field effect, with a charge carrier concentration up to 10^{13} cm^2. Furthermore, at carrier densities below 5×10^9 cm^2, suspended graphene exhibits low-temperature mobility approaching 200 000 cm^2 V^{-1} s^{-1}, which is not observed in semiconductors (figure 4.2) [44].

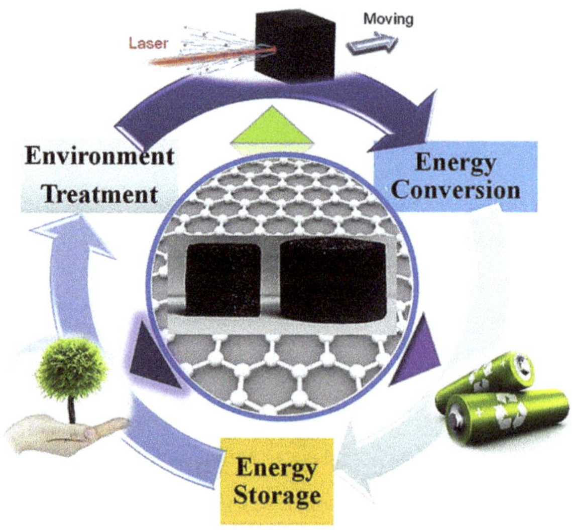

Figure 4.2. Three-dimensional graphene-based materials are used for various applications, particularly in the energy conversion/storage and environmental domains. (Reproduced with permission from [45]. Copyright 2018 Elsevier.)

Additionally, an unusual half-integer quantum Hall effect (QHE) for both electron and hole carriers in graphene has been discovered, which modifies the chemical potential utilizing the electric field effect [46, 47]. Using atomic force microscopy (AFM), the intrinsic mechanical characteristics of free-standing monolayer graphene membranes were investigated based on nano dimensions [39]. The mechanical properties of zigzag graphene and armchair graphene nanoribbons have been studied theoretically. The critical mechanical loads for failure and buckling of armchair ribbons appear lower than those for zigzag ribbons [48]. Graphene can also be modified chemically or physically using other components to provide advanced functionalities and better performance. However, graphene's poor processability, unique structure, and tough synthesis process have restricted its use as a bulk material. In particular, aggregation or re-stacking of graphene sheets reduces its intrinsic characteristics drastically [49, 50]. One potential alternative for preserving graphene's distinctive features in bulk and expanding its practical uses is to engineer graphene sheets to form three-dimensional (3D) architectures (e.g. foam, sponges, networks, aerogels, hydrogels) instead of re-stacking individual sheets [42, 43, 51–53]. More importantly, these 3D graphene-based materials retain graphene's intrinsic properties and add new collective physicochemical properties such as high porosity, low density, large specific surface area, excellent mechanical properties, unique electrochemical performance, and so on. Graphene and its variants have been manufactured, grown, and synthesized using a variety of processes (figure 4.3) [54].

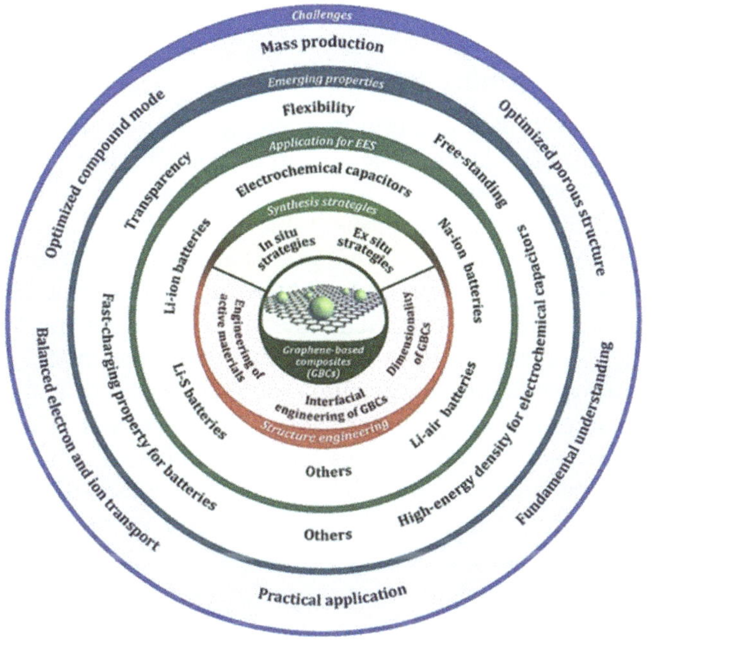

Figure 4.3. Three-dimensional graphene-based materials are used for various applications. (Reproduced with permission from [55]. Copyright 2019 Elsevier.)

Strong van der Waals forces bind adjacent graphene layers in graphite. As a result, immaculate graphene can be created by mechanically exfoliating graphite with adhesive tape, and graphene can be produced. Alternatives to the mechanical exfoliation of bulk graphite include the bottom-up creation of graphene sheets, and single- and few-layer graphene sheets have been grown on metal surfaces, such as Ni and Cu, using the chemical vapor deposition (CVD) process [33, 40, 54, 56–59].

4.2 Synthesis procedure

Significant effort has gone into developing synthesis methods for producing graphene-based composites with various morphologies, dimensions, structures, and characteristics over the last few years. To date, enormous effort has gone into developing synthetic methods for graphene and its derivatives, not only to reach a high yield of production but also to realize the solution or thin film-based process. Bottom-up and top-down methodologies have been utilized to synthesize graphene-based materials. Generally, chemists prefer a bottom-up approach for synthesizing graphene-based composites (GCs) with variable morphologies. The bottom-up strategy involves synthesizing graphene materials directly from carbon sources, such as chemical vapor deposition (CVD). The synthesis protocol or route for GCs can be categorized into four types: (i) one-step *in situ* formation of nanomaterials on graphenes; (ii) *in situ* growth of the nanocomposites or nanomaterials on graphene in a multi-step process; (iii) assemble of graphene and other nanomaterials in *ex situ* pathways; and (iv) formation of graphenes on low-dimensional nanomaterials in an *in situ* process. The synthesis protocols for graphene-based composites are described in what follows.

4.2.1 Chemical vapor deposition (CVD) process

Chemical vapor deposition is a common, scalable, and cost-effective approach for synthesizing single and few-layer, large-area graphene-based composites on a variety of substrates, opening a new path to large-scale manufacturing of high-quality graphene films for practical applications (table 4.1).

The most popular method for growing graphene with regulated layers and size is CVD. The same method may be used to synthesize 3D graphene-based composites by employing a porous carbon as the template and a transition metal as the catalyst, such as nickel (Ni) or copper (Cu) foam. After removing metal templates, free-standing 3D graphene-based composites were synthesized with predetermined morphologies. Using the template-assisted CVD approach, Cheng's team first developed a generic method and synthesized 3D graphene foam [63]. High-quality 3D GCs can also be produced by utilizing the template-assisted CVD approach inspired by the aforementioned study by employing various carbon sources or reaction conditions [64, 65].

4.2.2 Chemical oxidation of graphite

Chemical oxidation of graphite is the most common process to synthesize graphene, and is generally known as the bottom-up approach [66]. Using this technique,

Table 4.1. Some preparation methods of graphene and graphene-based composite materials by the CVD method.

Sl. No.	Synthesis method	Condition	Yield and properties	References
1.	Organic synthesis	High-temperature fusion of polycyclic aromatic hydrocarbons (temperature 1100 °C)	i. 30 nm thick films ii. Conductivity: 20 600 S m^{-1}	[60]
2.	CVD	• Carbon source: CH_4, H_2 • Substrate: Ni, Ru, Cu • Temperature: 1000 °C	Sheet size up to a few tens of micrometers.	[33, 58, 61]
3.	PECVD	• Carbon source: CH_4, H_2 • Substrate: Cu • Temperature: 650 °C	A large area of more than 1 cm of monolayer graphene.	[62]
4.	CVD	• Carbon source: pyrolysis of CH_4, poly(methyl methacrylate) • Substrate: Ni foam • Temperature: 1000 °C	i. Free-standing 3D graphene foam. ii. Specific surface area up to 850 m^2 g^{-1}.	[63]

graphite is first oxidized chemically and forms graphite oxide (GO), and is then reduced chemically to form graphene oxide. Raw graphite can be oxidized chemically using concentrated acids such as sulfuric, nitric, and phosphoric acid as well as highly oxidizing substances such as potassium permanganate and potassium perchlorate [67]. Hummers' method has been utilized as a process to synthesize the graphene/graphene oxide/reduced graphene oxide from raw graphite. In this process, potential oxidizing agents and concentrated acids are required to prepare graphene [68]. Dimanganeseheptoxide is been produced from $KMnO_4$ in the presence of a strong acid [69]. The reactions are as follows:

$$KMnO_4 + 3H_2SO_4 \rightarrow K^+ + MnO_3^+ + H_3O^+ + 3HSO_4^-$$
$$MnO_3^+ + MnO_4^- \rightarrow Mn_2O_7$$

4.2.3 Graphene-based materials

It is challenging to evenly disperse hydrophobic graphite flakes in water without dispersing agents. By reducing graphene oxide with hydrazine and dialysis while keeping the pH of the solution at around 10, chemically reduced graphene oxide (rGO), also known as chemically modified graphene, can be created since it contains oxygen-containing groups [70]. The resulting solubility of rGO in water, with a value of 0.5 mg ml^{-1}, is extremely low. However, organic solvents such as dimethyl

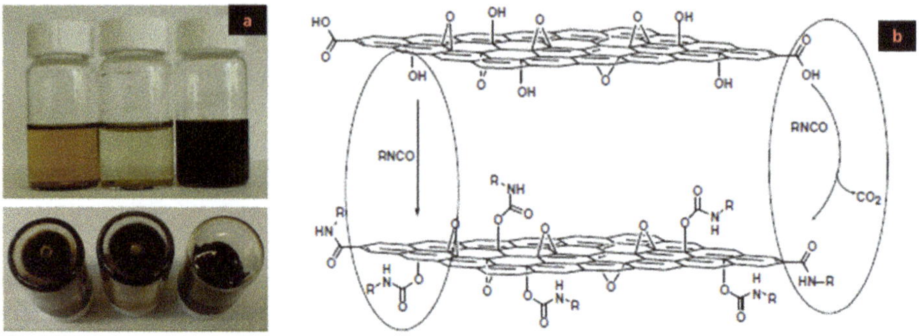

Figure 4.4. (a) Dispersion of GO in vials in different organic solvents such as DMF (left), phenyl isocyanate in water (middle), and isocyanate treated GO in DMF (right). (b) Proposed reaction mechanism during isocyanate treatment of GO. (Reproduced with permission from [72]. Copyright 2006 Elsevier.)

formamide (DMF), ethanol, acetone, dimethyl sulfoxide (DMSO), and tetrahydrofuran (THF) are useful to produce a suspension of highly reduced GO sheets (figure 4.4) [71].

rGO flakes can be modified by metal ions as well as functional groups. K-modified graphene (hKMG) flakes have been produced through the sequential chemical reaction of GO sheets in water with KOH and subsequently with hydrazine. This process provides a homogeneous aqueous suspension of graphene oxide flakes [73]. The treatment of GO with organic isocyanates could lead to the derivatization of both the edge carbonyl and surface hydroxyl functional groups via the formation of amides or carbamate esters (RnCO). The fabrication of GO–polymer composite sheets also led to the esterification of the carboxylic groups in GO with hydroxyl groups in poly(vinyl alcohol). Some organic reagents such as 3-amino-propyltriethoxysilane and 1-(3-aminopropyl)-3-methylimidazolium bromide, an example of amine-terminated organic molecules, can be used to functionalize the basal surface of GO [74]. Additionally, diazonium salts, such as sodium dodecyl benzenesulfonate (SDBS), were used to functionalize rGO platelets covalently, and the resulting rGO was easily dispersed in some polar organic solvents [75]. Noncovalent functionalization, which is based on the van der Waals force or the interaction between rGO and stabilizers, has a less detrimental effect on the structure of graphene and its derivatives than covalent functionalization. It also makes it possible to modify the solubility and electronic properties of these materials.

4.2.4 Graphene–metal-oxide-based composites

Graphene and its derivatives are good templates for synthesizing metal nanoparticles or metal oxides including noble metals or transition metals [76–90]. Depending on the type of the attached metal-based nanoparticles, graphene-based metal nanocomposites have been used in a wide range of applications, including surface-enhanced Raman scattering (SERS), catalysis, electrochemical sensing, etc [82, 90–93]. Graphene-based metal nanocomposites have been synthesized by various processes such as chemical reduction, microwave synthesis, hydrothermal synthesis, thermal oxidation, reduction, photochemical synthesis, gas-phase deposition, sol–gel processing, electroless metallization, CVD, etc [83, 93–96]. Among them the hydrothermal or solvothermal methods have been chosen to fabricate graphene-based metal nanocomposite due to their simplicity. Without post-synthetic annealing or calcination, these one-pot methods can produce nanostructures with great crystallinity. Typically, a single hydrothermal step was used for *in situ* generated SnO_2 nanorods on graphene sheets.

4.3 Application

4.3.1 Photovoltiac cells

Due to its simplicity of manufacture, light weight, and compatibility with flexible substrates, organic photovoltaic (OPV) cells have been proposed as a method of achieving low-cost energy through solar energy harvesting. The transparent conducting electrode through which light couples into the device is a crucial component of this kind of optoelectronic device. Transparent indium oxide (ITO) or fluorine-doped tin oxide (FTO) are frequently used as such electrodes in conventional OPVs. However, there are significant restrictions on the use of these materials for applications where cost, physical conformation, and mechanical flexibility are crucial due to the dearth of indium sources, demanding processing requirements, and the very brittle nature of metal oxides [97, 98].

4.3.2 Biomedical applications

Since the initial study on the use of graphene for biomedical applications in 2008, a great deal of impressive work has been done to investigate its potential for a variety of biomedical applications, including drug and gene delivery, biological sensing and imaging, antibacterial materials, and biocompatible scaffolds for cell culture [99].

4.3.3 Fuel cells

An electrochemical device called a fuel cell uses the oxidation of fuel at the anode and the reduction of oxygen from air at the cathode to generate electricity. The creation of effective fuel cells with high energy densities, low operating temperatures, and minimal environmental impact has received a lot of attention [100]. The best option for improved nanoelectrocatalysts for fuel cells is still carbon nanomaterials that support noble-metal-based nanocomposites. Due to its large surface area, superior conductivity, and outstanding chemical stability, graphene oxide-based nanocomposite as a 2D support should offer a favorable opportunity for the dispersion of metal-based nanocomposite to boost their electrocatalytic activity in fuel cells. Biofuel cells have been developed for the downsizing and portability of computing and communication devices. Currently, nanostructured electrode materials have tremendous potential for immobilizing enzymes in order to improve electron transfer between the enzyme and electrode while maintaining the enzyme's bioactivity. For instance, Li *et al* used a graphene-based nanocomposite as an electrode material in biofuel cells for the first time (figure 4.5) [101].

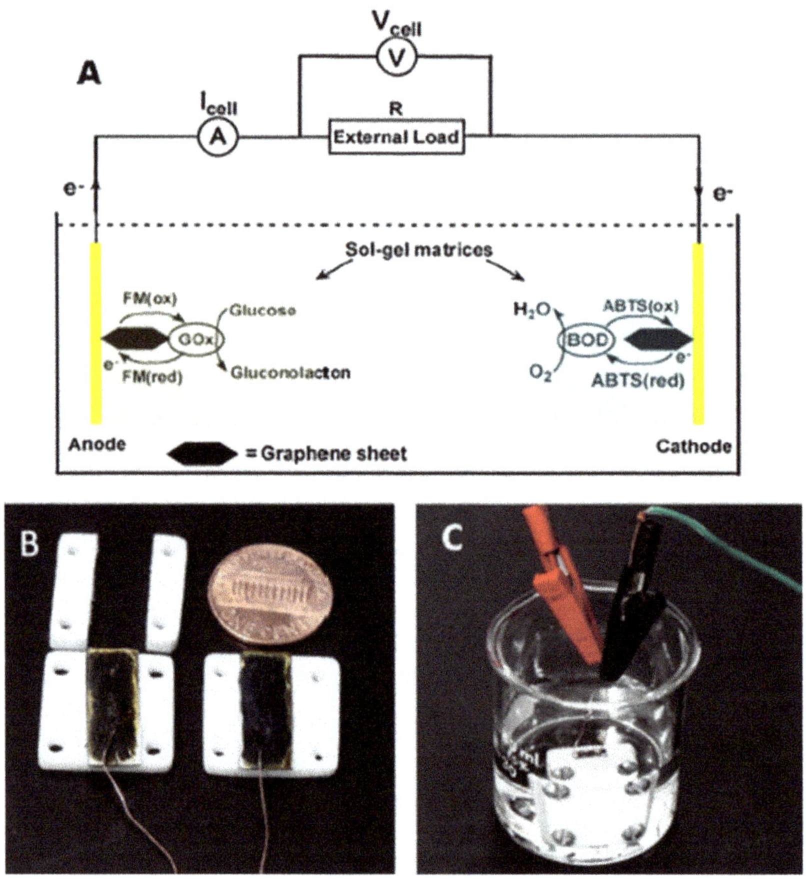

Figure 4.5. (a) Membraneless graphene-based enzymatic biofuel cell components, (b) enzymatic biofuel cell primary test set-up, and (c) practical representation of graphene-based enzymatic biofuel cell set-up for biological oxygen demand functionalized electrodes. (Reproduced with permission from [101]. Copyright 2009 Elsevier.)

4.3.4 Graphene-based FRET biosensors

The use of graphene in the creation of fluorescence resonance energy transfer (FRET) biosensors is receiving more and more attention. FRET is one of the more sophisticated technologies available for monitoring nanometer-scale distance and changes, both *in vivo* and *in vitro*. It entails the transfer of energy from a donor fluorophore to an acceptor fluorophore [102].

4.3.5 Energy storage devices

In order to meet both present and future energy needs, supercapacitors have received a lot of attention. Supercapacitors can be used in place of or in addition to batteries because of their quick charge–discharge rates, extremely long life, straightforward geometry, and environmentally benign design. Additionally, this cutting-edge technology can be produced on flexible and wearable electronics [103–105]. Supercapacitors can be used in hybrid electric vehicles since they charge quickly. This technology can be used in a variety of industries, including aerospace, manufacturing, and electronic devices, in addition to the realm of road transportation. Supercapacitors fall into two categories: pseudocapacitors, which store energy by redox reactions, and electric double-layer capacitors, which store energy electrostatically. However, the hybrid electrodes typically store energy using both the electrochemical and electrostatic methods. The thinnest nanoplate in the world, graphene exhibits exceptional electrical conductivity, a highly tunable surface area, chemical stability, and excellent mechanical behavior. There has also been a lot of interest in graphene with a metal oxide framework. Due to their complementary effects, a solid nanocomposite made of several oxides has better capacitance and cyclic efficiency. Distinct metal oxide manufacturing techniques result in different nanostructures, which in turn provide various electrochemical characteristics. The particular capacitance ranges of graphene produced by various extraction and production techniques vary. For example, a MnO_2 decorated graphene oxide nanocomposite displays a specific capacitance of 759 F g^{-1} which has been synthesized using a hydrothermal or solvothermal method with a durability of 88% after 3500 consecutive cycles [106]. Similarly, a graphene decorated $CuCo_2O_4$ composite has been prepared by solution phase along with thermal treatment which shows a specific capacitance of 1813 F g^{-1} with 96% cycle stability (figure 4.6) [107].

We present a range of graphene oxide-based electrode materials for supercapacitor applications in table 4.2.

4.4 Conclusion

In conclusion, a review of the significant advancements in graphene oxide-based nanocomposite research during the past few years has been performed. These fascinating research contributions include finding efficient methods for the large-scale production of high-quality graphene-oxide-based nanocomposites for band gap engineering and potential tunable new properties, sophisticated assembly methods for 2D and 3D functional thin films or architectures, nanohybrids with improved functionality, and high-performance devices. High-performance supercapacitor devices based on graphene oxide composites have been compared and

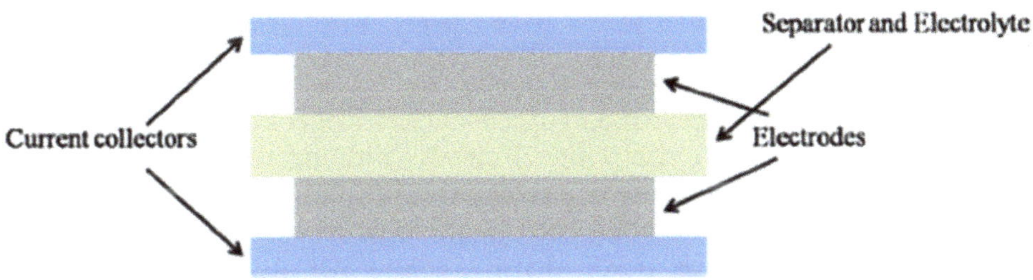

Figure 4.6. Internal structure of a supercapacitor. (Reproduced with permission from [108]. Copyright 2021 Elsevier.)

Table 4.2. Some recent reports on graphene oxide-based electrode materials for supercapacitors.

Active electrode materials	Supporting electrolyte	Capacitance (F g^{-1})	Durability	Reference
Graphene aerogel/CeO_2 nanocomposite	KOH	376.2	91% after 4000 cycles	[109]
MoS_2/graphene nanocomposite	Na_2SO_4	401	95.1% after 10 000 cycles	[110]
Graphene/lanthanum nanocomposite	KOH	156	78% after 400 cycles	[111]
MnO_2/rGO composite	Na_2SO_4	759	88% after 3500 cycles	[112]
$CuCo_2O_4$/graphene hollow spheres	KOH	1813	95% after 6000 cycles	[113]
NiO/Co_3O_4 decorated rGO	KOH	910	90% after 2000 cycles	[114]
$NiCo_2O_4$ nanorod decorated graphene	KOH	2147	94% after 2000 cycles	[115]
$MnFe_2O_4$ decorated nitrogen doped graphene sheet	KOH	905	83% after 65 000 cycles	[116]
Bismuth–iron oxide decorated graphene oxide	KOH	306	60% after 1000 cycles	[117]
Conducting polymer coated graphene oxide	Organic electrolyte	63 mF cm^{-1}	88% after 10 000 cycle	[118]
Polyaniline coated graphene oxide	H_2SO_4	299	89% after 2000 cycles	[119]
Polyaniline coated graphene oxide	H_2SO_4	257	98% after 1000 cycles	[120]
Strontium ferrite decorated graphene oxide	KOH	681	96% after 10 000 cycles	[121]
Ni–Fe decorated graphene oxide	Na_2SO_4	207	95% after 1000 cycles	[122]
Mn–Fe decorated graphene oxide	KOH	537	88% after 20 000 cycles	[123]
Ag decorated graphene oxide	NaOH	426	100% after 1000 cycles	[124]
Graphene oxide decorated on silicon	KOH	2 mF cm^{-2}	81% after 150 000 cycles	[125]

reviewed. The two most important requirements for a high-performance supercapacitor are a larger surface area and excellent electrical conductivity. Graphene is the ideal material to prepare composite electrodes with metal oxides or conducting polymers in this situation. Because of its comparatively large surface area and strong electrical conductivity, graphene has been found to be crucial for improving overall qualities such as capacitance, power density, energy density, and durability. The performance of supercapacitors is also affected by different electrolytes and synthesis techniques. To provide a greater surface area to the electrolyte ions, efforts are being made to generate vertical graphene sheets on the electrode. The capacitance values of the various graphene-based composite electrodes ranged from 73 to 3553 F g^{-1}. More research is still needed to address the need for commercial use, despite the 3D graphene-based nanocomposite's relatively good performance at a laboratory scale. Therefore, future studies should focus on, but not be limited to, the following areas. A large-scale, cost-effective, and eco-friendly technique to produce 2D or 3D graphene-oxide-based metal nanocomposites is needed. The majority of the fabrication techniques used today, including 3D printing, self-assembly, and utilizing graphene oxide as their primary building elements. The performance of graphene oxide-based nanocomposites is influenced greatly by the pore diameters, volumes, conductivity etc. Therefore, the graphene oxide-based nanocomposites must be constructed using precision fabrication techniques. It is possible to employ specifically created molecules with specific crosslinking sites in a subsequent stage to form graphene-oxide-based nanocomposites with customized structures, drawing inspiration from the DNA assembling process.

References

[1] Bi Z, Kong Q, Cao Y, Sun G, Su F, Wei X, Li X, Ahmad A, Xie L and Chen C M 2019 Biomass-derived porous carbon materials with different dimensions for supercapacitor electrodes: a review *J. Mater. Chem.* A **7** 16028–45

[2] Balahmar N, Mitchell A C and Mokaya R 2015 Generalized mechanochemical synthesis of biomass-derived sustainable carbons for high performance CO_2 storage *Adv. Energy Mater.* **5** 1500867

[3] Wu Z Y, Hu B C, Wu P, Liang H W, Yu Z L, Lin Y, Zheng Y R, Li Z and Yu S H 2016 Mo_2C nanoparticles embedded within bacterial cellulose-derived 3D N-doped carbon nanofiber networks for efficient hydrogen evolution *NPG Asia Mater.* **8** e288

[4] Zhang J W, Zhang H T, Du Z Y, Wang X, Yu S H and Jiang H L 2014 Water-stable metal–organic frameworks with intrinsic peroxidase-like catalytic activity as a colorimetric biosensing platform *Chem. Commun.* **50** 1092–4

[5] Cheng W, Zhang J, Wang Z, Wang M and Xie S 2014 Bacterial communities in sediments of a drinking water reservoir *Ann. Microbiol.* **64** 875–8

[6] Zhou W, Chen J, Chen M, Xu X, Tian Q, Xu J and Wong C P 2019 Rod-like anhydrous V_2O_5 assembled by tiny nanosheets as a high-performance cathode material for aqueous zinc-ion batteries *RSC Adv.* **9** 30556–64

[7] Leng C, Sun K, Li J and Jiang J 2017 From dead pine needles to O, N codoped activated carbons by a one-step carbonization for high rate performance supercapacitors *ACS Sustain. Chem. Eng.* **5** 10474–82

[8] Li M, Liu C, Cao H, Zhao H, Zhang Y and Fan Z 2014 KOH self-templating synthesis of three-dimensional hierarchical porous carbon materials for high performance supercapacitors *J. Mater. Chem.* A **2** 14844–51

[9] Raymundo-Piñero E, Leroux F and Béguin F 2006 A high-performance carbon for supercapacitors obtained by carbonization of a seaweed biopolymer *Adv. Mater.* **18** 1877–82

[10] Hu W, Peng C, Luo W, Lv M, Li X, Li D, Huang Q and Fan C 2010 Graphene-based antibacterial paper *ACS Nano* **4** 4317–23

[11] Zheng C, Zhang J, Zhang Q, You B and Chen G 2015 Three dimensional Ni foam-supported graphene oxide for binder-free pseudocapacitor *Electrochim. Acta* **152** 216–21
[12] Lu M M, Cao M S, Chen Y H, Cao W Q, Liu J, Shi H L, Zhang D Q, Wang W Z and Yuan J 2015 Multiscale assembly of grape-like ferroferric oxide and carbon nanotubes: a smart absorber prototype varying temperature to tune intensities *ACS Appl. Mater. Interfaces* **7** 19408–15
[13] Liu Y, Shi Z, Gao Y, An W, Cao Z and Liu J 2016 Biomass-swelling assisted synthesis of hierarchical porous carbon fibers for supercapacitor electrodes *ACS Appl. Mater. Interfaces* **8** 28283–90
[14] Madhu R, Sankar K V, Chen S M and Selvan R K 2014 Eco-friendly synthesis of activated carbon from dead mango leaves for the ultrahigh sensitive detection of toxic heavy metal ions and energy storage applications *RSC Adv.* **4** 1225–33
[15] Luo C, Zuo X, Wang L, Wang E, Song S, Wang J, Wang J, Fan C and Cao Y 2008 Flexible carbon nanotube–polymer composite films with high conductivity and superhydrophobicity made by solution process *Nano Lett.* **8** 4454–8
[16] Linares N, Silvestre-Albero A M, Serrano E, Silvestre-Albero J and García-Martínez J 2014 Mesoporous materials for clean energy technologies *Chem. Soc. Rev.* **43** 7681–717
[17] Zhang L L and Zhao X S 2009 Carbon-based materials as supercapacitor electrodes *Chem. Soc. Rev.* **38** 2520–31
[18] Wang Z, Li F, Ergang N S and Stein A 2006 Effects of hierarchical architecture on electronic and mechanical properties of nanocast monolithic porous carbons and carbon–carbon nanocomposites *Chem. Mater.* **18** 5543–53
[19] You B, Jiang J and Fan S 2014 Three-dimensional hierarchically porous all-carbon foams for supercapacitor *ACS Appl. Mater. Interfaces* **6** 15302–8
[20] You B, Yang J, Sun Y and Su Q 2011 Easy synthesis of hollow core, bimodal mesoporous shell carbon nanospheres and their application in supercapacitor *Chem. Commun.* **47** 12364–6
[21] You B, Wang L, Li N and Zheng C 2014 Improving the energy storage performance of graphene through insertion of pristine CNTs and ordered mesoporous carbon coating *Chem. Electro. Chem.* **1** 772–8
[22] Yuan C Z, Gao B, Shen L F, Yang S D, Hao L, Lu X J, Zhang F, Zhang L J and Zhang X G 2011 Hierarchically structured carbon-based composites: design, synthesis and their application in electrochemical capacitors *Nanoscale* **3** 529–45
[23] Wahid M, Parte G, Phase D and Ogale S 2015 Yogurt: a novel precursor for heavily nitrogen doped supercapacitor carbon *J. Mater. Chem.* A **3** 1208–15
[24] Schwenke A M, Hoeppener S and Schubert U S 2015 Synthesis and modification of carbon nanomaterials utilizing microwave heating *Adv. Mater.* **27** 4113–41
[25] Gao S, Chen Y, Fan H, Wei X, Hu C, Luo H and Qu L 2014 Large scale production of biomass-derived N-doped porous carbon spheres for oxygen reduction and supercapacitors *J. Mater. Chem.* A **2** 3317–24
[26] Sun K, Li J, Wu D and Jiang J 2020 Green synthesis of porous honeycomblike carbon materials for supercapacitor electrodes *Ind. Eng. Chem. Res.* **59** 14288–95
[27] Abbasi E, Akbarzadeh A, Kouhi M and Milani M 2016 Graphene: synthesis, bio-applications, and properties *Artif. Cells Nanomed. Biotechnol.* **44** 150–6
[28] Singh V, Joung D, Zhai L, Das S, Khondaker S I and Seal S 2011 Graphene based materials: past, present and future *Prog. Mater. Sci.* **56** 1178–271

[29] Nair R R, Blake P, Grigorenko A N, Novoselov K S, Booth T J, Stauber T, Peres N M R and Geim A K 2008 Fine structure constant defines visual transparency of graphene *Science* **320** 1308

[30] Service R F 2009 Carbon sheets an atom thick give rise to graphene dreams *Science* **324** 875–7

[31] Hall E O *et al* 2009 Control of graphene's properties by reversible hydrogenation: evidence for graphane *Science* **323** 610–4

[32] Li X, Wang X, Zhang L, Lee S and Dai H 2008 Chemically derived, ultrasmooth graphene nanoribbon semiconductors *Science* **319** 1229–32

[33] Kim K S, Zhao Y, Jang H, Lee S Y, Kim J M, Kim K S, Ahn J H, Kim P, Choi J Y and Hong B H 2009 Large-scale pattern growth of graphene films for stretchable transparent electrodes *Nature* **457** 706–10

[34] Geim A K 2009 Graphene: status and prospects *Science* **324** 1530–4

[35] Stankovich S, Dikin D A, Dommett G H B, Kohlhaas K M, Zimney E J, Stach E A, Piner R D, Nguyen S B T and Ruoff R S 2006 Graphene-based composite materials *Nature* **442** 282–6

[36] Tung V C, Allen M J, Yang Y and Kaner R B 2009 High-throughput solution processing of large-scale graphene *Nat. Nanotechnol.* **4** 25–9

[37] Geim A and Novoselov K 2007 The rise of graphene *Nat. Mater.* **6** 183–91

[38] Balandin A A, Ghosh S, Bao W, Calizo I, Teweldebrhan D, Miao F and Lau C N 2008 Superior thermal conductivity of single-layer graphene *Nano Lett.* **8** 902–7

[39] Lee C, Wei X, Kysar J W and Hone J 2008 Measurement of the elastic properties and intrinsic strength of monolayer graphene *Science* **321** 385–8

[40] Novoselov K S, Geim A K, Morozov S V, Jiang D, Zhang Y, Dubonos S V, Grigorieva I V and Firsov A A 2004 Electric field in atomically thin carbon films *Science* **306** 666–9

[41] Stoller M D, Park S, Yanwu Z, An J and Ruoff R S 2008 Graphene-based ultracapacitors *Nano Lett.* **8** 3498–502

[42] Cao X, Yin Z and Zhang H 2014 Three-dimensional graphene materials: preparation, structures and application in supercapacitors *Energy Environ. Sci.* **7** 1850–65

[43] Jiang L and Fan Z 2014 Design of advanced porous graphene materials: from graphene nanomesh to 3D architectures *Nanoscale* **6** 1922–45

[44] Du X, Skachko I, Barker A and Andrei E Y 2008 Approaching ballistic transport in suspended graphene *Nat. Nanotechnol.* **3** 491–5

[45] Wu Y, Zhu J and Huang L 2019 A review of three-dimensional graphene-based materials: synthesis and applications to energy conversion/storage and environment *Carbon* **143** 610–40

[46] Zhang Y, Tan Y W, Stormer H L and Kim P 2005 Experimental observation of the quantum Hall effect and Berry's phase in graphene *Nature* **438** 201–4

[47] Novoselov K S, Geim A K, Morozov S V, Jiang D, Katsnelson M I, Grigorieva I V, Dubonos S V and Firsov A A 2005 Two-dimensional gas of massless Dirac fermions in graphene *Nature* **438** 197–200

[48] Gao Y and Hao P 2009 Mechanical properties of monolayer graphene under tensile and compressive loading *Physica* E **41** 1561–6

[49] Si Y and Samulski E T 2008 Synthesis of water soluble graphene *Nano Lett.* **8** 1679–82

[50] Ma Y and Chen Y 2015 Three-dimensional graphene networks: synthesis, properties and applications *Nat. Sci. Rev.* **2** 40–53

[51] Wu Y *et al* 2015 Three-dimensionally bonded spongy graphene material with super compressive elasticity and near-zero Poisson's ratio *Nat. Commun.* **6** 6141

[52] Chowdhury S and Balasubramanian R 2017 Three-dimensional graphene-based macrostructures for sustainable energy applications and climate change mitigation *Prog. Mater Sci.* **90** 224–75

[53] Li C and Shi G 2012 Three-dimensional graphene architectures *Nanoscale* **4** 5549–63

[54] RozpŁoch F, Patyk J and Stankowski J 2007 Graphenes bonding forces in graphite *Acta Phys. Pol.* A **112** 557–62

[55] Wang B, Ruan T, Chen Y, Jin F, Peng L, Zhou Y, Wang D and Dou S 2020 Graphene-based composites for electrochemical energy storage *Energy Storage Mater.* **24** 22–51

[56] Li X *et al* 2010 Graphene films with large domain size by a two-step chemical vapor deposition process *Nano Lett.* **10** 4328–34

[57] Kalbac M, Reina-Cecco A, Farhat H, Kong J, Kavan L and Dresselhaus M S 2010 The influence of strong electron and hole doping on the Raman intensity of chemical vapor-deposition graphene *ACS Nano.* **4** 6055–63

[58] Sutter P W, Flege J I and Sutter E A 2008 Epitaxial graphene on ruthenium *Nat. Mater.* **7** 406–11

[59] Gomez De Arco L, Zhang Y, Schlenker C W, Ryu K, Thompson M E and Zhou C 2010 Continuous, highly flexible, and transparent graphene films by chemical vapor deposition for organic photovoltaics *ACS Nano* **4** 2865–73

[60] Wang X, Zhi L, Tsao N, Tomović Ž, Li J and Müllen K 2008 Transparent carbon films as electrodes in organic solar cells *Angew. Chem. Int. Ed.* **47** 2990–2

[61] Li X *et al* 2009 Large-area synthesis of high-quality and uniform graphene films on copper foils *Science* **324** 1312–4

[62] Wang Y, Xu X, Lu J, Lin M, Bao Q, Özyilmaz B and Loh K P 2010 Toward high throughput interconvertible graphane-to-graphene growth and patterning *ACS Nano* **4** 6146–52

[63] Chen Z, Ren W, Gao L, Liu B, Pei S and Cheng H M 2011 Three-dimensional flexible and conductive interconnected graphene networks grown by chemical vapour deposition *Nat. Mater.* **10** 424–8

[64] Liu Z, Tu Z, Li Y, Yang F, Han S, Yang W, Zhang L, Wang G, Xu C and Gao J 2014 Synthesis of three-dimensional graphene from petroleum asphalt by chemical vapor deposition *Mater. Lett.* **122** 285–8

[65] Pettes M T, Ji H, Ruoff R S and Shi L 2012 Thermal transport in three-dimensional foam architectures of few-layer graphene and ultrathin graphite *Nano Lett.* **12** 2959–64

[66] Dreyer D R, Ruoff R S and Bielawski C W 2010 From conception to realization: an historical account of graphene and some perspectives for its future *Angew. Chem. Int. Ed.* **49** 9336–44

[67] Adar F, Gouterman M and Aronowitz S 1976 Fluorescence, resonance Raman, and radiationless decay in several hemoproteins *J. Phys. Chem.* **80** 2184–91

[68] Su C Y, Xu Y, Zhang W, Zhao J, Tang X, Tsai C H and Li L J 2009 Electrical and spectroscopic characterizations of ultra-large reduced graphene oxide monolayers *Chem. Mater.* **21** 5674–80

[69] Koch K R and Krause P F 1982 Oxidation by Mn_2O_7: an impressive demonstration of the powerful oxidizing property of dimanganeseheptoxide *J. Chem. Educ.* **59** 973–4

[70] Li D, Müller M B, Gilje S, Kaner R B and Wallace G G 2008 Processable aqueous dispersions of graphene nanosheets *Nat. Nanotechnol.* **3** 101–5

[71] Park S, An J, Jung I, Piner R D, An S J, Li X, Velamakanni A and Ruoff R S 2009 Graphene oxide in a wide organic solvent *Nano Lett.* **9** 1593–7

[72] Stankovich S, Piner R D, Nguyen S B T and Ruoff R S 2006 Synthesis and exfoliation of isocyanate-treated graphene oxide nanoplatelets *Carbon* **44** 3342–7

[73] Park S, An J, Piner R D, Jung I, Yang D, Velamakanni A, Nguyen S B T and Ruoff R S 2008 Aqueous suspension and characterization of chemically modified graphene sheets *Chem. Mater.* **20** 6592–4

[74] Yang H, Shan C, Li F, Han D, Zhang Q and Niu L 2009 Covalent functionalization of polydisperse chemically-converted graphene sheets with amine-terminated ionic liquid *Chem. Commun.* 3880–2

[75] Lomeda J R, Doyle C D, Kosynkin D V, Hwang W and Tour J M 2008 Diazonium functionalization of surfactant-wrapped chemically converted graphene sheets *J. Am. Chem. Soc.* **130** 16201–6

[76] Gong J, Zhou T, Song D and Zhang L 2010 Monodispersed Au nanoparticles decorated graphene as an enhanced sensing platform for ultrasensitive stripping voltammetric detection of mercury(II) *Sensors Actuators* B **150** 491–7

[77] Fu C, Kuang Y, Huang Z, Wang X, Du N, Chen J and Zhou H 2010 Electrochemical co-reduction synthesis of graphene/Au nanocomposites in ionic liquid and their electro-chemical activity *Chem. Phys. Lett.* **499** 250–3

[78] Mastalir Á, Király Z, Patzkó Á, Dékány I and L'Argentiere P 2008 Synthesis and catalytic application of Pd nanoparticles in graphite oxide *Carbon* **46** 1631–7

[79] Kamat P V 2010 Graphene-based nanoarchitectures. Anchoring semiconductor and metal nanoparticles on a two-dimensional carbon support *J. Phys. Chem. Lett.* **1** 520–7

[80] Shi M, Shen J, Ma H, Li Z, Lu X, Li N and Ye M 2012 Preparation of graphene–TiO_2 composite by hydrothermal method from peroxotitanium acid and its photocatalytic properties *Colloids Surf.* A **405** 30–7

[81] Lu G, Mao S, Park S, Ruoff R S and Chen J 2009 Facile, noncovalent decoration of graphene oxide sheets with nanocrystals *Nano Res.* **2** 192–200

[82] Goncalves G, Marques P A A P, Granadeiro C M, Nogueira H I S, Singh M K and Grácio J 2009 Surface modification of graphene nanosheets with gold nanoparticles: the role of oxygen moieties at graphene surface on gold nucleation and growth *Chem. Mater.* **21** 4796–802

[83] Muszynski R, Seger B and Kamat P V 2008 Decorating graphene sheets with gold nanoparticles *J. Phys. Chem.* C **112** 5263–6

[84] Osváth Z *et al* 2010 The structure and properties of graphene supported on gold nanoparticles *Int. J. Electrochem. Sci.* **114** 1822–6

[85] Xu C, Wang X and Zhu J 2008 Graphene–metal particle nanocomposites *J. Phys. Chem.* C **112** 19841–5

[86] Huang X, Zhou X, Wu S, Wei Y, Qi X, Zhang J, Boey F and Zhang H 2010 Reduced graphene oxide-templated photochemical synthesis and *in situ* assembly of Au nanodots to orderly patterned Au nanodot chains *Small* **6** 513–6

[87] Jasuja K and Berry V 2009 Implantation and growth of dendritic gold nanostructures on graphene derivatives: electrical property tailoring and Raman enhancement *ACS Nano* **3** 2358–66

[88] Huang X, Li S, Huang Y, Wu S, Zhou X, Li S, Gan C L, Boey F, Mirkin C A and Zhang H 2011 Synthesis of hexagonal close-packed gold nanostructures *Nat. Commun.* **2** 292

[89] Carara S S, Batista R J C and Chacham H 2009 Modifications in graphene electron states due to a deposited lattice of Au nanoparticles: density functional calculations *Phys. Rev.* B **80** 115435

[90] Wang Z, Wu S and Zhang J *et al* 2012 Comparative studies on single-layer reduced graphene oxide films obtained by electrochemical reduction and hydrazine vapor reduction *Nanoscale Res. Lett.* **7** 161

[91] Wu H, Wang J, Kang X, Wang C, Wang D, Liu J, Aksay I A and Lin Y 2009 Glucose biosensor based on immobilization of glucose oxidase in platinum nanoparticles/graphene/chitosan nanocomposite film *Talanta* **80** 403–6

[92] Guo S, Wen D, Zhai Y, Dong S and Wang E 2010 Platinum nanoparticle ensemble-on-graphene hybrid nanosheet: one-pot, rapid synthesis, and used as new electrode material for electrochemical sensing *ACS Nano* **4** 3959–68

[93] Khan M S, Naqvi S A, Nisar N and Nawab F 2018 Medical students perspective about adverse effects of caffeine consumption *Rawal Med. J.* **43** 156–60

[94] Kim Y K, Na H K, Lee Y W, Jang H, Han S W and Min D H 2010 The direct growth of gold rods on graphene thin films *Chem. Commun.* **46** 3185–7

[95] Kim Y K, Na H K and Min D H 2010 Influence of surface functionalization on the growth of gold nanostructures on graphene thin films *Langmuir* **26** 13065–70

[96] Nayak A K, Das A K and Pradhan D 2017 High performance solid-state asymmetric supercapacitor using green synthesized graphene–WO_3 nanowires nanocomposite *ACS Sustain. Chem. Eng.* **5** 10128–38

[97] Peumans P, Yakimov A and Forrest S R 2003 Small molecular weight organic thin-film photodetectors and solar cells *J. Appl. Phys.* **93** 3693–723

[98] De Arco L G, Zhang Y and Zhou C 2011 Large scale graphene by chemical vapor deposition: synthesis, characterization and applications *Graphene – Synthesis, Characterization, Properties and Applications* (London: InTechOpen)

[99] Guo S and Dong S 2011 Graphene nanosheet: synthesis, molecular engineering, thin film, hybrids, and energy and analytical applications *Chem. Soc. Rev.* **40** 2644–72

[100] Mazumder V, Lee Y and Sun S 2010 Recent development of active nanoparticle catalysts for fuel cell reactions *Adv. Funct. Mater.* **20** 1224–31

[101] Liu C, Alwarappan S, Chen Z, Kong X and Li C Z 2010 Membraneless enzymatic biofuel cells based on graphene nanosheets *Biosens. Bioelectron.* **25** 1829–33

[102] Selvin P 2000 The renaissance of fluorescence resonance energy transfer *Nat. Struct. Mol. Biol.* **7** 730–4

[103] Kao H, Ling, Chuang C H, Chang L C, Cho C L and Chiu H C 2019 Inkjet-printed silver films on textiles for wearable electronics applications *Surf. Coat. Technol.* **362** 328–32

[104] Qiu Q, Zhu M, Li Z, Qiu K, Liu X, Yu J and Ding B 2019 Highly flexible, breathable, tailorable and washable power generation fabrics for wearable electronics *Nano Energy* **58** 750–8

[105] Xiang C, Li M, Zhi M, Manivannan A and Wu N 2013 A reduced graphene oxide/Co_3O_4 composite for supercapacitor electrode *J. Power Sources* **226** 65–70

[106] Jadhav S, Kalubarme R S, Terashima C, Kale B B, Godbole V, Fujishima A and Gosavi S W 2019 Manganese dioxide/reduced graphene oxide composite an electrode material for high-performance solid state supercapacitor *Electrochim. Acta.* **299** 34–44

[107] Kaverlavani S K, Moosavifard S E and Bakouei A 2017 Designing graphene-wrapped nanoporous $CuCo_2O_4$ hollow spheres electrodes for high-performance asymmetric supercapacitors *J. Mater. Chem.* A **5** 14301–9

[108] Lakra R, Kumar R, Sahoo P K, Thatoi D and Soam A 2021 A mini-review: graphene based composites for supercapacitor application *Inorg. Chem. Commun.* **133** 108929

[109] Mazloum-Ardakani M, Mohammadian-Sarcheshmeh H, Naderi H, Farbod F and Sabaghian F 2019 Fabrication of a high-performance hybrid supercapacitor using a modified graphene aerogel/cerium oxide nanoparticle composite *J. Energy Storage* **26** 100998

[110] Fu M, Zhu Z, Chen W, Yu H and Liu Q 2020 Microwave-assisted synthesis of MoS_2/graphene composites for supercapacitors *J. Mater. Sci.* **55** 16385–93

[111] Zhang J, Zhang Z, Jiao Y, Yang H, Li Y, Zhang J and Gao P 2019 The graphene/lanthanum oxide nanocomposites as electrode materials of supercapacitors *J. Power Sources* **419** 99–105

[112] Jadhav S, Kalubarme R S, Terashima C, Kale B B, Godbole V, Fujishima A and Gosavi S W 2019 Manganese dioxide/reduced graphene oxide composite an electrode material for high-performance solid state supercapacitor *Electrochim. Acta.* **299** 34–44

[113] Kaverlavani S K, Moosavifard S E and Bakouei A 2017 Designing graphene-wrapped nanoporous CuCo2O4 hollow spheres electrodes for high-performance asymmetric supercapacitors *J. Mater. Chem.* A **5** 14301–9

[114] Kumar R, Youssry S M, Soe H M, Abdel-Galeil M M, Kawamura G and Matsuda A 2020 Honeycomb-like open-edged reduced-graphene-oxide-enclosed transition metal oxides (NiO/Co_3O_4) as improved electrode materials for high-performance supercapacitor *J. Energy Storage* **30** 101539

[115] Li Y, Zhang M, Pan D, Wang Y, Xie J, Yan Z and Jing J 2017 Ultrathin hollow-structured $NiCo_2O_4$ nanorod supported on improved N-doped graphene for superior supercapacitor applications *J. Alloys Compd.* **722** 903–12

[116] Su L, Lei S, Liu L, Liu L, Zhang Y, Shi S and Yan X 2018 Sprinkling MnFe2O4 quantum dots on nitrogen-doped graphene sheets: The formation mechanism and application for high-performance supercapacitor electrodes *J. Mater. Chem.* A **6** 9997–10007

[117] Pattananuwat P, Khampuanbut A and Haromae H 2021 Novel electrode composites of mixed bismuth-iron oxide/graphene utilizing for photo assisted supercapacitors *Electrochim. Acta.* **370** 137741

[118] Zhang H, Tang X, Zhao D, Zheng N, Huang L, Sun T, Gu C and Ma Y 2020 Suppressing charge trapping effect in ambipolar conducting polymer with vertically standing graphene as the composite electrode for high performance supercapacitor *Energy Storage Mater.* **29** 281–6

[119] Macherla N, Singh K, Santosh M S, Kumari K and Lekkala R G R 2021 Heat assisted facile synthesis of nanostructured polyaniline/reduced crumbled graphene oxide as a high-performance flexible electrode material for supercapacitors *Colloids Surfaces A Physicochem. Eng. Asp.* **612** 125982

[120] Li Z F, Zhang H, Liu Q, Sun L, Stanciu L and Xie J 2013 Fabrication of high-surface-area graphene/polyaniline nanocomposites and their application in supercapacitors *ACS Appl. Mater. Interfaces* **5** 2685–91

[121] Fu M, Zhang Z, Zhu Z, Zhuang Q, Chen W, Yu H and Liu Q 2021 Facile synthesis of strontium ferrite nanorods/graphene composites as advanced electrode materials for supercapacitors *J. Colloid Interface Sci.* **588** 795–803

[122] Soam A, Kumar R, Sahoo P K, Mahender C, Kumar B, Arya N, Singh M, Parida S and Dusane R O 2019 Synthesis of Nickel Ferrite Nanoparticles Supported on Graphene Nanosheets as Composite Electrodes for High Performance Supercapacitor *ChemistrySelect* **4** 9952–8

[123] Fu M, Zhu Z, Zhuang Q, Zhang Z, Chen W and Liu Q 2020 In situ growth of manganese ferrite nanorods on graphene for supercapacitors *Ceram. Int.* **46** 28200–5

[124] Karami Z, Youssefi M, Raeissi K and Zhiani M 2021 An efficient textile-based electrode utilizing silver nanoparticles/reduced graphene oxide/cotton fabric composite for high-performance wearable supercapacitors *Electrochim. Acta.* **368** 137647

[125] Aradilla D, Delaunay M, Sadki S, Gérard J M and Bidan G 2015 Vertically aligned graphene nanosheets on silicon using an ionic liquid electrolyte: Towards high performance on-chip micro-supercapacitors *J. Mater. Chem.* A **3** 19254–62

Chapter 5

Synthesis and applications of graphane

Rajashree Sahoo, Paritosh Chaudhuri and Arpan Kumar Nayak

Since the development of graphene, a hexagonal crystal carbon monolayer structure, its atomically thin crystals have drawn a lot of interest. The chemically synthesized graphanes have generated a great deal of curiosity in this emerging material class. Despite the fact that bulk graphite is a chemically stable substance, it was shown that graphene with its single layer is an inert substance, yet its surface is quite reactive to individual atoms. Numerous graphene derivatives have so far been synthesized and described, such as fluorographene, chlorographene, and hydrogenated graphene (graphane). Using computational tools, the stability of graphene coated with bromine and iodine was predicted. Among these derivatives, the reversibly adjustable crystal structure, simple synthesis, and insulating electrical properties of graphane are unique for use in ultra-thin devices. The structural, electrical, magnetic, vibrational, and mechanical properties of graphane are covered in this review. Additionally, current developments in synthesis methods and potential uses for graphane are discussed.

5.1 Introduction

New material predictions and discoveries are always intriguing because they hold the possibility of new uses and qualities. Such a subject is even more engaging when the material consists just of carbon and hydrogen and exhibits certain unique topological properties such as reduced dimensionality. The report introduces a material called "graphane" and discuses the potential for easy synthesis. [1, 2].

The organic substances with only carbon and hydrogen atoms are called hydrocarbons. Some of these are found naturally as natural gas and crude oil. Polyethylene and other plastics are examples of those that are synthesized. They often make good fuels because they easily oxidize to produce CO_2 and H_2O along with the release of some amount of energy. Until now, all hydrocarbons were molecules with a carbon backbone and connected hydrogen atoms. A ring, a linear chain, or a combination of the two can serve as the backbone. In contrast, the

doi:10.1088/978-0-7503-5177-5ch5

compound discussed in this chapter, graphane, is predicted to be a two-dimensional extended covalently bound hydrocarbon. Graphene, its fully dehydrogenated equivalent, has been a hot topic in numerous recent investigations because of its peculiar transport properties.

In many respects, graphene is a special kind of substance. Despite being the thinnest crystal imaginable, it is also the strongest, stiffest, and most conductive substance known to exist. Additionally, it has a fascinating electrical spectrum that resembles a massless object, Dirac particles that cause a variety of intriguing phenomena, a typical number quantum Hall outcome, and Klein tunneling. The lack of an energy gap in graphene's electrical spectrum could be seen as a negative trait that prevents the use of graphene directly in electronics. Consequently, the research based on graphene has also been committed to producing an electronic energy gap. There are several ways to accomplish this: graphene can be reduced in dimensions by cutting it into flakes or ribbons, it can have periodic potentials or substrates that disrupt sublattice symmetry, and it can be achieved using functionalization, which destroys the bond network.

Graphene can be functionalized in a variety ways. It is feasible to replace part of the carbon (C) atoms in the layer-like structure of graphene with non-native atoms such as boron (B) and nitrogen (N). Doping charge can be acquired in this manner and the band gap can be opened slightly. The adsorption of atoms or other small particles is another method of functionalization. Physisorption and chemisorption are two categories of adsorption that can be distinguished based on the degree of adsorption strength. The graphene structure is essentially unaffected by physisorption, however, it can cause some slight electrical alterations like charge doping. Chemisorption is associated with powerful covalent connections between the adsorbate and the C atoms in graphene. The aromatic network is disrupted by these covalent linkages which induces significant structural changes in graphene. Among the adsorbates that can result in these modifications are, typically, OH groups or atomic radical species. The changed characteristics of graphene can be observed when the covering of chemisorbed adsorbates is dense enough. As in the case of oxygenation [5, 6], there is a significant alteration of properties. It is anticipated that a metal to insulator transition will occur in the wake of covalent functionalization [7].

5.2 Properties of graphane

Graphane is a collection of almost completely covered hydrogenated graphene structures. Each carbon atom and each hydrogen atom in the graphene layer ought to be bound together by covalent bonds. These C–H bonds can only happen if the C atoms in graphene switch from sp^2 to sp^3 hybridization, which causes the graphene sheet to bend.

No set method was found to pair every carbon to a hydrogen atom owing to the fact that hydrogen atoms can bind to the layered structured graphene from either below or above. Although *no priori* reason is found why it would be carried out in such an orderly manner, one typically assumes that the graphane structure is crystalline to make calculations and interpretations easier. The H atoms are thought to be chemisorbed following a straightforward pattern and, during the past ten years, numerous different patterns have been proposed. The many conformations, configurations, or (stereo) isomers of graphane are how they are commonly described. Because they are not connected by simple bond rotations, the various CH structures should be noted as not being conformers. We examine the composition and physical characteristics of several graphane isomers in more detail. We cover their electrical, vibrational, and optical characteristics, and we make some observations about how stable they are.

5.3 Atomic structure

The chair, stirrup, and boat configurations of the three most popular and stable graphane isomers—along with two more isomers—are shown schematically in figure 5.1. Before graphene was isolated in 2003, the first three configurations were originally put forth by Sluiter and Kawazoe [8]. However, it was not until 2007 that Sofo *et al*'s rediscovery of the chair and boat forms gained graphane, belonging to the graphene community, a great deal of attention [9]. The twist-boat [10], twist-boat-chair [11], armchair [12], a-tricycle [13], and many other structures have since been proposed. Most are, however, far less stable than some. The known arrangements are therefore not particularly noteworthy, apart from the tricycle arrangement, which is essentially a chair and bicycle combined, and the stirrup.

Table 5.1 compares the calculated stability of several graphane configurations in the literature. The chair configuration has consistently been found to be the most stable in computational simulations. As seen in figure 5.2(a), the hydrogen atoms are

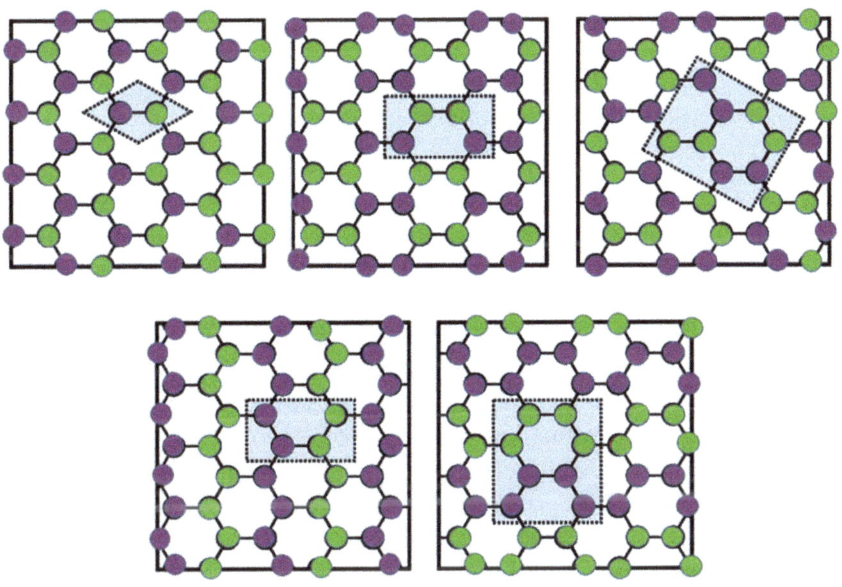

Figure 5.1. Five graphane isomers where each C atom is equal. H has been shown to adsorb above and below the graphene layer, in black and gray, respectively [14].

Table 5.1. The chair isomer's position in relation to the relative stability of numerous graphane isomers in the literature. The predicted total energies for each unit cell are provided in electronvolts.

Chair	Boat	Stirrup	Armchair	TB-chair	Twist boat	Tricycle	Reference
0.000	0.103	0.056	—	—	—	—	[8]
0.00	0.12	—	—	—	—	—	[9]
0.00	0.17	—	—	0.25	—	—	[10]
0.000	0.103	0.053	0.128	—	—	—	[12]
0.000	0.103	0.055	—	—	—	—	[15]
0.00	0.10	0.05	0.13	0.19	—	—	[16]
0.00	0.09	0.05	—	—	0.16	—	[11]
0.00	0.10	0.04	—	—	—	—	[17]
0.000	0.102	0.056	0.134	—	0.150	0.024	[13]

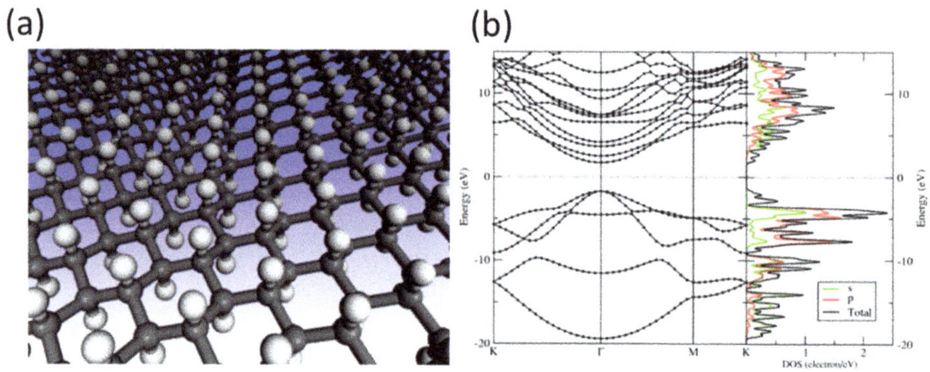

Figure 5.2. (a) Chair-shaped graphane crystal structure. (b) Graphane band structure. (Reproduced with permission from [9]. Copyright 2007 the American Physical Society.)

deposited alternately below and above the sheet like the structure of graphene, causing all of one sublattices' carbon atoms to travel upward while those of the other sublattice move downward. The stirrup configuration (also known as the wash-board, or zigzag pattern), which is prevalent in the literature, compared to the so called boat shape, is more stable. As shown in figure 5.1, the bottom left of the stirrup isomer is represented roughly as alternating zigzag H-atom-pointing chains. The stirrup configuration of graphane isomer is not as well- known as the boat configuration because the influential paper by Sofo *et al* [9] only examined the boat configuration and did not include the stirrup configuration. When demonstrating how attributes are dependent, four isomers covered by Leenaerts *et al* [12] are used as illustrations of the type of conformation. These isomers, according to Wen *et al*, are statistically more stable than benzene molecules [16].

We discovered two useful conformations for graphane: one with alternating pairs of hydrogen atoms in a boat-shaped conformation, and the other with a chair-like conformation on both sides of the plane. The space groups, bond lengths, atomic locations, and lattice parameters are shown in table 5.2. There is additional information available online that contains a representation of the unit cell for both conformers. In the chair conformer, each C–C bond joins C atoms with H atoms linked on the opposing sides of the plane. The computed length of the C–C bond, 1.52 Å, is similar to the sp^3 bond length of 1.53 Å in diamond and is significantly longer than that characteristic of sp^2 carbon in graphene of 1.42 Å. The bonds connecting C atoms to H atoms on opposite sides of the plane have a bond length of 1.52, while those connecting C atoms to H atoms on the same side of the plane have a bond length of 1.56, which is a little bit longer due to H–H repulsion. Similar in both conformations, the C–H bond length of 1.1 is typical of hydrocarbon molecules [9].

The chair conformation of graphane has an estimated binding energy of 6.56 eV/atom, making it just as stable as the related fluorinated molecule CF. We calculate the binding energy of CF to be 6.53 eV/atom using the same method. Even lower than the calculated value for other hydrocarbons with a 1:1 C:H ratio, such as benzene at 6.49 eV/atom and acetylene at 5.90 eV/atom, is the binding energy of graphane in the chair conformation. The most stable substance we discovered for this composition ratio is graphane. The chair conformation of graphane has a binding energy advantage of 0.055 eV/atom over the boat conformation. This energy difference is comparable to that in the fluorinated instance, where analogous computations of the electronic structure likewise produced a chair conformation advantage of 0.0725 eV/atom. Due to the attraction between the two H atoms that are connected to the first neighboring C atoms on the same side of the sheet, the boat structure is less stable than the chair conformation. As was previously indicated, this

Table 5.2. The crystal structure and binding energies in graphane's chair and boat conformations. The difference between the combined energies of the isolated atoms and the compound constitutes the binding energy.

	Chair	Boat
• Space group • A • B • C	• Unit cell • P-3m1(164) • 2.516 • 2.516 • 4.978	• *Pmmn* (59) • 4.272 • 2.505 • 4.976
• C • H	• Atomic positions • (2d) (1/3,2/3, 0.5419) • (2d) (1/3,2/3,0.7479)	• (4f) (0.8178, 0.5636) • (4f) (0.7439, 0.7717)
• C–C • C–H • Binding energy (eV/atom)	• Bond length (Å) • 1.52 • 1.11 • 6.56	• 1.56, 1.52 • 1.10 • 6.50

repulsion lengthens the C–C bond to 1.56. Although a high binding energy is crucial and necessary for the compound's stability, the formation energy in relation to other rival hydrocarbons is more crucial for the synthesis.

5.4 Electronic structure

A significant direct electronic band gap is present in the semiconducting material known as graphane (see figure 5.2(b)). According to the calculation approach, the gap varies in size from roughly 3.5–4.4 eV for local density approximation (LDA) and gradient approximation (GGA), respectively, in a hybrid functional (HSE-06) [12]. Further the traditional LDA- or GGA-based approximations of the exchange-correlation (XC) density functional theory (DFT) have some drawbacks that are addressed by the hybrid functional (HSE) [12, 13]. The latter involves what is known as self-interaction, an ad hoc contact between an electron and itself that results in a systematic underbinding of tightly localized states. By blending a portion of the Hartree–Fock exchange, HSE partially corrects this erroneous self-interaction. As a result, the structural, thermochemical, and electronic properties, such as bulk moduli, equilibrium-lattice constants, cohesive energies, and temperatures, for the realization of semiconductors and insulators have been improved [15]. By incorporating many-body interactions, the electronic energy gap can be determined with even more precision (Green function effective Coulomb (GW) approximation). These many-body interactions further widen the electronic energy gap of layered structured graphane to around 5.2 eV [12, 18].

The band gap fluctuates by less than 5% and is nearly configuration-independent. Figure 5.3 shows graphane's total and expected density of states (DOS) in the chair

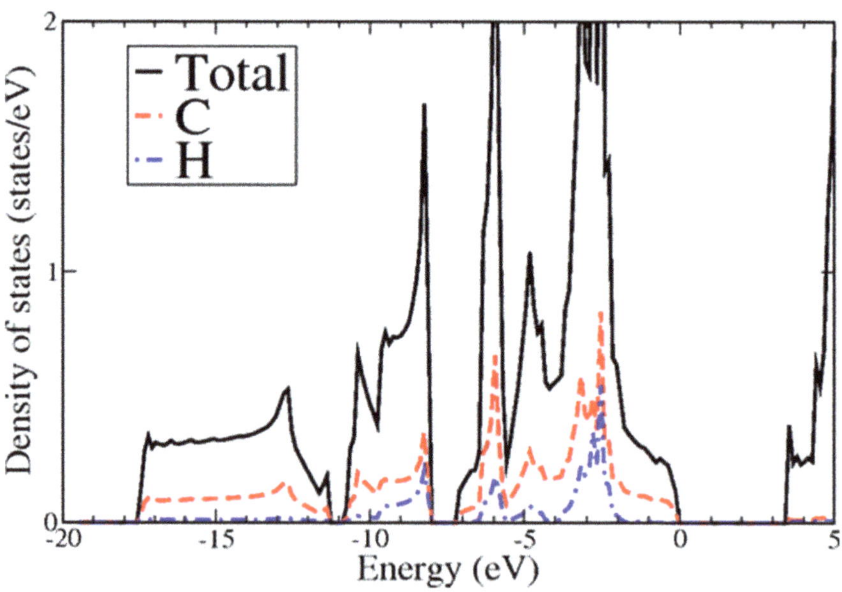

Figure 5.3. Graphane's total and expected density. (Reproduced with permission from [18]. Copyright 2009 the American Chemical Society.)

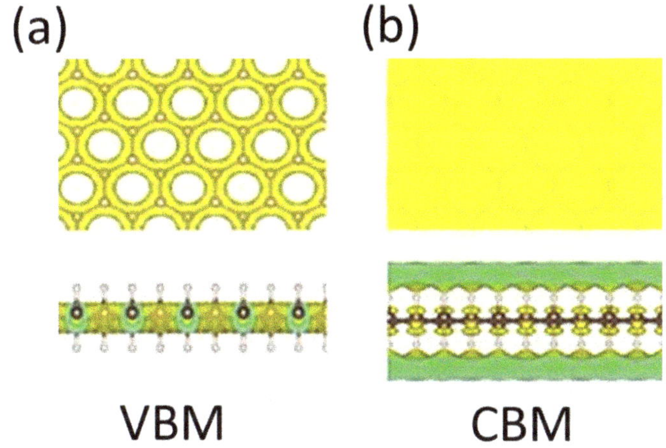

Figure 5.4. Top and side views of (a) VBM and (b) CBM states of graphane. (Reproduced with permission from [19]. Copyright 2013 American Chemical Society.)

formation. In the overall DOS, the electronic energy gap is easily discernible, but the projected DOS exhibits certain odd characteristics. But the contribution of C and H atoms to the conduction band states is negligile.

This becomes increasingly clear, as demonstrated in figure 5.4, when taking into account the density of the single-particle states at the valence band maximum (VBM) and conduction band minimum (CBM). The VBM's and CBM's properties and symmetry are very unlike. Although the conduction band states have plane-wave properties, the valence band states are made up of the C p_x and p_y orbitals. Above the H atoms, the CBM state performs as a loosely bound, delocalized electron [19].

5.5 Optical properties

Graphane's 2D structure affects its optical characteristics substantially and decreases the optical band gap. In 2D materials there is less screening, which results in significant hole–electron interaction [20], resulting in bound exciton states being found much below the CBM. According to calculations, the exciton binding energy for graphane is 1.6 eV [21], resulting in an optical gap of roughly 3.8 eV [20, 22, 23]. Figure 5.5(a) displays an exciton spectrum that is more comprehensive. The increase in the electronic energy gap brought on by electron–electron (e–e) interaction is almost completely canceled out by the electron–hole (e–h) interaction. As shown in figure 5.5(b), the absorption spectra display three distinct approximation levels: (i) without e–e interaction, (ii) with e–e and no e–h interaction, and (iii) with e–e and e–h interaction, respectively. Similar band gaps are produced by single-particle calculations (DFT) and simulations using the e–h and e–e interactions (G_0W_0+BSE).

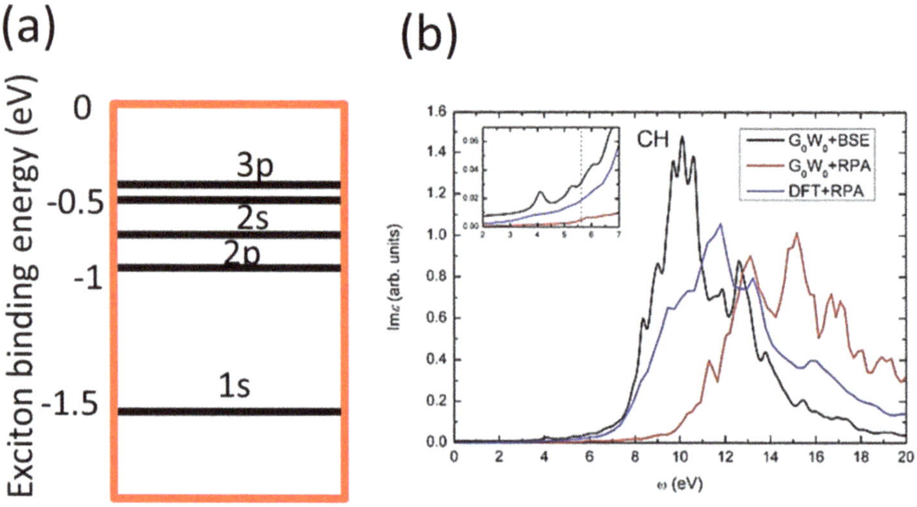

Figure 5.5. (a) The graphane bound state excitonic spectrum. (b) The imaginary portion of graphane's dielectric function calculated using DFT (DFT+RPA), with e–e interactions (G_0W_0+RPA), and with e–e and e–h interactions (G_0W_0+BSE). (Reproduced with permission from [22]. Copyright 2013 American Chemical Society.)

5.6 Vibrational properties

Graphane phonon dispersion in the chair configuration is a stable isomer, since there are no imaginary frequencies. There are three categories of phonons that make up graphane's phonon modes: low, moderate, and high frequency [24]. The high-frequency modes are found to be predominantly H modes from the projected phonon DOS, as would be anticipated given the CH-stretching modes. While both types of atoms contribute to the phonons' intermediate frequency group, the low-lying (acoustic) modes are predominately C modes.

The optical phonon spectrum of graphane at intermediate frequencies was proven by Savini *et al* who also demonstrated that the phonon intermediate frequency spectrum (optical) of graphane undergoes a massive Kohn anomaly upon hole doping [25]. It was predicted that this action will turn graphane to an electron–phonon (e–p) superconductor at a threshold temperature higher than 90 kelvins (K).

Due to the low dimensionality of graphane and the strong chemical interactions between the carbon atoms, there is a significant density of electronic states at the Fermi energy. Additionally, discrete breathers (DBs), vibrational modes of large-amplitude in defect free non-linear lattices which are limited spatially, were studied by Chechin *et al* to determine their presence and distinguishing features [26]. The DB frequency has been found to decrease with increasing amplitude and can also take on any value inside the phonon gap, even going into the phonon area of low-frequency.

5.7 Magnetic properties

Only four bonding partners for sp^3 hybridized C atoms are present in fully hydrogenated graphane (three C and one H). Therefore, it is not anticipated that the ideal graphane isomers will exhibit any intriguing magnetic properties. Only defects (missing hydrogen atoms [27, 28]) and partially hydrogenated graphane exhibit magnetism. Below, we go into greater depth about this. In graphane, substitutional flaws such as boron (B) atoms could also cause a magnetic reaction.

The tremendous rise in spin–orbit coupling [29, 30] is another significant result of hydrogenating graphane. The spin–orbit is enhanced by two orders of magnitude when the hybridization is switched from sp^2 to sp^3. H adsorption can transform the very weak second-order atomic spin–orbit coupling that exists in graphane into a first-order process. When $\Delta_{SO}^{at} \approx 10$ meV, the spin–orbit coupling in graphene reaches the order of the atomic spin–orbit coupling and is similar to that of diamond [27].

5.8 Mechanical properties

Graphene is the strongest material with extraordinary stiffness. The strength of graphene is reduced due to hydrogenation. This is due to the strong aromatic bond being substituted by single σ-bonds, which causes graphene to lose strength. Graphane's calculated 2D Young's modulus is 245 N m^{-1}, which is significantly lower than graphene's estimated [12, 31, 32] and measured value [33] of 340 N m^{-1}. Experimental values of graphane's Young's modulus are not yet available, however, because of the presence of flaws, it is anticipated that they will be less than the theoretical value. In the case of fluorinated grapheme a similar situation also happens, where the theoretical value is roughly twice as high as the experimental result [12, 34]. Highly anisotropic behavior can be seen in other graphane isomers, such as the stirrup and boat configurations [12, 35]. In some directions, Y (Young's modulus) might drop its value to half, and similarly highly anisotropic Poisson ratios are observed [35].

Although perfect graphane exhibits elastic behavior at pressures of at least 30%, minor flaws can significantly lower this limit [31]. Larger strains cause irreversible changes in graphane, and faults begin to manifest. Higher surface roughness and thermal contraction result from graphane's slightly lower out-of-plane stiffness compared to graphene [36].

Costamagna *et al* studied the thermal changes of single-layer graphane up to temperatures of at least 900 K [37]. By evaluating the height–height correlation function and the height–height mean square value for various system sizes and temperatures, they established that graphane was an unrippled system, in comparison to graphene, which subsequently adopts the membrane theory. This unexpected behavior can be explained by the fact that, in graphane, thermal energy is absorbed through in-plane bending modes, which involve changes in the angles between the carbon-carbon bonds within the buckled carbon layer, instead of causing significant fluctuations in the layer's out-of-plane orientation, as observed in graphene.

5.9 Synthesis

In 2009 Elias *et al* published a study on the experimental production of graphane [38]. Graphite was micromechanically cleaved to extract samples of graphene. These samples were either employed as free-standing membranes for transmission electron-microscopy (TEM) or were placed on top of an oxidized Si substrate. To remove any contaminates, the samples were first annealed at high temperature. Graphene samples were exposed for 2 h using a cold hydrogen plasma at 0.1 mbar pressure and a hydrogen–argon combination (10% H_2).

A useful tool for observing the changes brought about by hydrogenated graphene is Raman spectroscopy. The evolution of the Raman spectra for graphene, hydrogenated, and annealed graphene samples was explained by Sahin *et al.* In comparison to disordered and nanostructured carbon-based materials, hydrogenation also produces peaks at 1342 cm^{-1} (distinct D) and at 1610 cm^{-1} (D′) and a combination of both D + D′ peaks at roughly 2950 cm^{-1}. After the creation of the C–H sp^3 bonds, the translational symmetry of the C–C sp^2 bonds' translational motion was broken, giving rise to the D-peak of graphane. On account of the production of two additional bonds of C–H, the D-peak of both-sided hydrogenated graphene is twice as large as that for single-sided hydrogenated graphene.

Graphene's metallic properties come from annealing graphane, one can recover graphane in 24 h at 450 °C in an Ar environment. Increasing the annealing temperature could cause harm by injecting structural disturbance within the sample, which increases the bands' overall intensity to show that hole vacancies are added to the annealed graphene. all of (D, D′) defect-related peaks, and (D + D′) were strongly suppressed.

By using an Ar–H_2 plasma generated by a reactive ion etching (RIE) technique, Wojtaszek *et al* synthesized hydrogenated single and bilayer graphene [39]. They demonstrated that the selected plasma conditions can stop the graphene sheet from being harmed. The observed hydrogenation level of 0.05% can still be raised. Instead of being caused by the fragmentation of water adsorbates on the graphene surface by highly accelerated plasma electrons, the hydrogenation is caused by the hydrogen ions from the plasma.

In order to better understand how atomic hydrogen (H) dangling bond passivation at the interface between the silicon-terminated SiC and single-layer graphene might significantly affect the electrical characteristics of the graphene overlayer, Poh *et al* investigated graphite oxide exfoliation under high pressure (at 60–150 bar) and high temperature (at 200 °C–500 °C) conditions to produce partly hydrogenated graphene [40]. They discovered that the most effective reaction setting was at around 100 bar (H_2 pressure) under 500 °C. Because this process does not require a plasma source (unlike other research [38]), a lot of hydrogenated graphene can be created.

Guisinger *et al* employed room-temperature STM. Since hydrogen could not diffuse through the monolayer of graphene, this indicated that the graphene's edge was chemically bonded to the surface of the reconstructed SiC (0001) [41].

Even at saturated hydrogen coverage, Luo *et al* demonstrated that graphene layer hydrogenation via hydrogen plasma is reversible [42]. The hydrogen coverage is affected by several factors, including process length and plasma power. The hydrogenation energy barriers, which exhibit a pronounced layer dependence, regulate the rate of hydrogenation of graphene layers. The hydrogenation of bilayer and multilayer graphene on a SiO_2/Si substrate is shown to be significantly more practicable than that of single-layer graphene. The amount of hydrogen covering and the number of graphene layers both significantly influence dehydrogenation.

A graphane-like film produced via plasma-enhanced chemical vapor deposition (PE-CVD) was dehydrogenated by Wang *et al* to produce high-quality, monolayer graphene [43]. The benefit of the plasma deposition technique is that it may be utilized for wafer-scale processing, lithographic patterning, and deposition on silicon substrates with metal coatings while operating at temperatures (at least 350 °C) that are lower than those of thermal CVD procedures. One step toward designing an integrated circuit based on all-carbon electronics is the fabrication of graphane-like ribbons and squares using laser writing.

By electron irradiating graphene that has chemisorbed H_2O and NH_3 on the layer, Jones *et al* have created partially hydrogenated graphene on both sides and one side of the material [44]. The idea of hydrogenation was suggested because backscattered and secondary electrons, which fragment H_2O and NH_3 adsorbates, produce H^+ ions and H radicals.

By using electron-induced hydrogen silsesquioxane (HSQ) dissociation, Ryu *et al* successfully hydrogenated graphene [45]. Due to the increased chemical reactivity of a single sheet of graphene, hydrogenation occurred more quickly for single layers than for double layers. At normal temperature, the chance of hydrogen atom chemisorption on single layers is at least 0.03, which is 15 times higher than the likelihood for bilayers.

Theoretically, numerous authors suggested various methods for hydrogenating graphene.

Zhou *et al* proposed semi-hydrogenated graphene using density functional theory [46, 47] by putting an external electric field on completely hydrogenated graphene, which can remove H atoms from one side of graphane. Semi-hydrogenated graphene, also referred to as 'graphone', is an indirect band gap ferromagnetic semiconductor with a value of 0.43 eV. Graphone that stands alone, however, has been demonstrated to roll up [48]. Additionally, Ao *et al* demonstrated that hydrogenation of graphene layers is facilitated by the application of an electric field that catalyzes the dissociative adsorption of H_2 [49].

Ab initio calculations using density functional theory were carried out by Leenaerts *et al* to study the hydrogenation of a graphene bilayer [50]. If the hydrogen atoms are permitted to adsorb on both sides of the bilayer, 50% hydrogen coverage was attainable. Strong chemical bonds between the graphene layers have taken the place of the weak van der Walls forces in this instance. Due to the same hybridization, a bilayer of graphane is created at maximal coverage and exhibits electrical properties that are comparable to those of a single layer of graphane. Such a bilayer, which was recently achieved experimentally, can be thought of as the thinnest layer of diamond [51].

In order to study the electrical properties of hydrogenated bilayer graphene, Samarakoon *et al* used DFT simulations to apply a perpendicular electric bias, which causes a change from a semiconducting to a metallic state [52]. A ferromagnetic semiconductor with an adjustable band gap is also produced when hydrogen from one layer of bilayer graphane is discharged.

By imposing decorated hydrogen chemisorption onto the graphene produced moiré superlattice positions on an Ir substrate in the (111) plane, Balog *et al* showed that an energy gap may be opened in graphene. Results from STM and ARPES were combined to show that hydrogen adsorbate structures were stable and superior to ambient temperature. The band gap was not very sensitive to the exact hydrogen adsorbate structure. The outcomes of these experiments also confirm the notion of periodic lattice disturbances causing confinement-induced gap opening. The opening was created at the Fermi energy and was large enough that it may be used at room temperature for electronic applications. STM investigations were provided by Balog *et al* to demonstrate atomic hydrogen's local adsorbate structure on the graphene's base plane on a SiC support [53].

5.10 Applications of graphane

Graphane materials can be employed to create bipolar transistors, according to Gharekhanlou *et al* [54]. A precise analytical form of the two-dimensional filtered potential was presented by Cudazzo *et al* [55]. Gharekhanlou *et al* [56] created a 2D p–n junction using graphane that had a hydrogen deficiency. It was done to reduce the bandgap in an efficient manner. Their analysis, using the Shockley law of junctions, indicated that an exponential ideal I–V characteristic could be anticipated. The quality is anticipated. The graphane is widened by this or a transistor device using a graphene-like application.

P-doped graphane was utilized by Savini *et al* to create a prototype high-T_c electron–phonon superconductor with T_c as high as 150 K for a 1 nm nanowire, surpassing copper oxides [57].

By taking into account the order parameter variations, Loktev and Turkowski [58] and Kristoffel and Rägo [59] examined the superconducting characteristics of multilayer graphane. The outcome shows that, in the case of a single layer, the Berezinskii–Kosterlitz–Thouless (BKT) critical temperature, which corresponds to the vortex superconductivity (SC), is equal to the mean field (MF) temperature at a doping level of less than 0.01 and starts at a relatively low value. Additionally, they predicted that the critical temperature could be as high as 150 K, which is much higher than the maximum temperature in cuprates at ambient pressure.

Nechaev asserts that high-density hydrogen carrier intercalation in nanostructures resembling graphane can be used to power fuel cell vehicles [60]. With a 12.9 weight percent hydrogen storage capacity, polylithiated (OLi_2) functionalized graphane was used by Hussian *et al* as a possible hydrogen storage material [61, 62].

5.11 Conclusions

In this review, we have covered graphane from synthesis to applications. We have demonstrated that graphene can be hydrogenated one atom at a time, yielding graphane, a persistent direct band gap insulator with robust C–H bonds. There are various methods for the synthesis of graphanes, including thermal exfoliation of graphene and hydrogen plasma exposure to graphene oxides, graphene hydrogenation aided by STM, electron-induced HSQ dissociation, and plasma-enhanced CVD on graphene. Additionally, we have discussed graphene's electrical and magnetic characteristics. At the end, we demonstrated that hydrogenated graphenes may also find use in a variety of industries, including hydrogen storage, piezo-electricity, explosive detection, biosensing, and thermo-electricity devices. Future research on functionalized graphane, in our opinion, will elucidate a variety of fascinating features of this substance.

References

[1] Novoselov K S, Geim A K, Morozov S V, Jiang D, Zhang Y, Dubonos S V, Grigorieva I V and Firsov A A 2004 Electric field effect in atomically thin carbon films *Science* **306** 666–9

[2] Novoselov K S, Geim A K, Morozov S V, Jiang D, Katsnelson M I, Grigorieva I V, Dubonos S V and Firsov A A 2005 Two-dimensional gas of massless Dirac fermions in graphene *Nature* **438** 197–200

[3] Lee C G, Wei X D, Kysar J W and Hone J 2008 Measurement of the elastic properties and intrinsic strength of monolayer graphene *Science* **321** 385–8

[4] Balandin A A, Ghosh S, Bao W, Calizo I, Teweldebrhan D, Miao F and Lau C N 2008 Superior thermal conductivity of single-layer graphene *Nano Lett.* **8** 902–7

[5] Jung I, Dikin D A, Piner R D and Ruoff R S 2008 Tunable electrical conductivity of individual graphene oxide sheets reduced at low temperatures *Nano Lett.* **8** 4283–7

[6] Jia-An Y, Lede X and Chou M Y 2009 Structural and electronic properties of oxidized graphene *Phys. Rev. Lett.* **103** 086802

[7] Ezawa M 2013 Metal–insulator transition from graphene to graphane *Nanomater. Nanotechnol.* **3** 10

[8] Sluiter M H F and Kawazoe Y 2003 Cluster expansion method for adsorption: application to hydrogen chemisorption on graphene *Phys. Rev.* B **68** 085410

[9] Sofo J O, Chaudhari A S and Barber G D 2007 Graphane: a two-dimensional hydrocarbon *Phys. Rev.* B **75** 153401

[10] Samarakoon D K, Chen Z, Nicolas C and Wang X Q 2011 Structural and electronic properties of fluorographene *Small* **7** 965–9

[11] Samarakoon D K and Wang X Q 2009 Chair and twist-boat membranes in hydrogenated graphene *ACS Nano* **3** 4017–22

[12] Leenaerts O, Peelaers H, Hernández-Nieves A D, Partoens B and Peeters F M 2010 First-principles investigation of graphene fluoride and graphane *Phys. Rev.* B **82** 195436

[13] He C, Sun L Z, Zhang C X, Jiao N, Zhang K W and Zhong J 2012 Structure, stability and electronic properties of tricycle type graphane *Phys. Status Solidi RRL* **6** 427–9

[14] Sahin H, Leenaerts O, Singh S and Peeters F M 2015 Graphane *WIREs Comput. Mol. Sci.* **5** 255–72

[15] Artyukhov V I and Chernozatonskii L A 2010 Structure and layer interaction in carbon monofluoride and graphane: a comparative computational study *J. Phys. Chem.* A **114** 5389–96

[16] Wen X D, Hand L, Labet V, Yang T, Hoffmann R, Ashcroft N W, Oganov A R and Lyakhov A O 2011 Graphane crystals under pressure *Proc. Natl Acad. Sci. USA* **180** 6833–7

[17] Bhattacharya A, Bhattacharya S, Majumder C and Das G P 2011 The third conformer of graphane: a first principles DFT based study *Phys. Rev.* B **83** 033404

[18] Lebègue S, Klintenberg M, Eriksson O and Katsnelson M I 2009 Accurate electronic band gap of pure and functionalized graphane from GW calculations *Phys. Rev.* B **79** 245117

[19] Amini M N, Leenaerts O, Partoens B and Lamoen D 2013 Graphane and fluorographene-based quantum dots *J. Phys. Chem.* C **117** 16242–7

[20] Cudazzo P, Attaccalite C, Tokatly I V and Rubio A 2010 Strong charge-excitonic effects and the Bose–Einstein exciton condensate in graphane *Phys. Rev. Lett.* **104** 226804

[21] Huang S, Liang Y and Yang L 2013 Exciton spectra in two-dimensional graphene derivatives *Phys. Rev.* B **88** 075441

[22] Karlicky F and Otyepka M 2013 Band gaps and optical spectra of chlorographene, fluorographene and graphane from G0W0, GW0 and GW calculations on top of PBE and HSE06 orbitals *J. Chem. Theory Comput.* **9** 4155–64

[23] Reshak A H and Auluck S 2014 Electronic properties of chair-like and boat-like graphane *RSC Adv.* **4** 37411–8

[24] Peelaers H, Hernández-Nieves A D, Leenaerts O, Partoens B and Peeters F M 2011 Vibrational properties of graphene fluoride and graphane *Appl. Phys. Lett.* **98** 051914–6

[25] Savini G, Ferrari A C and Giustino F 2010 First-principles prediction of doped graphane as a high-temperature electron–phonon superconductor *Phys. Rev. Lett.* **105** 037002

[26] Chechin G M, Dmitriev S V, Lobzenko I P and Ryabov D S 2014 Properties of discrete breathers in graphane from *ab initio* simulations *Phys. Rev.* B **90** 045432

[27] Sahin H, Ataca C and Ciraci S 2009 Magnetization of graphane by dehydrogenation *Appl. Phys. Lett.* **95** 222510

[28] Berashevich J and Chakraborty T 2010 Sustained ferromagnetism induced by H-vacancies in graphane *Nanotechnology* **21** 355201

[29] Castro Neto A H and Guinea F 2009 Impurity-induced spin–orbit coupling in graphene *Phys. Rev. Lett.* **103** 026804

[30] Gmitra M, Kochan D and Fabian J 2013 Spin–orbit coupling in hydrogenated graphene *Phys. Rev. Lett.* **110** 246602

[31] Topsakal M, Cahangirov S and Ciraci S 2010 The response of mechanical and electronic properties of graphane to the elastic strain *Appl. Phys. Lett.* **96** 091912–5

[32] Muñoz E, Singh A K, Ribas M A, Penev E S and Yakobson B I 2010 The ultimate diamond slab: graphane versus graphene *Diam. Relat. Mater.* **19** 368–73

[33] Koenig S P, Boddeti N G, Dunn M L and Bunch J S 2010 Ultrastrong adhesion of graphene membranes *Nat. Nanotechnol.* **6** 543–6

[34] Nair R R *et al* 2010 Fluorographene: a two-dimensional counterpart of Teflon *Small* **6** 2877–84

[35] Colombo L and Giordano S 2011 Nonlinear elasticity in nanostructured materials *Rep. Prog. Phys.* **74** 116501–35

[36] Neek-Amal M and Peeters F M 2011 Lattice thermal properties of graphane: thermal contraction, roughness, and heat capacity *Phys. Rev.* B **83** 235437

[37] Costamagna S, Neek-Amal M, Los J H and Peeters F M 2012 Thermal rippling behavior of graphane *Phys. Rev.* B **86** 041408(R)

[38] Elias D C *et al* 2009 Control of graphene's properties by reversible hydrogenation: evidence for graphane *Science* **323** 610–3

[39] Wojtaszek M, Tombros N, Caretta A, van Loosdrecht P H M and van Wees B J 2011 A road to hydrogenating graphene by a reactive ion etching plasma *J. Appl. Phys.* **110** 063715

[40] Poh H L, Šanek F, Soferb Z and Pumera M 2012 High-pressure hydrogenation of graphene: towards graphane *Nanoscale* **4** 7006–11

[41] Guisinger N P, Rutter G M, Crain J N, First P N and Stroscio J A 2009 Exposure of epitaxial graphene on SiC(0001) to atomic hydrogen *Nano Lett.* **9** 1462–6

[42] Luo Z Q, Yu T, Kim K J, Ni Z H, You Y M, Lim S, Shen Z X, Wang S Z and Lin J Y 2009 Thickness-dependent reversible hydrogenation of graphene layers *ACS Nano* **3** 1781–8
[43] Wang Y, Xu X, Lu J, Lin M, Bao Q, Özyilmaz B and Loh K P 2010 Toward high throughput interconvertible graphane-to-graphene growth and patterning *ACS Nano* **4** 6146–52
[44] Jones J D, Mahajan K K, Williams W H, Ecton P A, Mo Y and Perez J M 2010 Formation of graphane and partially hydrogenated graphene by electron irradiation of adsorbates on graphene *Carbon* **48** 2335–40
[45] Ryu S, Han M Y, Maultzsch J, Heinz T F, Kim P, Steigerwald M L and Brus L E 2008 Reversible basal plane hydrogenation of graphene *Nano Lett.* **8** 4597–602
[46] Zhou J, Wang Q, Sun Q, Chen X S, Kawazoe Y and Jena P 2009 Ferromagnetism in semihydrogenated graphene sheet *Nano Lett.* **9** 3867–70
[47] Zhou J, Wu M M, Zhou X and Sun Q 2009 Tuning electronic and magnetic properties of graphene by surface modification *Appl. Phys. Lett.* **95** 103108
[48] Neek-Amal M, Beheshtian J, Shayeganfar F, Singh S K, Los J H and Peeters F M 2013 Spiral graphone and one-sided fluorographene nanoribbons *Phys. Rev.* B **87** 075448
[49] Ao Z M and Peeters F M 2010 Electric field: a catalyst for hydrogenation of graphene *Appl. Phys. Lett.* **96** 253106
[50] Leenaerts O, Partoens B and Peeters F M 2009 Hydrogenation of bilayer graphene and the formation of bilayer graphane from first principles *Phys. Rev.* B **88** 245422
[51] Rajasekaran S, Abild-Pedersen F, Ogasawara H, Nilsson A and Kaya S 2013 Interlayer carbon bond formation induced by hydrogen adsorption in few-layer supported graphene *Phys. Rev. Lett.* **111** 085503
[52] Samarakoon D K and Wang X Q 2010 Tunable band gap in hydrogenated bilayer graphene *ACS Nano* **4** 4126–30
[53] Balog R, Jørgensen B, Wells J, Lægsgaard E, Hofmann P, Besenbacher F and Hornekær L 2009 Atomic hydrogen adsorbate structures on graphene *J. Am. Chem. Soc.* **131** 8744–5
[54] Gharekhanlou B, Tousaki S B and Khorasani S 2010 Bipolar transistor based on graphane *J. Phys.: Conf. Ser.* **248** 012061
[55] Cudazzo P, Tokatly I V and Rubio A 2011 Dielectric screening in two-dimensional insulators: implications for excitonic and impurity states in graphane *Phys. Rev.* B **84** 085406
[56] Gharekhanlou B and Khorasani S 2010 Current–voltage characteristics of graphane p–n junctions *IEEE Trans. Electron Devices* **57** 209
[57] Savini A, Ferrari C and Giustino F 2010 Doped graphane: a prototype high-T_c electron–phonon superconductor *Phys. Rev. Lett.* **105** 037002
[58] Loktev V M and Turkowski V 2011 Possible high-temperature superconductivity in multi-layer graphane: can the cuprates be beaten? *Low Temp. Phys.* **164** 264
[59] Kristoffel N and Rägo K 2011 On the interband pairing in doped graphane *Phys. Lett.* A **375** 2246
[60] Nechaev Y S 2011 The high-density hydrogen carrier intercalation in graphane-like nano-structures, relevance to its on-board storage in fuel-cell-powered vehicles *Open Fuel Cells J.* **4** 16
[61] Hussain T, Maark T A, De Sarkar A and Ahuja R 2012 Polylithiated (OLi_2) functionalized graphane as a potential hydrogen storage material *Appl. Phys. Lett.* **101** 243902
[62] Hussain T, De Sarkar A and Ahuja R 2012 Strain induced lithium functionalized graphane as a high capacity hydrogen storage material *Appl. Phys. Lett.* **101** 103907

IOP Publishing

Nanocarbon Allotropes Beyond Graphene
Synthesis, properties and applications
Arpan Kumar Nayak and Santosh K Tiwari

Chapter 6

Synthesis and applications of graphyne

V Priyanka, Kuldeep Singh, Siladitya Laha and Tapas Ghatak

Ongoing efforts are being made to synthesize and identify emerging carbon allotropes with enhanced stability, innovative bonding features, unique properties, and inventive uses. The 'magical material' graphene has led the way for the discovery of a large number of additional two-dimensional compounds, such as graphyne, graphdiyne, graphone, and graphane. Graphyne is one of these materials whose characteristics have been investigated well. Graphyne, a similar 2D carbon allotrope to graphene containing double- and triple-bonded carbon atoms, has been shown to possess the same remarkable electrical characteristics as graphene. By substituting ⅓ of graphite's carbon–carbon bonds with ethyne units, a network made entirely of phenyl rings and C≡C is formed. It is expected that the highly unsaturated network would exhibit a number of useful qualities, including resistance to degradation from exposure to high temperatures, a wide band gap semiconductor, and the ability to intercalate Li^+ and Na^+ ions without causing interlayer disruption. The characteristics of these atypical systems are very pertinent to the search for organic conductors, electrochromic display materials, liquid crystals, synthetic ferromagnets, and nonlinear optical materials. This chapter provides a short overview of their characteristics, including their synthesis and applicability in addition to their structural, mechanical, physical, and chemical characteristics.

List of abbreviations

Sl. No.	Abbreviations	Full form
1	DFT	density functional theory
2	DOS	density of states
3	LS	least squares
4	DBA	dehydrobenzoannulene
5	CVD	chemical vapor deposition
6	GNTs	graphyne nanotubes

doi:10.1088/978-0-7503-5177-5ch6

(*Continued*)

7	NTs	nanotubes
8	NLO	nonlinear optical
9	TM	transition metal
10	HER	hydrogen evolution reaction
11	CRR	CO_2 reduction reaction
12	NRR	N_2 reduction reaction
13	OER	O_2 evolution reaction
14	ORR	O_2 reduction reaction
15	RDS	rate determining step
16	MO	metal–oxygen

6.1 Introduction

Carbon allotropes with low dimensions, which include fullerenes [1], carbon nanotubes [2], and graphene [3], have been synthesized in the last thirty years, providing a wide range of novel structural motifs for the development of high-performance carbon materials. Graphene, which consists of a single sheet of sp^2-bonded carbon atoms, is one of these carbon compounds and has been at the forefront of research on carbon materials for about a decade. Graphene is unique in numerous ways, which is attributable to its one-atom thickness and extremely symmetric network. Its low-energy excitation spectrum, for instance, differs from the quadratic correlation among both energy and momentum observed in normal semiconductors [4, 5].

Since the discovery of graphene-based 2D carbon compounds, graphdiyne and graphyne have captivated much research interest [6–8]. The structure and properties of graphyne were first accurately predicted by Baughman *et al* in 1987. [6]. Graphyne is a highly unsaturated network that is anticipated to have a variety of attractive physical properties [6, 9–11]. The band gap (1.2 eV) is less than that of polyacetylene, and Li^+ and Na^+ ions may penetrate deeper without causing interlayer deformation. The carbon allotrope that is composed of alkyne units is the one that is the most stable. Interest in graphynes and graphdiynes has surged lately because these sp^2 and sp hybridized carbon species provide networks with abundant π-conjunction, uniformly spaced holes, and tuneable electronic characteristics [12, 13]. Graphyne, like graphene, consists of sheets of carbon atoms one atom thick, however, unlike graphene the graphyne carbon atoms are hybridized with both sp/sp^2 atoms. Graphyne, which is similar to graphene but contains C=C and C≡C bonds, has been revealed to have the same extraordinary electronic properties as graphene [14]. Hexagonally graphyne is altered because of the presence of sp carbon atoms, which disrupts the original structure. As a direct consequence of this, numerous lattice types of graphyne with distinct geometries are produced. For instance, four different varieties of grapheme, namely α, β, γ, and 6,6,12-graphynes have been discovered so far, each with varying percentages of C≡C links, i.e. 100%, 66.67%, 33.33%, and 41.67%, respectively [6, 15]. Graphene and graphynes have similar topologies, but graphynes have different physical properties than graphene because they have C≡C links.

When acetylenic links are present in graphynes, both the fracture stress and Young's modulus are greatly decreased; this reduction is proportional to the fraction of linkages present [16]. The mechanical properties of 6,6,12-graphyne are the most directionally anisotropic among the four distinct graphynes. Due to the lack of freestanding graphynes in the laboratory, there have been fewer experimental studies of graphynes than theoretical studies. Thus far, only a limited number of synthetic investigations of graphdiyne films and nanoribbons have been published.

6.2 Properties

A sheet of sp and sp^2 bonded carbon atoms arranged in a crystal lattice to form graphyne has a thickness of precisely one atom. One-third of the C–C bonds in graphite may be replaced by acetylene units, resulting in a network consisting solely of phenyl rings and triple bonds, which have been termed graphyne. Apart from graphene, other two-dimensional planar carbon structures have been predicted to exist, such as graphyne [6, 15, 17, 18], although only molecular fragments have been produced so far [19, 20]. It is plausible that many different graphynes could exist, each with slightly different configurations of the double and triple bonds. Prior to 1960, it was merely assumed that graphyne existed, but after fullerenes were discovered people began to pay attention [21].

Graphyne is a kind of graphene in which the hexagonal carbon atoms are connected by C≡C bonds. It is of special interest to study the electrical structure of graphyne because, in contrast to the structures of other carbon-based materials such as diamond and graphite, graphyne's electrical structure is unique. There are various variants of graphyne, with γ-graphyne being the most prevalent form. However, additional varieties of graphyne have also been investigated [15]. The unit cells of γ-graphyne are generated by the arrangement of 12 carbon atoms that are arranged in such a way that they form a closed structure.

Graphene's mechanical properties, present in both finite and infinite sheets, are among the material's most alluring characteristics [16, 22, 23]. It was possible to correctly measure the acetylenic system's high-frequency vibrations by using rapid time steps of only 0.2 fs. The constant strain rate technique revealed an armchair modulus of 532.5 GPa and a zigzag modulus of 700.0 GPa. The armchair modulus and zigzag modulus determined using the energy minimization technique, on the other hand, were significantly different, measuring 629.4 GPa and 772.0 GPa, respectively. However, it is not a true dynamic method, so it does not model dynamic events such as failure accurately. As a result, the final stresses computed are substantially more appropriate. It should be mentioned that the thickness value of 3.20 Å was utilized in the conversion of N m^{-1} to GPa since it was found to have the lowest system potential energy about the interlayer distance. Zhang *et al* demonstrated that the formation and dissociation of bonds of various orders can be adequately predicted using a hydrocarbon potential (AIREBO) [16].

Additionally, a great deal of research has been put into analyzing the electrical properties of graphyne. Band gaps in the semiconductor area, spanning from 0.59 to 1.25 eV for thicknesses of 1–8 repetition units, were discovered by performing first-principles calculations on finite-width graphyne ribbons [24]. To obtain a 1 eV jump in the band gap when the tensile strain was 0.15, the compressive strain was 0.1, resulting in a 3 eV drop.

Other than mechanical methods, chemical methods have been used to influence the electrical characteristics of graphyne. Studies in physical chemistry reveal that coating sheets of graphyne with transition metal adsorbates drastically modifies the material's overall behavior [25]. In the presence of adsorbed chromium and iron, graphyne transforms into a spin-polarized metal, not a semiconductor. Energy-favorable absorptions suggest that a graphyne sheet could be used to achieve these results. Additionally, graphyne has been found to have certain peculiar optical properties. The complicated dielectric function of graphyne was determined as a function of energy absorption in electric fields both parallel to and perpendicular to its sheet [26].

Li *et al* successfully produced a novel 2D-C allotrope graphdiyne in 2010 by executing a cross-coupling method on a copper surface [8]. Typically, the dimensionality of carbon compounds is determined by their hybridization [27]. So, for example, 2D graphene is produced by sp^2 hybridization, while 1D and 3D carbon compounds are produced by sp and sp^3 hybridization [28, 29]. In contrast, graphyne is the only known 2D carbon material to have both sp and sp^2 hybridization, which explains its malleable and versatile crystal morphologies [30]. Because of the C≡C link, some unusual electrical structures are created. The Brillouin zone [26, 31] of graphene has a tiny gap in comparison to graphene with a zero gap. The controllability of the band gap offers up a broad variety of possible applications [8]. The distinctive features of graphyne have attracted the interest of researchers [15, 32, 33].

Recently, graphynes have gained attention as promising materials for many different fields of study and practical use, including those involving protection against UV light [34], transistors [35], catalysts [36], and materials for energy storage [37, 38]. As a first step toward understanding these potential uses, first-principles studies of their electronic structures [24, 39–41], mechanical properties [38, 42, 43], optical properties [26, 44], and thermal conductivity at room temperature [45] were conducted.

However, the three materials' minimum thermal conductivities at high temperatures and the impacts of strain on the optical characteristics are still unknown. On their properties, not much has been reported consistently. Until now, there has not been a comprehensive study of how strain affects the band structure and optics of graphyne, given these earlier investigations.

All forms of graphyne models are being investigated for their supercell structures, which are displayed in figure 6.1 as the red rectangles indicating unit cells. There are 8, 18, and 12 carbon atoms in each of the three types of graphyne. The sp^2 hybridized carbon is represented by the red balls, while sp hybridized carbon is represented by the gray balls. Two carbon atoms with sp hybridization form C≡C linkage and the quantity and location of acetylenic bonds can change to produce a variety of graphyne arrangements, with varying cavity radii.

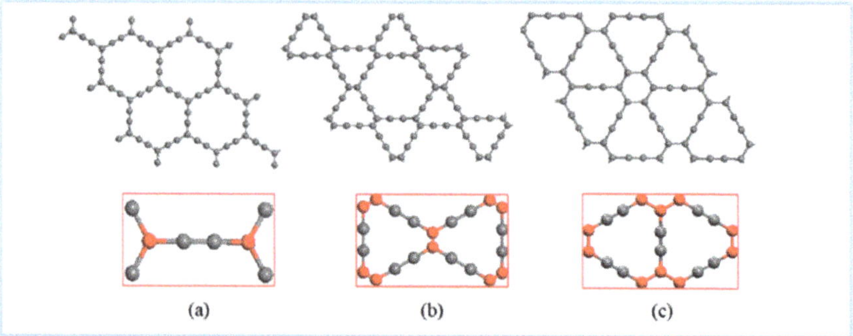

Figure 6.1. The solid-state structures of various forms of graphyne: (a) α-graphyne, (b) β-graphyne, and (c) γ-graphyne. (Reproduced from [131]. CC BY 4.0.)

The acetylenic bond is formed by sp^2 hybridization, while the σ and π bonds are formed by sp hybridization, both of which may be found in graphyne. α-graphyne does not have an sp–sp^2 hybridization bond. The cohesive energy per atom (echo) must be computed to determine the thermodynamic equilibrium of three different structures. The cohesive energy, defined as the energy required to dissociate graphyne's neutral atoms in the ground state at 0 K, is a measure of the compound's stability [46]. The energy released when atoms are bound to crystals is known as the cohesive energy of compounds. Greater stability of the compound is indicated by a lower negative value. It is calculated by using the formula [47]

$$E_{\text{coh}} = (E_{\text{total}} - n \times E_{\text{atom}})/n \tag{6.1}$$

n = number of C atoms/unit cells,
E_{atom} = energy of a single atom,
E_{total} = total energy.

That the cohesive energies are found to be negative for the three compounds demonstrates that these substances are all stable in their natural environments. The most stable member of this group is γ-graphyne because it has the largest negative cohesive energy. The order of the three different forms of graphynes in terms of their stability is as follows: γ-graphyne > β-graphyne > α-graphyne.

6.2.1 Electronic structures

The term 'band gap' refers to the gap that is created when the maximum of the valence band and the minimum of the conduction band are subtracted from each other. Figure 6.2 depicts three distinct graphyne structures together with the corresponding energy bands and densities of states for each one.

At the Fermi level, α-graphyne's DOS was very close to zero because it has a zero-band gap. Figures 6.2(b) and (c) demonstrate that the direct band gaps for β-graphyne and γ-graphyne are 0.028 eV and 0.447 eV, respectively.

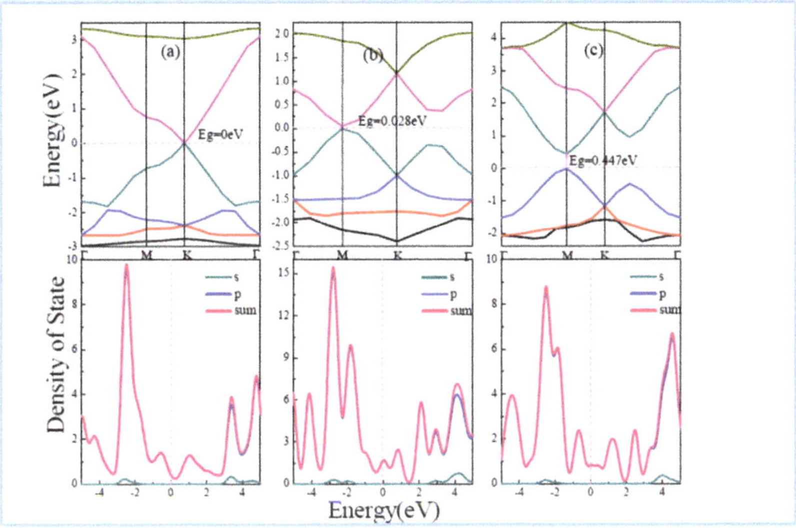

Figure 6.2. DOS of different graphynes (α, β, γ). (Reproduced from [131]. CC BY 4.0.)

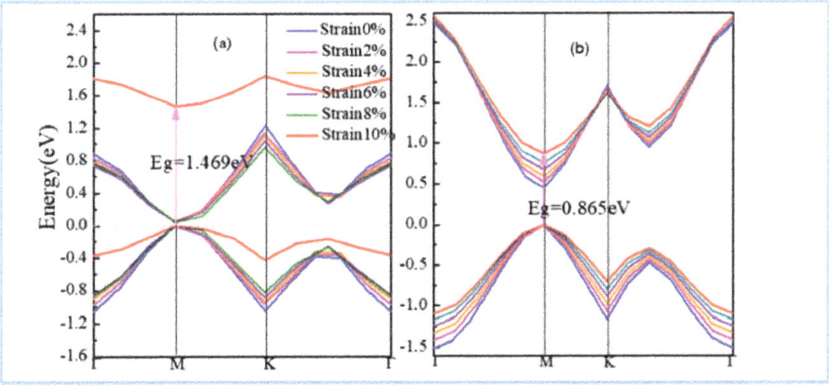

Figure 6.3. The graphical presentations of band gaps of strained (a) β and (b) γ graphyne. (Reproduced from [131]. CC BY 4.0.)

In every instance, there are broad energy bands with sharp ups and downs at the Fermi level. Furthermore, expanding p orbitals with highly delocalized carriers of low effective mass are shown in PDOS to dominate the energy band near the Fermi level.

Both strained β-graphyne and unstrained γ-graphyne have distinct energy band structures, which are depicted in figures 6.3(a) and (b), respectively. Before the strain hits 8%, the band gap of β-graphyne does rise marginally; but, beyond that point, it increases dramatically, reaching 1.469 eV when the strain is at 10%. The control of band gaps in semiconductor materials can improve their optoelectronics application.

Despite γ-graphyne's fixed band gap, this may be changed regularly. Because of this, γ-graphyne is an excellent optoelectronic material.

Stress causes displacement and defects in the lattice, preventing carriers from moving. This shows up in the material by making it less conductive and metallic. When stress reaches a certain level, the β-graphyne may become quite sensitive. The 10% strain produces a significant DOS peak at the Fermi level, indicating the localization of p orbitals. While DOS is essentially constant, there is an increase in the band gap of γ-graphyne, indicating the possibility of a simple strain-based regulatory mechanism. Graphene's imperfections make it less metallic and more like a semiconductor, according to all of the above evidence.

It is clear from the bond type and length that the metallic, strain-insensitive, zero-band-gap α-graphyne does not display sp^2–sp^2 hybridization. The bond length of β-graphyne, which is more susceptible to strain, is longer (1.456) than the bond length of γ-graphyne (1.423). The electronic structure of graphyne is stabilized by sp–sp hybridization, according to this study.

6.2.2 Elasticity

The elasticity of the material can be attributed to crystal defects and bonds. If there is external stress, the elastic limit of the lattice is indicated by its elastic constant [48]. For measuring C_{ij}, *ab initio* calculations are commonly utilized, with the least squares (LS) fitting of the coefficients of elasticity in the stressed unit cell that serves as a guide [49]. So, the [01] and [10] are the directions of the calculated elastic constants.

The tensile strength of a film may well be determined by analyzing its elastic modulus B^{2D}, shear modulus G^{2D}, and Young's modulus E^{2D}. The crystal stability under shear force is denoted by the Poisson ratio ν of the 2D material. More stability is present under shear force when the ratio is smaller. The bond strength and external-pressure resistance can be represented by all of these variables in some cases [50]. The shear anisotropy factor A is equal to 1 which means that the material is in elastic isotropy [51].

The elastic modulus B^{2D} of the 2D material is a formula for calculating the tensile resistance of layered material:

$$B^{2D} = (C_{11} + C_{22} + C_{12})/4. \tag{6.2}$$

The Young's modulus E^{2D} of the 2D material in the [10] and [01] directions (plane rigidity) are calculated using the equation [52]

$$E^{2D}{}_{[10]} = (C_{11}C_{22} - C_{12}{}^2)/C_{22},\quad E^{2D}{}_{[01]} = (C_{11}C_{22} - C_{12}{}^2)/C_{11}. \tag{6.3}$$

Poisson's ratio is given as follows [52]:

$$\begin{aligned} V^{2D}{}_{[10]} &= C_{12}/C_{22} \\ V^{2D}{}_{[01]} &= C_{12}/C_{22}. \end{aligned} \tag{6.4}$$

The shear modulus G^{2D} is given as follows [51]:

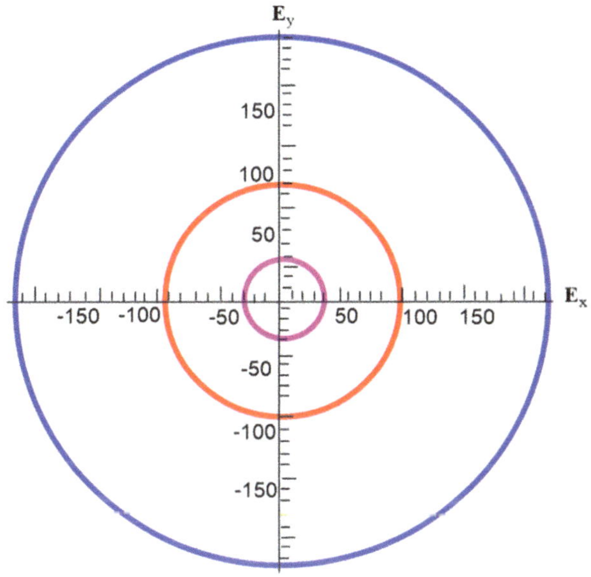

Figure 6.4. The graphical presentation of graphyne's 2D Young's modulus. The pink, red, and blue lines represent α-, β-, and γ-graphyne, respectively.

$$\begin{aligned} G^{2D}{}_{[10]} &= G^{2D}{}_{[01]} = C_{66} \\ A &= 4C_{66}/(C_{11} + C_{22} - 2C_{12}). \end{aligned} \tag{6.5}$$

In light of the aforementioned equations, it is clear that the mechanical properties of γ-graphyne are outstanding. Since the shear anisotropy factor A is equal to 1, this denotes that all three structures are in elastic isotropy. Figure 6.4 depicts two-dimensional diagrams of the Young's modulus E^{2D} to more accurately represent the isotropy of the three structures.

6.2.3 Thermal conductivity

Heat transport in 2D materials becomes more divergent with rising temperature and lower thermal conductivity [53, 54]. A good understanding of the factor that impacts thermal conductivity, including the lowest values at extreme temperatures, is necessary for understanding their applications. The minimum thermal conductivity (k_{min}) is measured using the Clark model [55]. The calculation of the average acoustic wave velocity v_m, Debye temperature θ_D [56], longitudinal wave velocity, and average acoustic transverse v_l and v_t [57] is as follows:

$$\text{Clark model: } k_{min} = 0.87 k_B M_a{}^{-2/3} E^{1/2} \rho^{1/6} \tag{6.6}$$

$$\Theta_D = h/k_B (3/4\pi M)^{1/3} v_m \tag{6.7}$$

$$V_m = (1/3(2/V_t)3 + 1/(V_l)^3)^{-1/3} \tag{6.8}$$

$$\nu_t = \sqrt{G/\rho}, \quad \nu_l = \sqrt{(3B + 4G)/3\rho} \tag{6.9}$$

E = Young's modulus,
ρ = density,
k_B = Boltzmann constant,
$M_a = [M/(n \times N_A)]$ (average mass of the atoms),
M = molar mass,
n = number of atoms,
N_A = Avogadro number.

The order of minimum thermal conductivity, Debye temperature, and acoustic wave velocities of different graphynes are as follows: γ-graphyne > β-graphyne > α-graphyne. All these quantities are derived from equations (6.6) and (6.9).

6.2.4 Optical properties

Light absorption gives insight into a material's optical characteristics and electrical structure, this takes place as a result of light's interaction with the electrons and atoms of the substance it penetrates. The dielectric constant, as shown in equation (6.10), may be used to describe it in most instances.

$$\varepsilon(\omega) = \varepsilon_1(\omega) + i\varepsilon_2(\omega). \tag{6.10}$$

Dielectric absorption power is intrinsically linked to the imaginary part ε_2 and may therefore be estimated simply using the Kramers–Krönig dispersion, the real portion ε_1, the molar extinction coefficient, and the reflectivity obtained from ε_2 [58]. For the three graphyne structures, the absorption spectra, reflectivity, ε_1, and ε_2 are shown in figures 6.5(a)–(d), respectively, as functions of energy.

Figures 6.5(a) and (c) exhibit absorption spectra that match the ε_2 graph in terms of trend and peaks. Sharp peaks with increasing intensities with photon energy can be found in both the near-infrared and visible light spectra. Substantial delocalization and lower electron effective mass are seen in the energy band diagram, which shows a dramatic change in band shape in the energy range (spectrum) (0–3.5 eV).

Figure 6.5(a) shows a high concentration of free carriers due to an absorption coefficient that increases uniformly as the energy range gets closer to visible light. Visible light and the neighboring energy range have a −ve real part of the dielectric constant, as seen in figure 6.5(d), which indicates high conductivity. Additionally, there is a small divergence between the degenerated band gap and the spectrum's peak energy, indicating that relaxation occurred during the transition.

Furthermore, it shows that the energy required for the transition of an electron is more complicated than just energy levels difference. The reflectivity and absorptivity possessed by all three types of graphynes are high in the IR to the visible region. Visible light absorption and reflectivity were both highest for β-graphyne (figure 6.6–6.8).

Graphynes are a new class of 2D materials that show significant potential in the field of optoelectronics. The optical properties of graphyne under strain were never

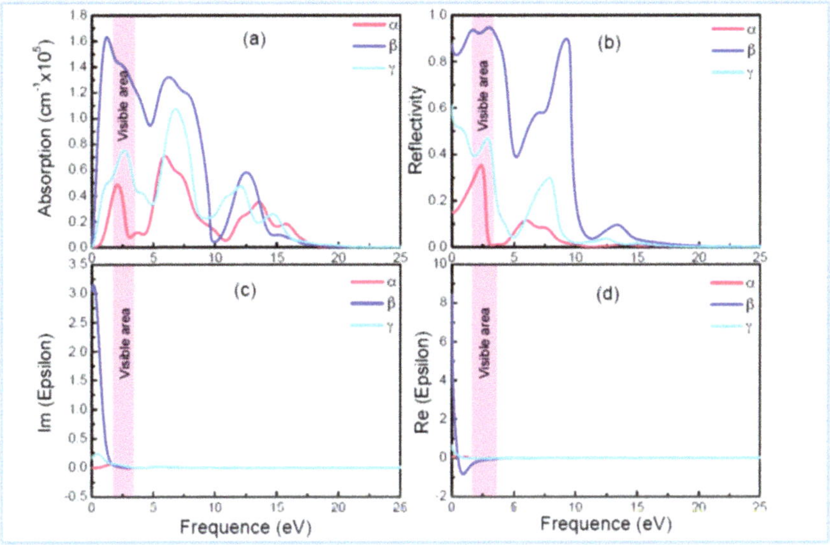

Figure 6.5. (a) Absorption spectra, (b) reflectivity, (c) imaginary, and (d) real part of the dielectric constants of the three graphyne structures. (Reproduced from [131]. CC BY 4.0.)

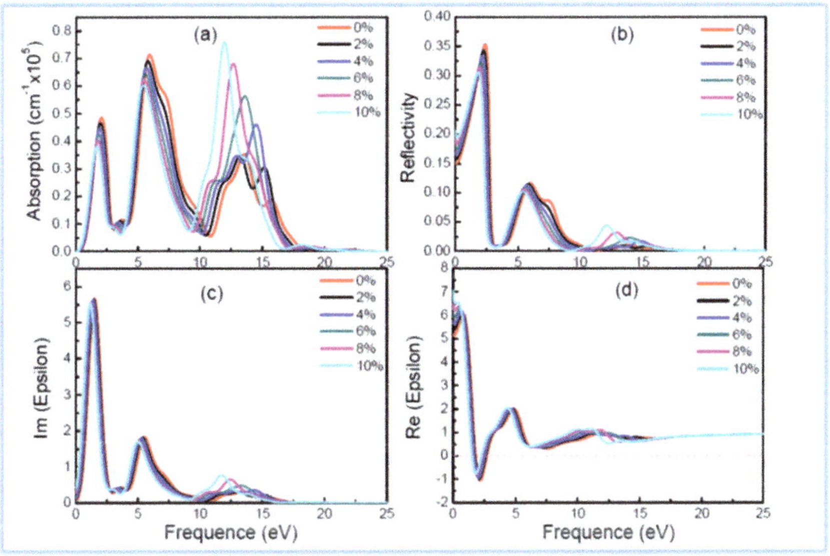

Figure 6.6. The graphical presentation of (a) absorption spectra, (b) reflectivity, (c) imaginary, and (d) real parts of the dielectric constants of α-graphyne. (Reproduced from [131]. CC BY 4.0.)

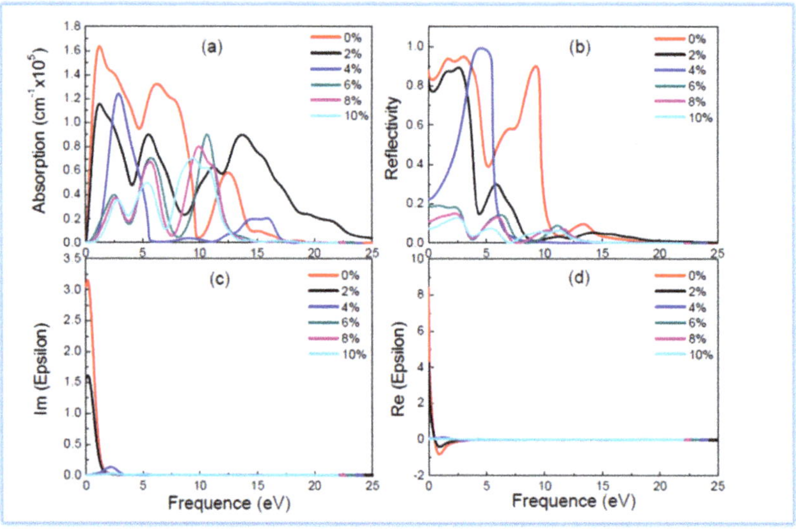

Figure 6.7. The graphical presentation of (a) absorption spectra, (b) reflectivity, (c) imaginary, and (d) real part of the dielectric constants of β-graphyne. (Reproduced from [131]. CC BY 4.0.)

examined. Different strains have an impact on the optical properties of three different graphyne materials, as shown in the following sections.

In combination with the electronic structures, the band gaps of α-graphyne remain zero from 0% to 10% strain. The absorption spectra seem to be closely related to this phenomenon. No matter how much strain is used, all absorption spectra start to rise at 0 eV. From the perspective of absorption, when strain grows, the peaks between 0 eV and 10 eV diminish, whereas the peak between 10 eV and 15 eV rises. The red-shift phenomenon was seen in the three main peaks. The general trend of reflectivity matches up nicely with the absorption spectrum, but the static reflectivity values (at 0 eV) increase from 2% to 10% strain. From 0 to 10 eV, the imaginary component of dielectric constants does not change significantly. Because of this, peaks located between 10 eV and 15 eV rise in value as strain increases, while the process of red-shift occurs. When the real component of dielectric constants is included, the value of static dielectric constants grows by a significant amount.

Before the strain reaches 8%, the band gap for β-graphyne rises slightly concerning strain; after that, it increases quickly, reaching 1.469 eV at 10% strain. This process is highly correlated to the absorption spectrum. It takes more energy for a bigger band gap to be transported by carriers. Figure 6.7 illustrates that β-graphyne is quite sensitive to strain, as shown. It only takes a tiny amount of strain to have a big impact on optics. When the strain is lowered to 2%, there is a drastic drop in the absorption peak, static reflectivity, and static dielectric constants. Something is interesting about the blue-shift effect in terms of absorption spectra and reflectivity. In particular, a large amount of force is required to open the β-graphyne band gap.

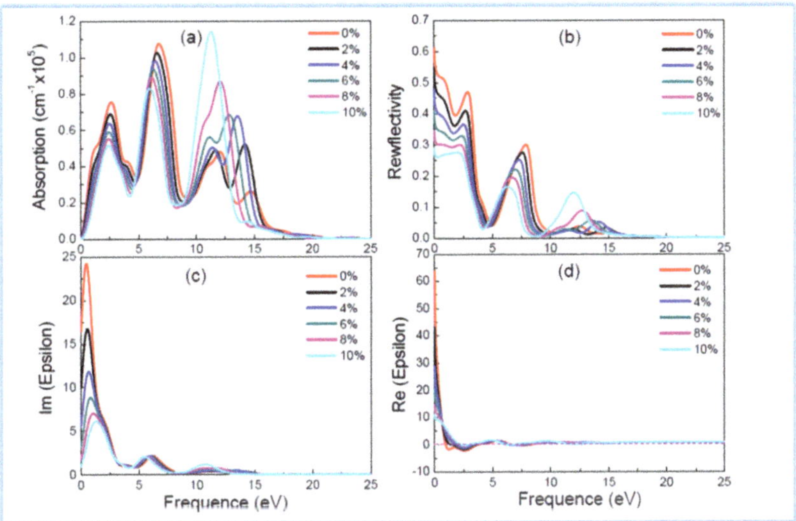

Figure 6.8. The graphical presentation of (a) absorption spectra, (b) reflectivity, (c) imaginary, and (d) real part of the dielectric constants of γ-graphyne. (Reproduced from [131]. CC BY 4.0.)

When strain is increased, all special values fall linearly, as illustrated in figure 6.8. According to figures 6.8(a) and (b), as tensile strain rises, the energy peak of the reflectivity curve shifts downward, and the peak height decreases. These findings imply that strain can influence the optical characteristics of γ-graphyne precisely in engineering applications.

6.3 Synthesis

In 1994 Diederich [19] reported that powerful new organometallic synthetic techniques, such as modern acetylene chemistry, can be used for high-quality organic synthesis. These play a huge role in the success of efforts to develop new carbon phases. In addition, to find unique and unusual synthesis approaches, it may be essential to be open to novel methodologies created in other fields.

In 2008 Haley came up with the idea of using synthesis methods to create substructures of non-natural, planar carbon networks of graphyne that are based on the dehydrobenzo[12]annulene framework [59].

6.3.1 Haley's work

The major obstacle to graphyne-related research is synthetic accessibility. For the synthesis of (**1**) controlled oligotrimerazation of cyclo[12]carbon (**2**) forms an accessible path (scheme 6.1). This strategy has been shown to be inadequate for the synthesis and detection of **2** in a mass spectrometer [60]. Synthetically accessible hexaethynylbenzene is the most basic subunit of the graphyne component; it can serve as a C≡C scaffold upon which graphyne mimics can be constructed. Vollhardt *et al* synthesized the parent hydrocarbon (**3**) in 1986 via a Sonogashira

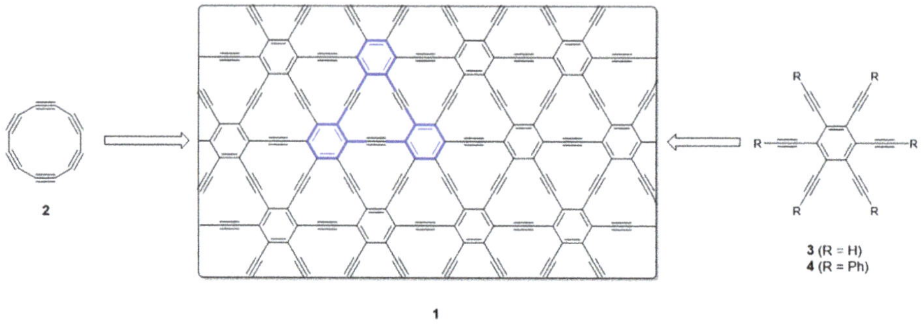

Scheme 6.1. Synthesis of graphyne from compounds **2**, **3**, and **4**.

cross-coupling process between hexabromobenzene and trimethylsilylacetylene, followed by protiodesilylation [61]. The resulting molecule was found to be extremely heat- and oxygen-sensitive, which is a frequent issue in permethylated π-systems. This instability has notably restricted research using **3**. Since the first report, substituted hexaethynylbenzenes with D_{6h} [62, 63], D_{3h} [64], and C_{2v} [65, 66] symmetries have been produced, however, the latter two systems need extremely complicated syntheses.

For the most part, the 'advanced' subunits of the graphyne family are hexakis (phenyl)benzenes (**4**). Molecule **4** was also synthesized from hexabromobenzene and revealed to exhibit a significant third-order nonlinear susceptibility [63].

Instead of using acetylenic scaffolding, macrocyclic segments can be created as network mimics and used as an alternative technique. A family of compounds called dehydrobenzoannulenes (DBAs) [18] is ideal for this use.

And DBAs are not simply the foundation for carbon allotropes; they are also ligands in organometallic chemistry, hosts for chelating guest molecules, probes for weakly induced ring currents, and raw materials for fullerenes, bucky tubes, bucky onions, and other carbon-rich compounds. As a result, annulene compounds are more resistant to heat, light, and oxygen.

Hexadehydrotribenzo[12]annulene ([12]DBA, **5**) is the smallest macrocyclic unit in the special case of graphyne. In the more than 40 years after its first preparation [67, 68], several methods have been developed for the synthesis of triyne **5**, with overall yields ranging from low to moderate [69, 70] (scheme 6.2). A modified route of the original method for Cu-mediated intermolecular cyclotrimerization still provides the easiest and most direct access to **5** in 47% yield, along with 8% yields of tetramer and a trace of the hexamer, it is still the fastest and most efficient method [71].

The simultaneous separation of tetramer and hexamer in addition to **5** reveals a significant shortcoming of intermolecular methods. Such syntheses are straightforward to carry out, but they tend to produce unwanted cyclooligomeric and/or polymeric products, resulting in lower yields of the material. Trimeric, tetrameric, or even extremely strained dimeric species might be predominant depending on the substituents used [69, 70]. A particular macrocycle might be difficult (or impossible)

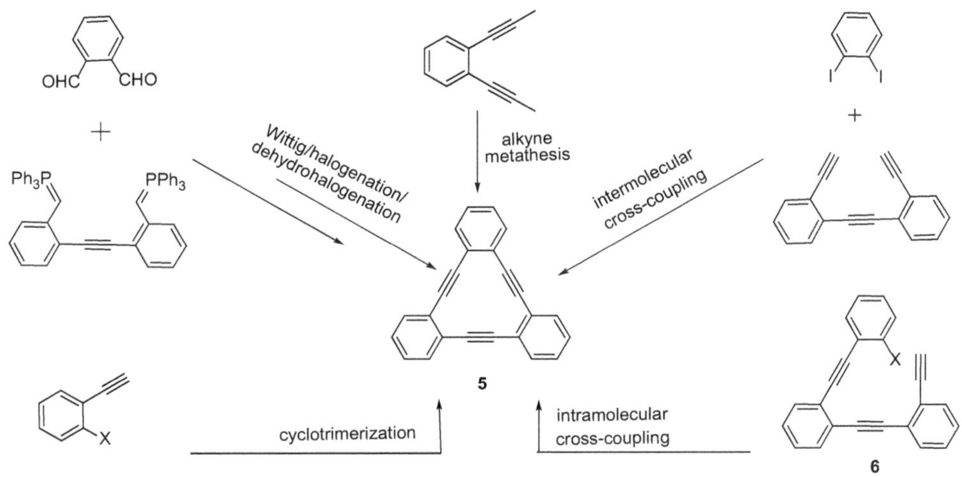

Scheme 6.2. Various methods employed for the preparation of triyne **5**.

to extract from a mixture because of its closeness in composition, structure, and solubility. A range of functionalized [12]DBA derivatives has been produced using the cyclotrimerization technique, despite its disadvantages [70]. Using an intramolecular method, DBA **5** may be built to reduce side products [72]. The most important part of the chemistry for synthesis of **6** includes consecutive Sonogashira cross-coupling. $Pd(DBA)_2$ was used for intramolecular alkynylation under high dilution conditions, and the only product obtained was **5**. Even though an additional six steps using easily available chemicals were required, a 35% yield was nevertheless achieved using this method (scheme 6.2). More crucially, the absence of bigger cyclooligomeric macrocycles made it easier to isolate and purify the product.

To build more complicated graphyne network substructures such as bis[12]DBA **7**, an intramolecular cyclization method is crucial (scheme 6.3). Compound **7** was prepared efficiently by the use of a double intramolecular Sonogashira reaction of **8** in the first attempt [72]. A relatively low isolated yield of **7a** resulted from the inadequate cyclization process. After 2000 publications, it has been shown that while intermolecular techniques have the potential to be used in the synthesis of the carbon-rich backbone of **7a–b**, the overall yields are quite low [73, 74].

We recently focused on the 'bow-tie' bis[12]DBA moiety again, using an intramolecular strategy where the alkyne metathesis was used as a ring-forming technique, as catalyst development and macrocyclization applications for alkyne metathesis have increased dramatically in the last several years [75, 76]. This novel approach, which combines alkyne metathesis efficiency with pre-organized propynyl groups, gives enhanced access to graphyne substructures, yielding **7c–d** with attached solubilizing alkyl groups in around 20% of the overall yield [77].

The Sonogashira reaction may be used to build both intra and intermolecular variants of the 'diamond' substructure **9** [78] (scheme 6.4). It is one of the few

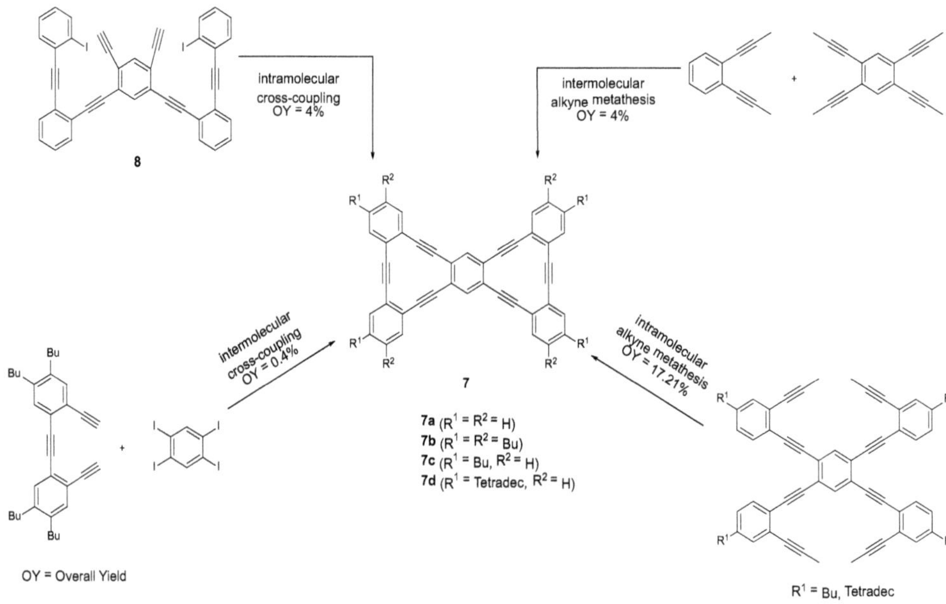

Scheme 6.3. Synthesis of complicated graphyne bis[12]DBA **7**.

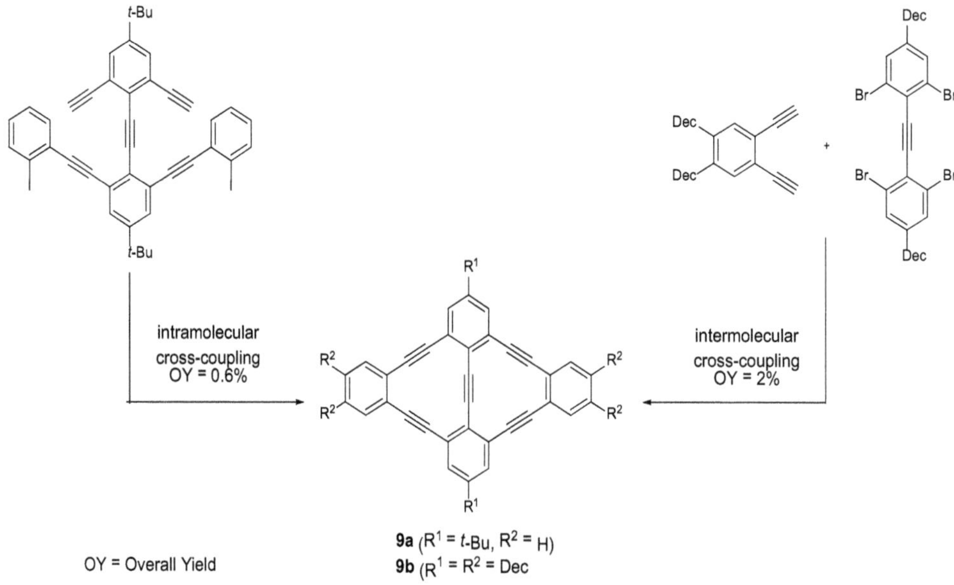

Scheme 6.4. Synthesis of 'diamond' structured graphyne **9**.

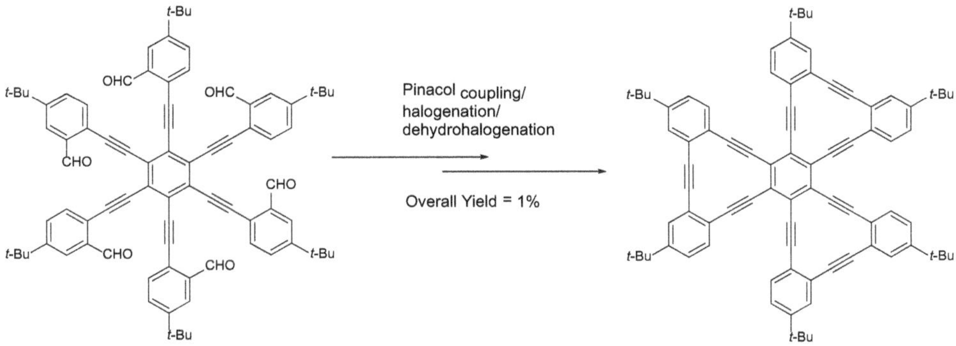

Scheme 6.5. Synthesis of 'trefoil' subunit **10**.

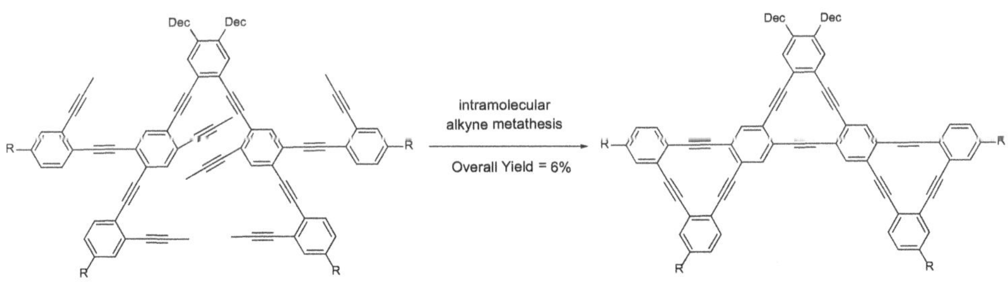

Scheme 6.6. Preparation of **11** through direct alkyne metathesis.

molecules for which the intramolecular synthesis pathway has a lower overall yield, although neither approach is appropriate for producing vast amounts of material.

The 'trefoil' subunit **10** is made up of three [12]DBAs joined by a benzene ring in the middle (scheme 6.5) [79]. Sonogashira's reaction with hexabromobenzene is the initial step in creating the carbon skeleton. Thereafter, the aldehydes undergo intramolecular pinacol coupling, the resultant hydroxyls are converted to chlorides, and the rings are formed by double HCl elimination using a strong base with an overall yield of 1%.

The isomeric tris[12]DBA (**11**), which possesses the longest linear chromophore of all the known graphyne subunits, may be readily achieved using directed alkyne metathesis (scheme 6.6), with a yield of about 6%. There have been unsuccessful attempts to synthesize the 'linear' tetrakis[12]DBA substructure by alkyne metathesis, with only three of the four rings being formed [77].

6.3.2 Synthesis of gamma graphyne

Li *et al* investigated the possibility that γ-graphyne may be formed by a cross-coupling mechanism involving sp- and sp^2-hybridized carbon. Cook *et al* have performed a Huisgen coupling reaction via the mechanochemical method by

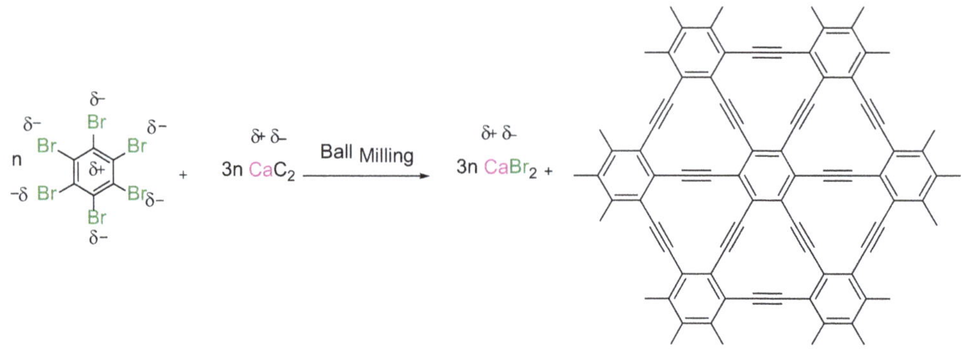

Scheme 6.7. Schematic presentation of the synthesis of γ-graphyne by the mechanochemical route.

performing azide-alkyne cycloaddition without the need for a copper cocatalyst [80]. The Suzuki–Miyaura cross-coupling process may be carried out via the mechanochemical reaction, according to Schneider *et al* used a mechanochemical technique to manufacture certain alkynyl carbon compounds [81, 82]. Cross-coupling reactions utilizing CaC_2 and 1,3,5-tribromobenzene have also recently yielded hydrogen substituted graphyne by Li *et al* [83]. These investigations motivate to synthesizing of γ-graphyne via a mechanochemical cross-coupling process. Mechanochemical synthesis of γ-graphyne is shown to be a straightforward and high-yielding process. Scheme 6.7 depicts the usage of CaC_2 and hexabromobenzene as precursors in a cross-coupling process driven by ball milling. In theory, the composition and lattice constant of the sample as it was made fit well with those of γ-graphyne.

This study is the first to reveal a straightforward, high-yield technique for making γ-graphyne. The EDX and XPS studies showed that the prepared γ-graphyne sample contains elemental carbon. Since there are an equal number of sp^2 and sp hybridized carbon atoms in elemental carbon, γ-graphyne was recognized in Raman and high-resolution C1s core-level XPS spectra. The structure of γ-graphyne was confirmed by XRD, SARD, and TEM. It has a lattice constant value of 0.69 nm with monocrystalline nanosheet morphology. The UV–vis DRS spectroscopy and Mott–Schottky plots were used to deduce the energy band structure. Li *et al* have concluded that their study not only explains the XRD pattern but also offers a method to synthesize γ-graphyne, and energy band structure to promote further studies.

Chemical vapor deposition of organic precursor molecules has recently been described as a method for synthesizing graphene on metal surfaces, and this might be used as a support and template in future attempts to synthesize the material on surfaces [84]. Whether or not it is worth attempting to synthesize extended graphyne materials is determined by their qualities, in particular their electrical properties. It is important to remember that no substantial amounts of graphyne have been synthesized yet in the laboratory, only traces have been created.

6.4 Potential applications of graphynes

Among the family of 2D carbon compounds, graphene stands out as one of the most intriguing carbon allotropes due to its potential use in many different scientific disciplines. The combination of its unusual sp- and sp^2-hybridization and flat structure makes it an intriguing candidate for use in cutting-edge nanotechnology research. Metal atoms may form strong bonds with graphyne because of the presence of the unusual P_x–P_y π/π^* states in the band structure. The excellent electronic transport properties and fascinating structure suggest that graphyne-based nanomaterials will have possible use as energy storage devices, gas separation, therapy, electronic devices, detector, biomedicine, solar cells, H_2 storage, magnetism, water desalination and purification, making battery, and catalysis application. In the following, we examine these applications briefly.

6.4.1 Electronic devices

The development of highly efficient energy-related devices with a tunable band gap in nanostructures has important significance [85]. The inherent nonzero band gap of graphynes, in contrast to graphene, has been measured to be from 0.44 to 1.22 eV, depending on the method of calculation used [26, 86]. However, the band gap may be altered in a variety of ways, including by altering the number of layers, applying pressure, strain [87], doping heteroatoms [88, 89], and adjusting the graphyne building blocks themselves [90].

Indeed, graphyne can also display different kinds of morphologies depending on the structural engineering, for example cages [91], nanoribbons [17], and nanotubes [92]. The value of the band gap may be dramatically impacted by tailoring 2D graphyne nanosheets to take on unusual shapes, such as an armchair or the zigzag edges of nanoribbons. The theoretical impact of N and B atomic doping on the electronic characteristics of graphynes was investigated by Gamallo *et al* [93]. Doping with nitrogen and boron converts graphynes into n-type and p-type species, allowing for the injection of holes into the valence band and the transfer of additional electrons to the conduction band. Indeed, the electronic characteristics of graphynes are transmitted from the semiconductor to the metal transition through substitution on sp and sp^2 carbon by varying amounts of impurities. A precisely tunable intrinsic nonzero band gap and doping of heteroatom of graphynes open the way towards its applications for designing infinitely flexible and wearable electronic devices.

6.4.1.1 Optoelectronic devices

Optoelectronics, often known as optronics, refers to electronic devices and systems used for detecting and manipulating light. These investigations are often categorized as a sub-field of photonics. Optoelectronic gadgets are simple operating phenomena such as optical-to-electrical or electrical-to-optical transducers. Because of their narrow, controllable natural band gap, graphynes have enhanced ambipolar carrier transport capabilities and a wider range of light absorption, making them promising candidates for use in photonics and optoelectronics. The band gap is substantial, more akin to a semiconductor (E_g = 1.2 eV) than metal or semimetal, as theoretically anticipated by Baughman *et al.* The substantial band gap and well-characterized nature of related conjugated polymers and linear structures suggest that graphynes will exhibit optical characteristics [6]. Because of their high carrier mobility and large operating frequency range (300–800 nm), materials based on graphynes may potentially be a promising component in photodetection systems [94]. Bhattacharya's group theoretically calculated the electronic and optical properties of the armchair and zigzag graphyne nanotubes (GNTs) by using the partial density of states and crystal orbital Hamilton population analysis. The effect of N and B atoms on the band gap was studied concerning the B/N substitution site and increasing diameter of the NTs. The dielectric function reveals that, in low-energy areas, perpendicular polarization is suppressed relative to parallel polarization. The low-energy region has been found responsible for anisotropy. In this study they also found that the optical properties of graphynes were mainly because of the tubular geometry rather than its parent planar structure [95]. The capability of light polarization could be useful for the tuning of the optical response of GNT-based optoelectronic devices. The value of the band gap can also be regulated by the bilayer mix with heteroatoms and could also be very useful for optoelectronic devices [96]. Introducing π-conjugated chains with donor/acceptor groups such as –$(CH{=}CH)_m$–NH_2/NO_2– enhance the second-order nonlinear optical responses. The graphyne and conjugated chain may undergo significant electron transfer with this sort of alteration. This electron transfer decreases the transition energy, which induces large static first hyperpolarizability (β_0) [97]. The value of hyperpolarizability depends monotonously upon the length and train of the conjugated chain. In 2022, Iqbal *et al* studied the role of Ag cluster doping on graphynes and its influence on optical and nonlinear optical (NLO) properties by following the TD-DFT/DFT techniques precisely. Doping of silver clusters at different positions on graphynes, such as center, side, and perpendicular, significantly improved the charge transfer properties, such as the enhanced λ_{max} in the range 368–536 nm and decreased the band gap in the range 2.58–4.73 eV compared to pure graphyne (GY) with 5.78 eV and 265 nm λ_{max} [98]. The current studies suggested that Ag cluster doping on two-unit cells on graphynes and its influence on NLO properties might be a very useful future development of nanoscale optoelectronic devices.

6.4.1.2 Magnetism and spintronic devices

Graphyne-based 2D carbon materials might be employed as magnetic and spintronic devices after changing their mechanical and chemical properties. Introducing sp^3-type defects in 2D carbon materials for breaking the delocalized π-electrons for avoiding sp^3-type functional clustering is a well-known methods to induce magnetism. Chemical adsorption of H atoms into the graphene surface introduces ferromagnetism. The absorption of H atoms on sp^2 carbon on alternative atoms forms a strong sigma bond between the carbon and hydrogen atoms. After hydrogenation, sp carbon atoms become sp^2, sp^2 becomes sp^3, and unhydrogenated C atoms leave the unpaired localized electrons to generate the maximum magnetic moment [99]. He *et al* calculated theoretically the doping effect of d-group metal

(TM) on the graphyne surface by using DFT+U calculations. Doping of transition metal atoms (TM = vanadium, chromium, iron, manganese, cobalt, and nickel) can efficiently change the magnetic properties of γ-GY films and triggered charge transfer between them [99]. The d-metal doping also modifies the band gap of graphynes and changes the magnetic properties depending on the metals, with V/Mn/Co@γ-GY behaving as spin-polarized half-semiconductors, while Cr/Fe@γ-GY behaves as a metal. The redistribution of electrons among the s, p, and d orbitals in the TM atom, caused by a charge transfer, is thought to be responsible for the material's improved magnetic characteristics. Like TM, non-metals, such as O, P, B, N, and S doping, also affect the magnetic properties of γ-GY films depending on the site of doping. Theoretical studies suggested that the boron and sulfur atoms desire to engage the sp^2-carbon atoms, even though oxygen, nitrogen, and phosphorous atoms prefer to be substituted on sp carbon atoms at the C≡C linkage. The polarized charges mostly appeared at the C–N bond along with the magnetic moments value of 0.58 and 0.31 μ_B for B-1 and B-2 while O-1/2, P-1/2, and S-1/2 doped graphyne appeared at the acetylene linkage [25]. Evidence from these investigations reveals that the chemical doping of graphyne films may affect their charge transfer and magnetic characteristics.

6.4.1.3 Anode materials for Li-ion batteries

Graphyne, among all the 2D carbon allotropes, has received enormous research interest owing to its excellent physical characteristics with tunable electronic properties and excellent promise for energy storage devices, in particular lithium-ion battery anodes. The mechanical properties of the anode have a significant impact on the battery's performance and lifespan during the charge/discharge cycle. High mobility and high ion storage capacity are essential for ideal anode materials for Li- or Na-ion batteries. Ji *et al* used molecular dynamics modeling methods to investigate the mechanical characteristics of multilayer graphyne sheets oriented in both the zigzag and armchair orientations. According to the Young's modulus, the mechanical properties of α-, β-, and γ-graphyne multilayer structures depend on the number of layers/sheets and acetylenic linkage. A higher number of acetylenic linkages in α-graphyne multilayer structures allows for larger elementary cell sizes and more charge carrier storage capacity than is seen in other graphynes [100]. The 3D Li diffusion with corresponding energy barriers 0.53 –0.57 eV could be achieved by multilayer α-graphynes with special atomic structures [101]. In graphene, out-plane diffusion of Li is prohibited due to a high barrier of over 8 eV [102]. Consequently, graphynes have quite high mobility of Li ions. Multilayer graphene-based nanomaterials have been shown to have distinct uses as anodes for energy storage systems such as lithium-ion batteries and can boost battery capacity when compared to conventional graphite anodes.

6.4.1.4 Other electronic devices

Graphyne-based 2D materials decorated with heteroatoms and transition metals can be used either as semiconductors or semimetals with Dirac cones. A highly tunable band gap also makes them very useful nanomaterials for various electronic devices [87]. Graphynes could be used for quantum transport, as predicted by many theorists. Structurally modified zigzag α-graphyne nanoribbons can be used as a molecule signal converter, dual-spin filter diode, and spin caloritronic device, as shown by first-principles quantum transport calculations by Yao's group [103]. Unique electronic, mechanical, chemical, as well as structural properties suggest that graphynes could also be used as interface modifiers or dopant materials in perovskite solar cells [104]. Photoelectrochemical and electrochemical studies suggest that γ-graphyne could be used as the predominant electrode material in supercapacitors. Galvanostatic charge/discharge measurements results indicate that the γ-graphyne supercapacitor delivers a maximum specific capacitance of 81 F g^{-1} at 0.2 A g^{-1} and a capacitance retention rate of 87.5% even after 5000 cycles at 3 A g^{-1} [105]. As we know, Li-ion batteries have some advantages, but also many disadvantages such as short lifetime, high cost, and safety issues. As a result of calcium's low toxicity, multivalent nature, and wide distribution, Ca-ion batteries have recently garnered the attention of scientists. Zhou *et al* evaluated the application of graphyne in the Ca-ion batteries by using the B3LYP-gCP-D3 method. Investigated results indicate that BN-doped graphyne has higher diffusion coefficient values 5.14×10^{-9} cm^2 s^{-1} than normal graphyne 3.27×10^{-11} cm^2 s^{-1} determined for the Ca^{2+} on the surface. based on studies, BN-analogous (BN-yne) might be good material for Ca-ion batteries [106]. Graphyne-based materials could be excellent candidates for thermoelectric equipment. These predictions have been made after theoretically calculating the zT values of 3.0 and 4.8 for p-type holes and n-type electrons, respectively, at room temperature by using DFT calculations and MD simulations [107]. Thus, the current zT values are superior to those of most of the existing thermoelectric materials, promoting to use of graphynes-based materials for high-powered thermoelectric devices.

6.4.2 Catalytic applications of graphynes

Currently, various 2D carbon allotropes are commonly used as catalysts for numerous chemical transformations because of their designable and flexible structures, such as $g-C_3N_4$, $Mo/Co/CeS_2$, and graphene. Usually, graphyne carbon atoms behave inactively as a catalyst. Several strategies, including the incorporation of functional groups and doping with various heteroatoms or metals, have been established to improve the catalytic activity of graphynes. After such modifications, graphyne-based materials might be used for a wide range of catalytic transformations such as the O_2 evolution reaction, H_2 evolution reaction (HER), selective hydrogenation, CO_2 reduction reaction (CRR), N_2 reduction reaction (NRR), and so on.

6.4.2.1 H_2/O_2 evolution and oxygen reduction reactions

The HER, O_2 evolution reaction (OER), and O_2 reduction reaction (ORR) employing multifunctional catalysts play critical roles in the advancement of electrochemical energy systems. There are two subcategories of graphyne-based materials utilized in HER/OER/ORR, namely metal and metal-free catalysts. The inert pristine graphyne-containing metal-free catalysts can be activated by doping heteroatoms and changing their architectures [108]. Co@GY and Ni@3B-GY were anticipated to be excellent multifunctional electrocatalysts for HER/OER/ORR based on experimental and DFT simulation results. Nitrogen doping at sp^2-N carbon not only increases the positive charge on adjacent carbon but could also boost the work functions of graphyne, thus promoting O_2 decomposition during the ORR process [109]. A systematic study has been carried out by the Chattopadhyay group for the mechanism of ORR in basic conditions on numerous N-doped graphynes, namely α-graphyne, β-graphyne, γ-graphyne, and 6,6,12-graphyne. Except for γ-graphyne, all other N-doped graphynes can perform ORR reactions [110]. Recently, in 2022 Xu *et al* synthesized a F-doped γ-graphyne/PtPd catalyst and presented their application for the oxygen reduction reaction. It showed almost similar activity compared to commercially available Pt/C with more rapid kinetics [111]. The modified γ-graphyne speeds up the diffusion of oxygen during catalysis by increasing its ability to adsorb oxygen and transport electrons to and from the surface of the catalyst.

6.4.2.2 Nitrogen reduction and fixation reactions

Ammonia is not only a nutritional source for most living things on Earth, but is also an essential raw material for various agrochemical, medicine, and chemical industries. Traditional methods such as the Haber–Bosch process face many disadvantages such as operating at high temperatures and pressure, low yield, high energy consumption, and severe environmental pollution. Hence, it is of great importance to design an efficient catalytic system for NRRs. Through structural and electronic tuning with the doping of transition metals and heteroatoms, graphynes could play a crucial role as catalysts for electrocatalytic NRRs. Xie *et al* demonstrated the catalytic properties of chromium (Cr) co-decorated graphyne towards the NRR using DFT calculations. The results of the study suggested that the inert N≡N bond is activated efficiently by Cr-graphyne and exhibits high activity, selectivity, stability, and feasibility [112]. In 2022 Li *et al.* investigated the γ-graphyne-like boron nitride (BN) sheet-supported single metal atom doped material as an electrocatalyst for nitrogen reduction reaction. They used a series of different metal atom doped graphyne-based materials for the N_2 reduction reaction. Mechanistic studies indicate that V/BN has the lowest energy barrier in the rate determining step (RDS). Further, studies of the mechanism in detail suggested that the excellent catalytic activity in the V/γBN sheet is mostly attributed to the electron donation and back-donation mechanism [113]. Sun *et al* studied monometallic and bimetallic transition metal atom decorated graphyne systematically for NRRs utilizing DFT studies. According to current research, bimetallic Ru decorated 2D graphyne has higher catalytic activity for NRRs, with a maximal free energy change (ΔG) of 0.43 eV [114]. Overall, after proper changes in the physical and chemical properties, graphyne-based nanomaterials could be used as excellent catalysts for NRRs.

6.4.2.3 Carbon dioxide reduction reaction

At present, scientists have major concerns about rising atmospheric carbon dioxide (CO_2) concentrations and their impact on the environment, resulting in global warming. Hence, it is an urgent requirement to develop techniques for efficient CO_2 capture and catalytic conversion into useful products. Atomically scattered catalysts tend to alleviate the greenhouse effect and generate useful products by electrochemical reduction of carbon dioxide. Lu and co-workers studied the role of earth-abundant triple transition metal (TM = Cu, Fe, and Co) embedded graphyne for CO_2 reduction into C_1 products through DFT calculations. According to the results, using the 3TM-graphyne as a catalyst could produce HCOOH, CH_3OH, and CH_4 from CO_2 depending on the applied potential [115]. Single metal-doped graphyne can also reduce CO_2, and theoretical studies have been performed by Kim and co-workers using a series of transition metals such as Ti, V, Cr, and Mn. The results of their study suggest that the most suitable candidate for selective reduction of carbon dioxide to CH_4 is Cr decorated graphyne. In chromium-based catalytic reduction of CO_2 to CH_4, the optimal reaction path followed is $^*CO_2 \rightarrow {}^*COOH \rightarrow {}^*CO \rightarrow {}^*CHO \rightarrow {}^*CH_2O \rightarrow {}^*CH_2OH \rightarrow {}^*CH_2 \rightarrow {}^*CH_3 \rightarrow CH_4$ [116]. The pore size of graphyne nanoparticles may influence their catalytic activity in the CO_2 reduction process. Chen *et al* studied the pore size effect of Cu-embedded graphyne for catalytic reduction of carbon dioxide. The effect of the skeleton structure of the graphynes on the attached Cu atoms was investigated using graphynes with a range of pore sizes [117].

6.4.2.4 Water splitting

The generation of hydrogen via photoelectrochemical water splitting has gained a great deal of attention in the last decade as it is a renewable source of green energy. The pioneering catalysts suffer from several drawbacks such as high cost, low efficacy, and selectivity. To overcome such drawbacks, the highly modifiable architecture and the electronic properties of graphynes make them promising candidates for water-splitting reactions. Cui *et al* studied a graphyne/TiO_2 (GY/TiO_2) p–n junction which contains p-type graphyne nanosheets surrounded by TiO_2 nanotube arrays in 3D space and found them to be 1.7 times more active than bare TiO_2 for water splitting. The modified heterojunction could extend solar light absorption, accelerate charge separation, and boost sunlight-driven water splitting with TiO_2 [118]. Shen *et al* applied strain on the nitrogen and cobalt anchored graphyne (Co@N1-GY) and studied its effect on catalytic activity and found that 0.5% tensile strain gives the best result.

6.4.2.5 Other catalytic reactions

Because of their unique electronic and structural properties, graphynes have the potential to be used as catalysts in a variety of different processes, including the oxidation of methane, carbon monoxide, and nitric oxide. Handling and proper use of methane gas, which is a major component of natural gas, is one of the biggest challenge facing the scientific community. Methane gas is highly flammable and inert, which makes it uneconomical and unsafe to use, thus it is mostly wasted in flares at remote sites. Density functional theory calculations done by the Sun and co-workers suggested that 4d-metal atom-loaded graphenes could be promising catalysts for transforming methane into methanol [119]. Graphyne-based materials, after being electronically tuned by doping of transition metals with anchored H4,4,4-graphyne, can also be used as promising catalysts for CO/NO oxidation reactions, as studied theoretically by Dai *et al* [120].

6.4.3 Applications in the field of energy

Their large surface area, excellent conductivity, strong affinity with external atoms, evenly distributed in-plane pores, and outstanding mechanical qualities allow graphynes to be a viable energy conversion material. In addition to their use in energy-related applications including batteries, hydrogen storage nanosystems, and high-performance catalysts, graphene-based nanomaterials are incredibly resilient in structure, and range from 1D nanotubes/nanoribbons/nanowires to 2D nanowalls/nanosheets and 3D nanostructure arrays.

6.4.3.1 Rechargeable batteries

Graphyne is an ideal material for energy storage devices due to its typical layered and porous structures, and its excellent electronic properties. Its vast surface area and regularly spaced porous channels provide ideal ionic confinement for metal ions. Sun *et al* [121] were the first to calculate Li atom adsorption and diffusion on graphyne monolayers. Graphynes have a storage capacity for Li atoms that can reach LiC_3. Intending to enhance their functionality as Li-ion battery anodes, Zhang *et al* looked into strained monolayer and bilayer graphynes [122]. The Li atom capacity can be increased by using applied strain, and monolayer graphynes can withstand up to 12% biaxial strain. Materials based on graphynes that have been doped with heteroatoms are expected to offer a versatile alternative to pure graphynes for use in rechargeable batteries. Theoretically, Singh *et al* investigated the possibility of using N-doped grapheme nanosheets as anode components for lithium, sodium, and magnesium ion batteries [123]. Li and Na atoms may be absorbed on the N-graphyne monolayer with a storage capacity of 623–2180 mAh g^{-1}, surpassing 2D phosphorene and borophane materials.

6.4.3.2 Storage of hydrogen

Graphynes' layered as well as porous structures are also excellent for hydrogen and methane storage systems. The ability of scandium-adorned holey graphyne, a recently synthesized carbon allotrope, to store hydrogen was investigated by Shukla *et al* using DFT and molecular dynamics simulations. It was demonstrated that holey graphyne's unit cell can absorb 6 scandium atoms which can absorb a total of 30 hydrogen molecules, with average binding energies and temperatures of -0.36 eV/H_2 and 464 K, respectively. The Bader charge analysis revealed that 1.9e charges are transferred from the 3d/4s orbital of the scandium atom to the C-2p orbital of graphyne. The structural stability of the scandium-decorated holey graphyne system is confirmed by *ab initio* molecular dynamics simulations at higher desorption temperatures.

6.4.4 Other applications of graphynes

Because of its flat sp^2 and sp hybridized framework, graphyne exhibits exceptional conjugated structures, pore distribution, and electrical properties that may be tuned. The intriguing architectures and electrical characteristics imply that graphyne-based compounds may find use in gas detection, gas separation, C1–C2 alcohol separation, methane oxidation, water purification, etc.

6.4.4.1 Water purification

Desalination of brackish water and seawater has received increasing attention due to the rapid depletion of numerous water resources. The pore size of graphyne permits the penetration of hydrogen and oxygen, but it is rather small for an H_2O molecule to pass through, making it ineffective for desalination and filtration of water. Water molecules easily pass through the membrane when the pore size is greater, as in the case of graphyne-3, and the permeability is approximately a hundred times more than that of industrial osmosis membranes [124]. The graphyne-3 membrane excludes all salts and pollutants with a rejection rate of 100%, including KCl, NaCl, C_6H_6, $CuSO_4$, and CCl_4 [125]. Therefore, it is a strong candidate for the filtration of contaminated water and the desalination of seawater. Graphyne-4 is recommended as an improved purification membrane when both the rejection rate and permeability are considered. Furthermore, it is claimed that negative charging might enhance water permeability and salt rejection even further [126]. Other than the graphyne-*n* structures, α-graphyne and β-graphyne also demonstrate outstanding water desalination capabilities with high water permeability and a 100% rejection rate.

6.4.4.2 Methane oxidation

Since collection and utilization are not financially feasible, methane, which makes up the majority of natural gas, is flared wastefully at isolated locations. One effective way to address this problem is the direct conversion of methane into liquid fuels such as methanol. Single metal atoms demonstrated reasonable stability at the pores of graphyne. Instead of expensive H_2O_2 or N_2O, abundant O_2 was used in this work. This thorough investigation shows that the dissociative adsorption of O_2 can produce catalytically active metal–oxygen (MO) moieties in both iron and cobalt single atoms scattered at adjacent graphyne pores [119]. Additionally, the Co–O moiety placed on graphyne showed adequate selectivity and good catalytic activity for the conversion of methane to methanol. Surprisingly, single metal atoms dispersed on a graphyne support were able to mimic the enzymatic formation of catalytically active M–O moieties. The methane-to-methanol conversion mechanisms and methanol production rates were shown to be directly influenced by the strength of the M–O bond by microkinetic analysis.

6.4.4.3 Separation of C1–C2 alcohols

The abundant nanopores in graphyne membranes have rendered them a great choice for alcohol purification. The issue, however, is in the rational design of GYs for effective (C1–C2) alcohol separation. By using molecular simulations, Jin *et al* created the first GYs for the recovery of methanol and the dehydration of ethanol [127]. Because preferred adsorption is the dominant mechanism for separating methanol from water, methanol is pre-selected in this process, and its flow rises with positive charges up to a certain point. The positively charged GY-M membrane achieves the best performance when this surface charge amendment is used for the separation of methanol and ethanol. Because of the influence of size sieving during the separation of water and ethanol, water is pre-selected. The GY-M is functionalized with polar groups to lower the permeation barrier and blocking effect, promoting water flux independent of aperture size. GY membranes may be used in organic solvent recovery and dehydration, as shown in the study.

6.4.4.4 Gas detection

Due to the serious issue of harmful gas emissions into the environment, numerous researchers are developing various sensors to identify these chemicals. Zhang *et al* used DFT simulations to examine BCl_3 adsorption on pure, sulfur-doped, and chromium-doped graphynes [128]. By switching out the transition metal Cr for the element C, the sheet's sensitivity and reactivity are significantly increased. Adsorption of BCl_3 reduces the HOMO–LUMO gap in Cr-doped graphene from 2.18 to 1.38 eV, enhancing the electrical conductivity greatly. DFT calculations were utilized by Heravi *et al* to investigate the Cl_2 and Br_2 interactions regarding Co-doped graphyne [129]. For Br_2 and Cl_2 adsorption, two vertical and parallel arrangements were found. Calculations revealed that Br_2 has a greater affinity for the graphyne nanosheet than Cl_2. The HOMO–LUMO gap (E_g) and electrical resistance of pristine sheets were not affected significantly by Br_2 or Cl_2. Its sensitivity and reactivity to bromine and chlorine gases increased significantly by adding a Co atom to the structure of pure graphyne. As a result, in the presence of Cl_2, the Co-doped graphyne may selectively detect Br_2 gas.

Derakhshande *et al* recently used DFT simulations to study the adsorption of cyanogen (C_2N_2) gas. The electrical characteristics of pure graphyne do not considerably change as C_2N_2 gas is faintly adsorbed on it with an adsorption energy of 11.0 kcal mol^{-1} [130]. Cu-doping enhances graphyne performance and increases its reactivity and sensitivity to C_2N_2, making it a potential candidate for use in C_2N_2 chemical sensors. Calculations show that the Cu-doped sheet's HOMO/LUMO gap (E_g) is reduced by about 37% because of the C_2N_2 adsorption, from 2.42 to 1.52 eV. This implies that the nanosheet's electrical conductivity has been enhanced.

6.5 Conclusion

In conclusion, we have discussed graphyne's chemical, physical, and mechanical characteristics as well as its uses in nanotechnology. Graphane, in contrast to graphone, has been synthesized effectively via a variety of distinct approaches.

Several fascinating and possibly useful properties of graphyne have been uncovered via research on this compound. Depending on the edge configuration, zigzag graphyne nanoribbons may be either antiferromagnetic semiconductors or non-magnetic semiconductors, but armchair graphyne nanoribbons always exhibit nonmagnetic semiconductor behavior. Large-scale synthesis of graphyne has not yet been achieved, while efforts are still being made to do so; this is necessary for providing experimental data to evaluate the theoretical predictions and resolve certain differences concerning mechanical characteristics. Transistor electronics and optoelectronics are two areas where graphyne's expected electronic characteristics show significant potential.

Acknowledgments

VP acknowledges VIT University for providing a fellowship. TG thanks VIT University for providing the 'VIT SEED GRANT' and Science and Engineering Research Board (SERB), India for providing Start-up Research Grant (SRG/2020/000760).

References

[1] Kroto H *et al* 1985 C_{60}: buckminsterfullerene *Nature* **318** 162–3
[2] Iijima S 1991 Helical microtubules of graphitic carbon *Nature* **354** 56–8
[3] Novoselov K S *et al* 2004 Electric field effect in atomically thin carbon films *Science* **306** 666–9
[4] Wallace P R 1947 The band theory of graphite *Phys. Rev.* **71** 622
[5] Bostwick A *et al* 2007 Quasiparticle dynamics in graphene *Nat. Phys.* **3** 36–40
[6] Baughman R, Eckhardt H and Kertesz M 1987 Structure–property predictions for new planar forms of carbon: layered phases containing sp^2 and sp atoms *J. Chem. Phys.* **87** 6687–99
[7] Haley M M, Brand S C and Pak J J 1997 Carbon networks based on dehydrobenzoannulenes: synthesis of graphdiyne substructures *Angew. Chem. Int. Ed. Engl.* **36** 836–8
[8] Li G *et al* 2010 Architecture of graphdiyne nanoscale films *Chem. Commun.* **46** 3256–8
[9] Narita N *et al* 1998 Optimized geometries and electronic structures of graphyne and its family *Phys. Rev.* B **58** 11009
[10] Narita N *et al* 2000 Electronic structure of three-dimensional graphyne *Phys. Rev.* B **62** 11146
[11] Narita N, Nagai S and Suzuki S 2001 Potassium intercalated graphyne *Phys. Rev.* B **64** 245408
[12] Ivanovskii A 2013 Graphynes and graphdyines *Prog. Solid State Chem.* **41** 1–19
[13] Li Y *et al* 2014 Graphdiyne and graphyne: from theoretical predictions to practical construction *Chem. Soc. Rev.* **43** 2572–86
[14] Wu W, Guo W and Zeng X C 2013 Intrinsic electronic and transport properties of graphyne sheets and nanoribbons *Nanoscale* **5** 9264–76
[15] Malko D *et al* 2012 Competition for graphene: graphynes with direction-dependent Dirac cones *Phys. Rev. Lett.* **108** 086804
[16] Zhang Y, Pei Q and Wang C 2012 Mechanical properties of graphynes under tension: a molecular dynamics study *Appl. Phys. Lett.* **101** 081909

[17] Coluci V *et al* 2003 Families of carbon nanotubes: graphyne-based nanotubes *Phys. Rev.* B **68** 035430
[18] Baughman R H *et al* 1993 Fullereneynes: a new family of porous fullerenes *Chem. Phys. Lett.* **204** 8–14
[19] Diederich F 1994 Carbon scaffolding: building acetylenic all-carbon and carbon-rich compounds *Nature* **369** 199–207
[20] Bunz U H, Rubin Y and Tobe Y 1999 Polyethynylated cyclic π-systems: scaffoldings for novel two and three-dimensional carbon networks *Chem. Soc. Rev.* **28** 107–19
[21] Balaban A, Rentia C C and Ciupitu E 1968 Chemical graphs. 6 Estimation of relative stability of several planar and tridimensional lattices for elementary carbon *Rev. Roum. Chim.* **13** 231
[22] Peng Q, Ji W and De S 2012 Mechanical properties of graphyne monolayers: a first-principles study *Phys. Chem. Chem. Phys.* **14** 13385–91
[23] Cranford S W and Buehler M J 2011 Mechanical properties of graphyne *Carbon* **49** 4111–21
[24] Pan L *et al* 2011 Graphyne- and graphdiyne-based nanoribbons: density functional theory calculations of electronic structures *Appl. Phys. Lett.* **98** 173102
[25] He J *et al* 2012 Magnetic properties of single transition-metal atom absorbed graphdiyne and graphyne sheet from DFT+ U calculations *J. Phys. Chem.* C **116** 26313–21
[26] Kang J *et al* 2011 Elastic, electronic, and optical properties of two-dimensional graphyne sheet *J. Phys. Chem.* C **115** 20466–70
[27] Yang B *et al* 2017 Dirac cones and highly anisotropic electronic structure of super-graphyne *Carbon* **113** 40–5
[28] Hu M *et al* 2016 Superhard sp^2–sp^3 hybrid carbon allotropes with tunable electronic properties *AIP Adv.* **6** 055020
[29] He C *et al* 2013 Two viable three-dimensional carbon semiconductors with an entirely sp^2 configuration *Phys. Chem. Chem. Phys.* **15** 680–4
[30] Zhang H *et al* 2012 Tunable hydrogen separation in sp–sp^2 hybridized carbon membranes: a first-principles prediction *J. Phys. Chem.* C **116** 16634–8
[31] Owens F J 2017 On the possibility of planar graphyne and graphdiyne chains *Solid State Commun.* **250** 75–8
[32] Puigdollers A R, Alonso G and Gamallo P 2016 First-principles study of structural, elastic and electronic properties of α-, β- and γ-graphyne *Carbon* **96** 879–87
[33] Jiang P *et al* 2017 Thermoelectric properties of γ-graphyne from first-principles calculations *Carbon* **113** 108–13
[34] Bhattacharya B, Singh N B and Sarkar U 2015 Pristine and BN doped graphyne derivatives for UV light protection *Int. J. Quantum Chem.* **115** 820–9
[35] Sarma J, Chowdhury R and Jayaganthan R 2014 Graphyne-based single electron transistor: *ab initio* analysis *Nano* **9** 1450032
[36] Wu P *et al* 2015 Graphyne-supported single Fe atom catalysts for CO oxidation *Phys. Chem. Chem. Phys.* **17** 1441–9
[37] Lu R *et al* 2013 Boron-substituted graphyne as a versatile material with high storage capacities of Li and H_2: a multiscale theoretical study *Phys. Chem. Chem. Phys.* **15** 16120–6
[38] Liu Y *et al* 2014 Hydrogen storage using Na-decorated graphyne and its boron nitride analog *Int. J. Hydrog. Energy* **39** 12757–64

[39] Li H *et al* 2020 The electronic structure and physicochemical property of boron nitridene *J. Mol. Graphics Modell.* **94** 107475
[40] Wang G *et al* 2014 Strain engineering of Dirac cones in graphyne *Appl. Phys. Lett.* **104** 213107
[41] Long M *et al* 2011 Electronic structure and carrier mobility in graphdiyne sheet and nanoribbons: theoretical predictions *ACS Nano* **5** 2593–600
[42] Srinivasu K and Ghosh S K 2012 Graphyne and graphdiyne: promising materials for nanoelectronics and energy storage applications *J. Phys. Chem.* C **116** 5951–6
[43] Rouhi H *et al* 2013 Young's modulus and Poisson's ratio of monolayer graphyne *J. Nanostruct.* **3** 303–7
[44] Shao Z-G and Sun Z-L 2015 Optical properties of α-, β-, γ-, and 6,6,12-graphyne structures: first-principle calculations *Physica* E **74** 438–42
[45] Zhang Y, Pei Q and Wang C 2012 A molecular dynamics investigation on thermal conductivity of graphynes *Comput. Mater. Sci.* **65** 406–10
[46] Dickman S, Senozan N and Hunt R 1970 Thermodynamic properties and the cohesive energy of calcium ammoniate *J. Chem. Phys.* **52** 2657–63
[47] Chiou W C Jr and Carter E A 2003 Structure and stability of Fe_3C–cementite surfaces from first principles *Surf. Sci.* **530** 88–100
[48] Jiao L *et al* 2016 Electronic, elastic, and optical properties of monolayer BC_2N *J. Solid State Chem.* **244** 120–8
[49] Le Page Y and Saxe P 2002 Symmetry-general least-squares extraction of elastic data for strained materials from *ab initio* calculations of stress *Phys. Rev.* B **65** 104104
[50] Wilhoit R C, Chao J and Hall K R 1985 Thermodynamic properties of key organic oxygen compounds in the carbon range C_1 to C_4. Part 1. Properties of condensed phases *J. Phys. Chem. Ref. Data* **14** 1–175
[51] Nika D *et al* 2009 Phonon thermal conduction in graphene: role of Umklapp and edge roughness scattering *Phys. Rev.* B **79** 155413
[52] Andrew R C *et al* 2012 Mechanical properties of graphene and boronitrene *Phys. Rev.* B **85** 125428
[53] Yang Z *et al* 2016 The thermal properties and thermoelectric performance of γ-graphyne nanoribbons *J. Phys. D: Appl. Phys.* **49** 145102
[54] Chen Z-Q *et al* 2015 Elastic properties, hardness, and anisotropy in baddeleyite IVTMO_2 (M = Ti, Zr, Hf) *Sci. China Mater.* **58** 893–905
[55] Clarke D R 2003 Materials selection guidelines for low thermal conductivity thermal barrier coatings *Surf. Coat. Technol.* **163** 67–74
[56] Bouhemadou A 2010 First-principles study of structural, electronic and elastic properties of Nb_4AlC_3 *Braz. J. Phys.* **40** 52–7
[57] Ravindran P *et al* 1998 Density functional theory for calculation of elastic properties of orthorhombic crystals: application to $TiSi_2$ *J. Appl. Phys.* **84** 4891–904
[58] Waters K R *et al* 2000 On a time-domain representation of the Kramers–Krönig dispersion relations *J. Acoust. Soc. Am.* **108** 2114–9
[59] Haley M M 2008 Synthesis and properties of annulenic subunits of graphyne and graphdiyne nanoarchitectures *Pure Appl. Chem.* **80** 519–32
[60] Tobe Y, Fujii T, Matsumoto H, Naemura K, Achiba Y and Wakabayashi T 1996 Generation of cyclocarbons with 4*n* carbon atoms (C_{12}, C_{16}, and C_{20}) by [2 + 2]

cycloreversion of propellane-annelated dehydroannulenes *Angew. Chem. Int. Ed. Engl.* **35** 1800–2

[61] Diercks R and Vollhardt K P C 1986 Cobalt-katalysierte Cyclisierung von Bis(2-ethinyl-phenyl)ethin-neue Synthese von gewinkeltem [3]Phenylen (Terphenylen), einem Molekül mit einer Cyclohexatrien-Einheit *Angew. Chem.* **98** 268–70

[62] Praefcke K, Kohne B and Singer D 1990 Hexaalkynyltriphenylene: a new type of nematic-discotic hydrocarbon *Angew. Chem. Int. Ed. Engl.* **29** 177–9

[63] Kondo K *et al* 1995 The third-order optical non-linearity of the phenylethynyl-substituted benzene system *J. Chem. Soc., Chem. Commun.* **1** 55–6

[64] Anthony J E, Khan S I and Rubin Y 1997 1,3,5/2,4,6-differentiated hexaalkynylbenzenes: absorption and fluorescence properties of a D_{3h}-symmetric donor-substituted system *Tetrahedron Lett.* **38** 3499–502

[65] Tobe Y, Kubota K and Naemura K 1997 Diels–Alder reactions of tetraethynylcyclopentadienones. an approach to differentially substituted hexaethynylbenzenes of C_{2v} symmetry *J. Org. Chem.* **62** 3430–1

[66] Tovar J D *et al* 1997 Synthesis and x-ray characterization of an octaalkynyldibenzooctadehydro[12]-annulene *J. Org. Chem.* **62** 5656

[67] Campbell I *et al* 1966 1,2;5,6;9,10-tribenzocyclododeca-1,5,9-triene-3,7,11-triyne and 1,2;5,6;9,10;13,14-tetrabenzocyclohexadeca-1,5,9,13-tetraene-3,7,11,15-tetrayne *Chem. Commun.* **1966** 87–9

[68] Staab H A and Graf F 1966 Zur Konjugation in makrocyclischen Bindungssystemen IV: Synthese und Eigenschaften von 1:2,5:6,9:10-tribenzo-cyclododeka-1.5.9-trien-3.7.11-triin *Tetrahedron Lett.* **7** 751–7

[69] Jones C 2005 *MM Haley in Acetylene Chemistry* (Weinheim: Wiley)

[70] Spitler E L, Johnson C A and Haley M M 2006 Renaissance of annulene chemistry *Chem. Rev.* **106** 5344–86

[71] Hara K *et al* 2004 Mono- and bis(tetrathiafulvaleno)hexadehydro[12]annulenes *Chem. Commun.* **18** 2042–3

[72] Kehoe J M *et al* 2000 Carbon networks based on dehydrobenzoannulenes. 3. Synthesis of graphyne substructures *Org. Lett.* **2** 969–72

[73] Miljanić O Š, Vollhardt K P C and Whitener G D 2003 An alkyne metathesis-based route to ortho-dehydrobenzannulenes *Syn. Lett.* **2003** 0029–34

[74] Iyoda M *et al* 2004 Copper-mediated simple and efficient synthesis of tribenzohexadehydro [12]annulene and its derivatives *Synthesis* **2004** 1527–31

[75] Schrock R R 1995 The alkoxide ligand in olefin and acetylene metathesis reactions *Polyhedron* **14** 3177–95

[76] Zhang W and Moore J S 2007 Alkyne metathesis: catalysts and synthetic applications *Adv. Synth. Catal.* **349** 93–120

[77] Johnson C A, Lu Y and Haley M M 2007 Carbon networks based on benzocyclynes. 6. Synthesis of graphyne substructures via directed alkyne metathesis *Org. Lett.* **9** 3725–8

[78] Sonoda M *et al* 2004 Convenient synthesis and photophysical properties of tetrabenzo-pentakisdehydro[12]annuleno[12]annulene *Chem. Lett.* **33** 972–3

[79] Yoshimura T *et al* 2006 Synthesis and properties of trefoil-shaped tris(hexadehydrotribenzo [12]annulene) and tris(tetradehydrotribenzo[12]annulene) *Org. Lett.* **8** 2933–6

[80] Cook T L, Walker J A and Mack J 2013 Scratching the catalytic surface of mechanochemistry: a multi-component CuAAC reaction using a copper reaction vial *Green Chem.* **15** 617–9

[81] Li Y *et al* 2017 Efficient destruction of hexachlorobenzene by calcium carbide through mechanochemical reaction in a planetary ball mill *Chemosphere* **166** 275–80

[82] Li Y *et al* 2017 Synthesis and supercapacitor application of alkynyl carbon materials derived from CaC_2 and polyhalogenated hydrocarbons by interfacial mechanochemical reactions *ACS Appl. Mater. Interfaces* **9** 3895–901

[83] Lee J *et al* 2018 Synthesis of hydrogen substituted graphyne through mechanochemistry and its electrocatalytic properties *Acta Phys. Chim. Sin.* **34** 1080

[84] Kim K S *et al* 2009 Large-scale pattern growth of graphene films for stretchable transparent electrodes *Nature* **457** 706–10

[85] Zuo Z *et al* 2017 A facile approach for graphdiyne preparation under atmosphere for an advanced battery anode *Chem. Commun.* **53** 8074–7

[86] Luo G *et al* 2011 Quasiparticle energies and excitonic effects of the two-dimensional carbon allotrope graphdiyne: theory and experiment *Phys. Rev.* B **84** 075439

[87] Zhang C *et al* 2018 First principles study of trirutile magnesium bismuth oxide: ideal bandgap for photovoltaics, strain-mediated band-inversion and semiconductor-to-semimetal transition *Comput. Mater. Sci.* **149** 158–61

[88] Xu J *et al* 2020 Bandgap opening of graphdiyne monolayer via B, N-codoping for photocatalytic overall water splitting: design strategy from DFT studies *J. Phys. Chem.* C **124** 6624–33

[89] Bu H *et al* 2012 Isoelectronic doping of graphdiyne with boron and nitrogen: stable configurations and band gap modification *J. Phys. Chem.* A **116** 3934–9

[90] Majidi R and Karami A 2015 Electronic properties of B- and N-doped graphyne nanotubes *Comput. Mater. Sci.* **97** 227–30

[91] Zhang S *et al* 2012 Theoretical investigations of sp–sp^2 hybridized zero-dimensional fullerenynes *Nanoscale* **4** 2839–42

[92] Karami A 2015 Acrolein adsorption on graphyne nanotube: a density functional theory study *Fuller. Nanotub. Carbon Nanostruct.* **23** 885–9

[93] Ruiz-Puigdollers A and Gamallo P 2017 DFT study of the role of N- and B-doping on structural, elastic and electronic properties of α-, β- and γ-graphyne *Carbon* **114** 301–10

[94] Li Y *et al* 2021 Graphdiyne visible-light photodetector with ultrafast detectivity *Adv. Opt. Mater.* **9** 2001916

[95] Bhattacharya B *et al* 2015 Electronic and optical properties of pristine and boron–nitrogen doped graphyne nanotubes *Phys. Chem. Chem. Phys.* **17** 19325–41

[96] Bhattacharya B, Sarkar U and Seriani N 2016 Electronic properties of homo- and heterobilayer graphyne: the idea of a nanocapacitor *J. Phys. Chem.* C **120** 26579–87

[97] Hou N, Feng R and Fang X H 2021 Effective enhancement of the second-order nonlinear optical responses of graphynes by introducing π-conjugated chains with donor/acceptor groups *ChemistrySelect* **6** 10806–16

[98] Zahid S *et al* 2022 Silver cluster doped graphyne (GY) with outstanding non-linear optical properties *RSC Adv.* **12** 5466–82

[99] Wang Y *et al* 2018 Tuning the electronic and magnetic properties of graphyne by hydrogenation *Appl. Surf. Sci.* **452** 181–9

[100] Momen R *et al* 2022 Evaluation of mechanical properties of multilayer graphyne-based structures as anode materials for lithium-ions batteries *Eur. Phys. J. Plus* **137** 360
[101] Zhang H *et al* 2011 High mobility and high storage capacity of lithium in sp–sp^2 hybridized carbon network: the case of graphyne *J. Phys. Chem.* C **115** 8845–50
[102] Khantha M *et al* 2008 Interaction and concerted diffusion of lithium in a (5, 5) carbon nanotube *Phys. Rev.* B **78** 115430
[103] Ni Y *et al* 2016 The spin-dependent transport properties of zigzag α-graphyne nanoribbons and new device design *Sci. Rep.* **6** 25914
[104] Ahmad S O A *et al* 2021 Application of two-dimensional materials in perovskite solar cells: recent progress, challenges, and prospective solutions *J. Mater. Chem.* C **9** 14065–92
[105] Ding W *et al* 2020 Ultrasound-promoted synthesis of γ-graphyne for supercapacitor and photoelectrochemical applications *Ultrason. Sonochem.* **61** 104850
[106] Wu L, Zhang H and Zhou J 2021 A simulation on the graphyne and its inorganic BN-like nanosheets as anode materials for Ca-ion batteries *Mol. Simul.* **47** 326–33
[107] Sun L *et al* 2015 Graphdiyne: a two-dimensional thermoelectric material with high figure of merit *Carbon* **90** 255–9
[108] Lv Q *et al* 2018 Selectively nitrogen-doped carbon materials as superior metal-free catalysts for oxygen reduction *Nat. Commun.* **9** 3376
[109] Kang B *et al* 2017 Revealing the importance of nitrogen doping site in enhancing the oxygen reduction reaction on β-graphyne *Carbon* **123** 415–20
[110] Das B K, Sen D and Chattopadhyay K 2022 Mechanism of oxygen reduction reaction in alkaline medium on nitrogen-doped graphyne and graphdiyne families: a first principles study *Chem. Phys. Chem.* **23** e202100900
[111] Wang W *et al* 2022 A cathode material of fuel cells: F-doped γ-graphyne/PtPd nanocomposite from plasma activation and gamma irradiation *ACS Appl. Energy Mater.* **5** 2036–44
[112] Song W *et al* 2021 Carbon-coordinated single Cr site for efficient electrocatalytic N_2 fixation *Adv. Theory Simul.* **4** 2100044
[113] Li T, Wu Y and Pei M 2022 Screening of transition metal single-atom catalysts doped on γ-graphyne-like BN sheet for efficient nitrogen reduction reaction *J. Alloys Compd.* **908** 164675
[114] Jasin Arachchige L *et al* 2020 Theoretical investigation of single and double transition metals anchored on graphyne monolayer for nitrogen reduction reaction *J. Phys. Chem.* C **124** 15295–301
[115] Zhou S *et al* 2021 Triple-atom catalysts 3TM-GYs (TM = Cu, Fe, and Co; GY = graphyne) for high-performance CO_2 reduction reaction to C_1 products *Appl. Mater. Today* **25** 101245
[116] Fu L *et al* 2021 Construction of Cr-embedded graphyne electrocatalyst for highly selective reduction of CO_2 to CH_4: a DFT study *Chem. Eng. J.* **414** 128857
[117] Ni Y *et al* 2020 Pore size effect of graphyne supports on CO_2 electrocatalytic activity of Cu single atoms *Phys. Chem. Chem. Phys.* **22** 1181–6
[118] Qiu D *et al* 2021 Assembling γ-graphyne surrounding TiO_2 nanotube arrays: an efficient p–n heterojunction for boosting photoelectrochemical water splitting *Dalton Trans.* **50** 15422–32
[119] Jasin Arachchige L *et al* 2021 Mechanistic insights into direct methane oxidation to methanol on single-atom transition-metal-modified graphyne *ACS Appl. Nano Mater.* **4** 12006–16

[120] Tang Y *et al* 2022 Theoretical insights into the CO/NO oxidation mechanisms on single-atom catalysts anchored $H_{4,4,4}$-graphyne and $H_{4,4,4}$-graphyne/graphene sheets *Fuel* **319** 123810

[121] Sun C and Searles D J 2012 Lithium storage on graphdiyne predicted by DFT calculations *J. Phys. Chem.* C **116** 26222–6

[122] Zhang Q, Tang C and Fu L 2019 The first-principle study on the performance of biaxial strained graphdiyne as the Li-ion battery anode *Appl. Surf. Sci.* **497** 143723

[123] Makaremi M *et al* 2018 Theoretical investigation: 2D N-graphdiyne nanosheets as promising anode materials for Li/Na rechargeable storage devices *ACS Appl. Nano Mater.* **2** 127–35

[124] Kou J *et al* 2013 Water permeation through single-layer graphyne membrane *J. Chem. Phys.* **139** 064705

[125] Xue M, Qiu H and Guo W 2013 Exceptionally fast water desalination at complete salt rejection by pristine graphyne monolayers *Nanotechnology* **24** 505720

[126] Wu B *et al* 2017 Water and salt permeability of monolayer graph-n-yne: molecular dynamics simulations *Carbon* **123** 688–94

[127] Liu Q *et al* 2022 Efficient separation of (C_1–C_2) alcohol solutions by graphyne membranes: a molecular simulation study *J. Membr. Sci.* **644** 120139

[128] Wu L and Zhang H 2022 A molecular modeling on the boron trichloride gas detection by S- and Cr-doped graphyne *J. Mol. Model.* **28** 38

[129] Heravi M R P 2021 Theoretical investigation of Br_2 and Cl_2 detection by the pristine and Co-doped graphyne *Research Square* 10.21203/rs.3.rs-534934/v1 27 May

[130] Alamri S, Rajhi A A and Derakhshande M 2022 Potential detection of C_2N_2 gas by the pure, Al, and Cu-doped graphynes: a DFT study *Mol. Simul.* **48** 574–83

[131] Hou X *et al* 2018 Study of electronic structure, thermal conductivity, elastic and optical properties of α, β, γ-graphyne *Materials* **11** 188

IOP Publishing

Nanocarbon Allotropes Beyond Graphene
Synthesis, properties and applications
Arpan Kumar Nayak and Santosh K Tiwari

Chapter 7

Synthesis and applications of graphdiyne

Sutripto Majumder

Graphdiyne (GDY) is a one-atom-thick carbon allotrope which consists of evenly distributed pores made up of sp^2- and sp-hybridized carbon atoms with high degrees of conjugation. Because of its unusual and interesting structural, physical, and chemical qualities, it has piqued the curiosity of both research and industry. In this chapter we will briefly review the fabrication and practical applications of GDY. This chapter starts with the preparation methods, which are divided into three categories: solid, liquid and gaseous. This is followed by discussion of the modification of GDY through doping and composite formation. We then discuss the prominence of GDY in different applications such as energy harvesting and storage, followed by electronic, biological, and environmental applications with the latest references. The main aim of this chapter is to provide a clear understanding regarding the synthesis and recent applications of the novel material GDY.

7.1 Introduction

Compounds made up of carbon possess outstanding physicochemical, electrical, mechanical, and even optical capabilities, which have played a key role in recent material advancements [1]. Researchers have paid close attention to the creation or manufacture and enriching of the variety of new carbon allotropes, including fullerenes, carbon nanotubes, as well as graphene [2–4]. In this context, researchers have used a variety of synthesis approaches to further change the structures and enhance the characteristics of existing carbon allotropes [5]. Carbon allotropes are known to be made up of three hybrid forms of carbon from the viewpoint of structure and content (sp, sp^2, and sp^3) [6, 7]. Given that all of the three kinds of hybrid carbon have distinctive characteristics, it might be both inspiring and meaningful, if newly designed carbon-based materials could be produced with high precision, this could allow the various hybridized forms of carbon to 'cooperate' or excellently display about their own unique benefits [8].

doi:10.1088/978-0-7503-5177-5ch7

The sp^2 type carbon, which has been used to make carbon nanotubes as well as graphene, has been shown to have excellent characteristics. Graphdiyne (GDY) is a one-atom-thick carbon allotrope made up of equally distributed pores made up of sp^2- and sp-hybridized carbon atoms with high degrees of conjugation. As a result, GDY has piqued researchers' curiosity [9]. Using a cross-coupling process, GDY film was initially created on large-area copper foil [10]. The presence of sp^2 and sp carbon furnishes GDY with high-conjunction, broad interplanar spacing, and superior chemical stability with adjustable electronic characteristics, according to theoretical predictions and observations [11, 12].

GDY seems to be the most durable diacetylene-bonded quasi-carbon allotrope [13], with a direct band gap (0.46 eV) and with very strong carrier mobility at ambient temperature (10^4–10^5 $cm^2\ V^{-1}\ s^{-1}$), suggesting that this material is a perfect candidate for forthcoming nanoelectronics [14]. Utilizing GDY with diverse morphologies (e.g. films, powders, and nanowalls) synthesized using distinctive preparation techniques has found useful applications [15–23]. Although significant progress has been made in the pursuit of good quality GDY manufacturing, including with specific features that render it acceptable for potential implementation, there is still a significant gap between the ideal and reality. Pristine GDY has nearly always been polycrystalline or amorphous, and its thickness control remains challenging. As a consequence of this, the attributes of immaculate GDY were outstanding in accordance with the hypothetical projected conclusions [24].

We examine the advancements of GDY synthesis methods extensively in this chapter, which may be divided into the three basic states of matter. Furthermore, doping directed reformations of GDY and GDY-based composites are investigated thoroughly. Finally, we discuss the viability of the aforementioned numerous synthesis techniques, before reviewing the numerous possible applications that are already available.

7.2 Synthesis of GDY

GDY is an excellent 2D carbon allotrope that is made up of sp^2 as well as sp carbon atoms with four different forms of carbon–carbon bonds: C_{sp2}–C_{sp2} bonds, C_{sp2}–C_{sp} bonds, and C_{sp}–C_{sp} triple as well as single bonds connecting contiguous C=C bonds. GDY provides functional materials with a different approach that introduces new areas for materials science as well as technological development. Manufacturing and designing of GDY not only includes different nanostructures but also unique features that can speed up its development and contribute to many new applications. Numerous initiatives to fabricate GDY with controllable geometries, including extendable GDY frameworks, have been pursued to date. In general, the synthesis of GDY has involved six major aspects: (i) solid phase, (ii) liquid, and (iii) gas phase fabrication, (iv) doping oriented reformations, (v) composite fabrications, and finally (vi) GDY analogizing. The following sections deal with the different synthesis methods.

7.2.1 Solid phase method

Few-layer GDYs are produced through the solid phase method, which is a simple synthesizing technique to generate GDY samples on a large scale solely in the solid phase and is suitable for industrial applications [25, 26]. The solid phase methods are divided into two basic types which are as follows.

7.2.1.1 Bottom-up method

The popular Bottom-up method used for the fabrication of GDY is the vapor–liquid–solid (VLS) method. As the name suggests, the VLS method is a type of chemical vapor deposition. VLS is a three step process. (i) Fabrication of a liquid alloy droplet to be sprayed onto the substrate surface from where the wire will be grown. (ii) The material to be produced is introduced as a vapor that soaks into the surface of the liquid as well as disperses inside the droplet. (iii) Along this liquid–solid interface, supersaturation as well as nucleation contribute towards radial crystallization. Zao *et al* describe a VLS growth technique for creating new GDY nanowires on a ZnO nanorod array as shown in figures 7.1(a) and (b) [27]. GDY

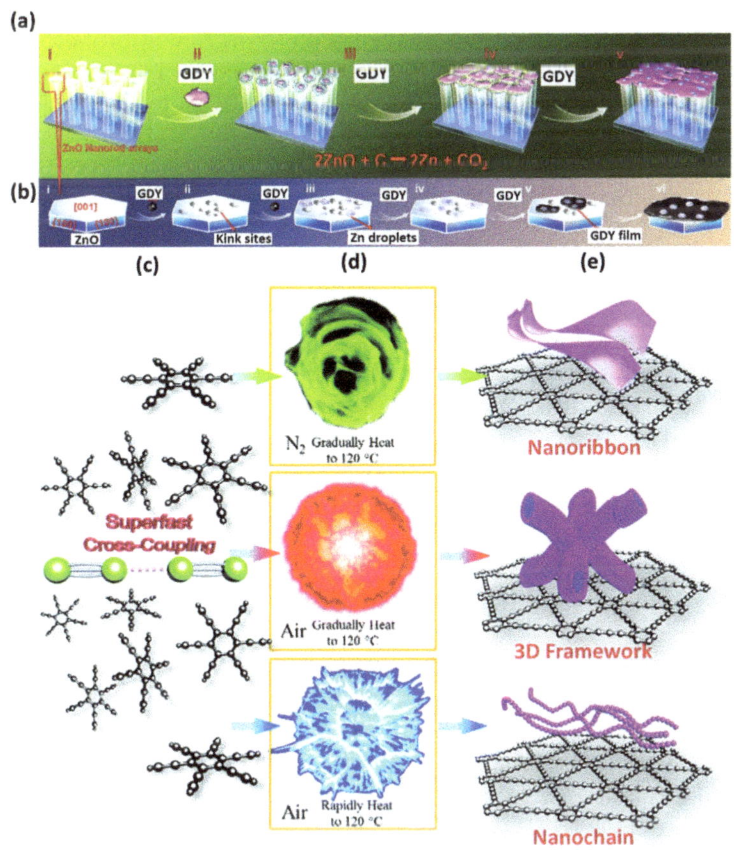

Figure 7.1. (a) Diagrammatic representation of the development of a GDY film incorporating a slight decrease of a self-catalyzed as well as saturated VLS model, mostly on the top surface of single ZnO nanoribbons. (b) The growth of GDY at the top surface of hexagonal ZnO nanoribbons. (Reproduced with permission from [27]. Copyright 2015 Springer Nature.) (c) The HB before treatment and (d) three different treatments of GDY to form (e) nanoribbons, a 3D framework, and a nanochain through the explosion method. (Reproduced with permission from [28]. Copyright 2017 The Royal Society of Chemistry.)

thin films were generated effectively by catalysis of Zn droplets after these GDY segments were coated on the surface ZnO nanorods under the influence of argon. A modified VLS growing technique was then used, where they used a mishmash of reduction in addition to self-catalyzed VLS growth of an ultra-thin GDY sheet with good conductivity and strong field effect mobility. The VLS method is applied due to the formation of one-dimensional nanostructures of the compound.

7.2.1.2 Explosion method

The explosion method deals with a bulk carbon substance devoid of any extra metal catalyst, which is a required prerequisite for the bottom-up method that was described earlier [29]. It can be understood simply as the formation of GDY produced over the surface of different substrates with a particular temperature through an explosion using monomers of the same group. As-grown GDY powders are obtained by the introduction of hexaethynylbenzene (HB) monomers that link uniformly with each counterpart in the solid phase in this approach, implying that practically every HB would react to produce GDY, resulting in a high yield of GDY (up to 98%) [28]. By simply adjusting the environment and reaction temperature of the HB precursors (at 120 °C), GDY powders with three various geometries, comprising GDY nanoribbons, 3D GDY structure, as well as GDY nanochains, can be created successfully, as shown in figures 7.1(c)–(e).

7.2.2 Liquid phase method

In particular compared to the solid phase approaches outlined above, the wet chemical procedure seems to be a more efficient way to manufacture large-area GDY films in real world operation [30, 31]. The liquid phase model of the fabrication of GDY depends on two kinds of techniques. The first is homo-coupling reactions and the second is the reaction over substrates. The following are some of the well-known methods which make the fabrication of GDY easy.

7.2.2.1 Acetylenic coupling in solution-phase method

For GDY growth, there are two types of acetylenic coupling processes [32].

7.2.2.1a Cu-catalyzed terminal alkyne coupling reaction

In such a process, a Cu-catalyzed alkyne coupling process, also known as the Glaser coupling reaction [33], incorporates the coupling of terminal alkynes to produce symmetric or cyclic bisacetylenes. With the influence of cuprous salts, alkaline solutions, and oxidants, these terminal alkynes produce diyne compounds. During the catalytic process the Cu(I) is found to reoxidized by oxygen within reaction media [34]. This makes the Glaser coupling process complicated. Therefore, as a solution to this problem, a method in which the Cu(II) catalyst is replaced with the Cu(I) Eglinton reaction has been developed [35]. Furthermore, the Eglinton coupling process typically utilizes a pyridine-like organic base ligand with methanol. This process is performed on room temperature (RT) which makes it more manageable.

7.2.2.1b Coupling reactions of alkynylsilanes mediated by a Cu (I) salt
Hiyama coupling was developed by Nishihara *et al* where the coupling reactions of alkynylsilanes are mediated through Cu(I) salt [36]. The GDY was found to be fabricated using trimethylsilyl (TMS) as the monomer over alkyne in a polar organic solvent such as *N,N′*-dimethylformamide (DMF) combined with dimethyl sulfoxide (DMSO).

Li *et al* started by making sizable GDY films on Cu foil surfaces [10]. The precursor HB had first been adsorbed over the Cu surface, then polymerized into a carbon film by an *in situ* homo-coupling reaction using Cu catalysis inside this procedure. Additional coupling processes have been added to certain surface-mediated liquid phase synthesis platforms ever since. By employing diverse surfaces/interfaces as well as precursor solution concentrations, GDY films exhibiting various morphologies as well as quality were produced through solid–liquid or liquid–liquid reaction methods. According to Gao *et al* the Eglinton coupling process, when paired only with a solution-phase van der Waals epitaxial technique, may produce extremely crystallized ultra-thin GDY films on graphene [37]. Following sections delves into the specific methods.

7.2.2.2 Substrate mediated method
Within this section we will discuss about the substrate mediated approach which can be subdivided into copper-substrate mediated and non-copper-substrate mediated.

7.2.2.2.1 Copper mediated method
Li's group developed GDY fabrication through a Cu-surface-mediated process [10]. The group prepared GDY through an *in situ* Glaser coupling reaction initiated through HB monomers over Cu foil. Here the substrate (i.e. Cu foil) acted as (i) the reservoir of catalyst and (ii) the pattern for the development of GDY. Pyridine was used as the solvent. The mechanism of the growth of GDY depends on the polymerization reaction which is limited by the controlled diffusion of catalyst ions in the solution from the substrate surface. The GDY that formed at the solid–liquid interface was not an ideal 2D structure. The chemical reaction that takes place for the formation of GDY is shown in figure 7.2(a). Figures 7.2(b) and (c) show an FESEM image and AFM of the GDY film.

In 2015 a modified Glaser–Hay coupling method was employed by Zhou *et al* to develop GDY nanowalls over Cu foils, as shown in figure 7.3(a) [38]. Here the group used acetone as the solvent. The concentration of the catalyst was controlled as result of which at the initial stage the Cu ions dispersed over the Cu foil which formed active sites that fundamentally initiated the coupling reaction. The formation of active sites the continued to extend over the entire substrate surface (due to the controlled dissolution of the Cu ions of the catalyst ions). This initiates the formation of nanosheets, which were found to be the building blocks for the formation of nanowall structures. Further as the reaction continues these nanowall structures result in the formation of a 3D honeycomb-like structure, which is shown in figure 7.3(b). Further, for the estimation of their thickness, the GDY nanowalls were exfoliated mechanically and transferred to a Si/SiO_2 substrate, after which AFM was performed, which showed that the well-formed nanowalls were 15.5 nm thick, as shown in figure 7.3(c).

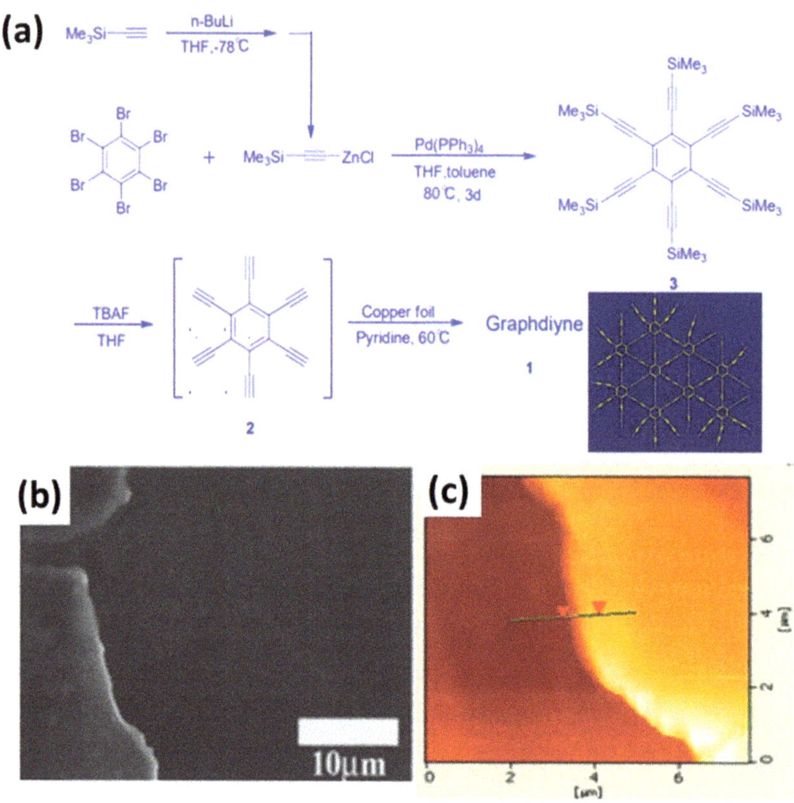

Figure 7.2. (a) Synthesis route for GDY film. (b) FESEM and (c) AFM of the GDY film. (Reproduced with permission from [10]. Copyright 2010 The Royal Society of Chemistry.)

Instead of 2D Cu foils, other substrates, for instance 1D Cu wires, 3D Cu mesh, as well as Cu foam, have been utilized for the preparation of GDY nanostructured thin films with similar *in situ* coupling of the HB method. Wang *et al* employed a grooved template to produce numerous microscale reaction channels when in close contact with flat copper foil, as shown in figure 7.4(a), which was then placed into the HB dissolved in pyridine [39]. Now, owning to the linear pillar structure over the template, the growth of the GDY took place according to the contact between the reaction solution and the microgrooved template. GDY grows when bubbles of air which are already present in the environment are allowed to expand. The author also suggested two possibilities which can be designated as lyophilic and superlyophilic templates, as shown in figure 7.4(a). In the lyophilic templates the air bubbles are restricted to the middle of the template whereas in the superlyophilic template the problem of the air pockets is avoided. As a result, the growth of GDY which formed on a lyophilic template shows uneven film formation at the end/edges of the Cu foil. On the other hand, uniform formation of GDY stripes takes place over Cu foil with a superlyophilic template structure, as shown in figure 7.4(b).

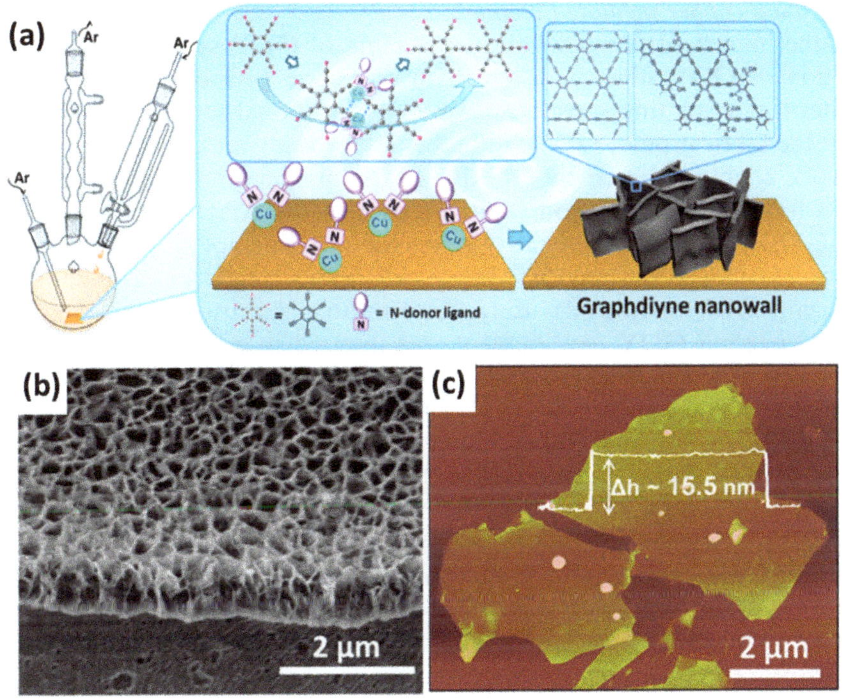

Figure 7.3. (a) Schematic illustration of the experimental set-up. SEM images of GDY nanowalls on Cu substrate. (b) A coss-sectional view SEM image and (c) an AFM image of an exfoliated sample on a Si/SiO_2 substrate. (Reproduced with permission from [38]. Copyright 2015 American Chemical Society.)

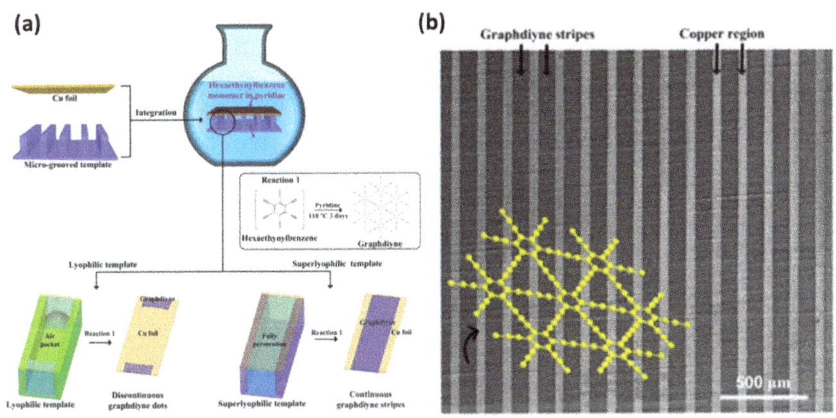

Figure 7.4. (a) Schematic showing the technique using both lyophilic and superlyophilic templates for the formation of continuous GDY stripes. (b) FESEM image of the subsequent formation of continuous GDY stripes made using the superlyophilic template. (Reproduced with permission from [39]. Copyright 2016 Wiley.)

7.2.2.2.2 Non-copper mediated method

One of the important issues within the domain of GDY synthesis is realizing the production of GDY films with a certain collection of morphologies on various substrates (i.e. insulating substrates as well as functional semiconductors). Li *et al* describe the fabrication of GDY nanotubes (NTs) using an aluminum oxide (AAO) framework enabling extremely effective field emission studies [40]. The scientists used Cu foil as a catalytic source at the base of the AAO pattern in this study, allowing GDY to develop *in situ* just on the AAO template. The 'Cu-envelope' synthesis technique, devised by Gao *et al*, is a straightforward approach for synthesizing GDY nanowalls using customizable substrates [41]. The intended substrate enabling GDY development was inserted in a Cu foil envelope, which was then deposited in a mixture solution of HB, acetone, and pyridine, with *N*,*N*,*N*′, *N*′-tetramethylethylenediamine (TMEDA), as shown in figure 7.5(a). A suitable concentration of catalyst might permeate through the interface between both the HB monomer solution as well as the target substrate under alkaline conditions, in which the Glaser–Hay coupling process of HB monomers supposedly occurred, resulting in *in situ* formation of GDY over the target substrate. Furthermore, they also reported the formation of GDY nanostructured tin films over 1D, 2D, and 3D substrates. Zhao *et al* employed a multipurpose method known as the 'controlled-release method' for the preparation of GDY without a Cu

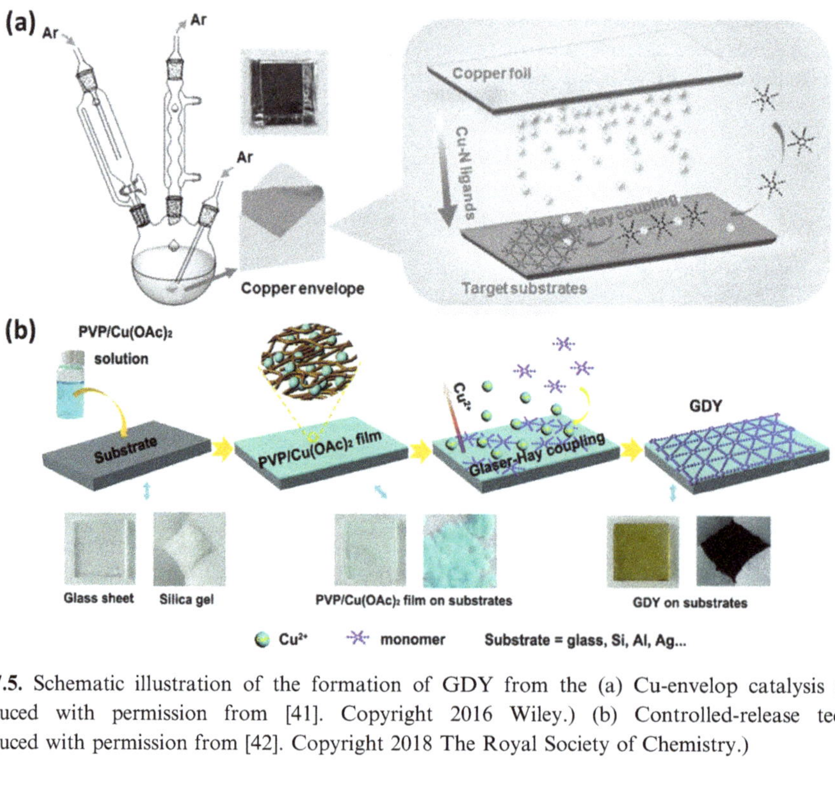

Figure 7.5. Schematic illustration of the formation of GDY from the (a) Cu-envelop catalysis method. (Reproduced with permission from [41]. Copyright 2016 Wiley.) (b) Controlled-release technique. (Reproduced with permission from [42]. Copyright 2018 The Royal Society of Chemistry.)

substrate [42]. In short, a high molecular weight polyvinylpyrrolidone was used with Cu(II)–acetate composite film, as shown in figure 7.5(b). The Cu(II) ions might migrate through the catalyst reservoir towards the target substrates including SiO_2, ZnO, Al, and others, in which the acetylenic coupling reaction (following Glaser–Hay coupling) might occur, causing GDY development.

7.2.2.3 Interface supported method

Matsuoka *et al* developed a new methodology for the development of GDY in 2017 [43]. The group produced GDY interfacial synthesis at liquid–liquid and gas–liquid interfaces, the result of which which may be described as a bottom-up covalent nanosheet with conjugated carbon–carbon bonding. They dispersed HB as a monomer and $Cu(OAc)_2$ as a catalyst in two different types of insoluble solvents, dichloromethane and water, respectively. The reaction process which took place was the Eglinton coupling process at the liquid–liquid boundary, producing a GDY thin film with an exceptionally low monomer concentration, as shown in figures 7.6(a) and (b). The reaction mechanism which was followed for the formation of GDY is shown in figure 7.6(c).

Additionally, when sprinkling a tiny quantity of HB solution just above this Cu acetate aqueous solution, the HB coupling process was limited towards the liquid–vapor interface. Again at the liquid–vapor interface, this technique results in an improved crystallographic hexagon shaped GDY nanostructure with just a uniform thickness of 3 nm as well as an average size of 1.5 m. Moreover, the GDY nanosheets were imperceptible to the human eye, but could still be transported to a variety of flat surfaces using the Langmuir–Schäfer process, as shown in figure 7.6(d).

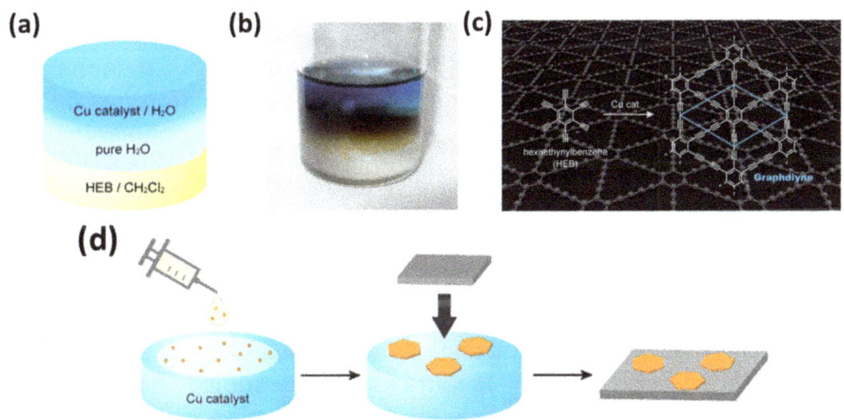

Figure 7.6. The liquid–liquid interfacial synthesis technique: (a) schematic and (b) photograph. (c) Reaction suggested for the formation of GDY. (d) Schematic of the gas–liquid interface synthesis as well as transfer process. (Reproduced with permission from [43]. Copyright 2017 American Chemical Society.)

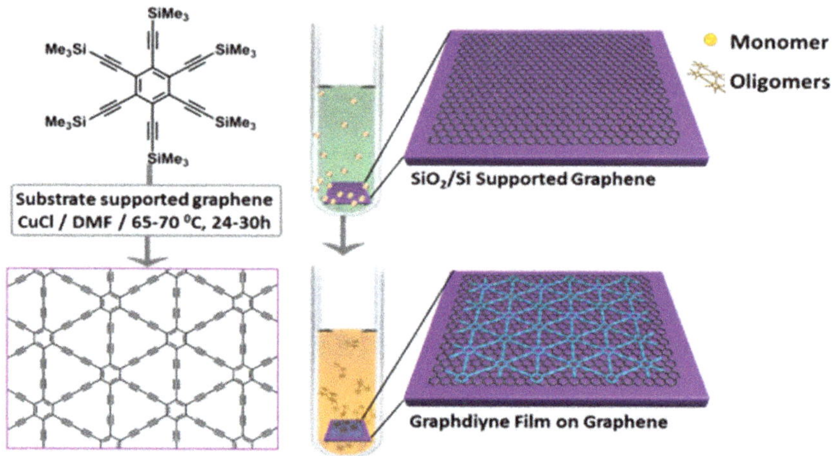

Figure 7.7. Hiyama-coupling reaction that produces a GDY coating over graphene's surface. (Reproduced with permission from [44]. Copyright 2018 American Chemical Society.)

7.2.2.4 Solution-phase van der Waals epitaxy

Apart from the above-mentioned techniques, GDY can also be synthesized through the solution-phase van der Waals epitaxy method. Typically, for this technique Gao *et al* involved three major aspects to create high-quality GDY thin films: monomer design, coupling reaction, as well as restricting the coupling reaction to the 2D plane [37]. Further, monomer HB was used for the Eglinton coupling reaction at room temperature. Finally, when the coupling processing took place, graphene was employed as the epitaxial platform limiting free rotation mostly over alkyne–aryl single bonds in monomers. Further, the authors suggested that the monomers have very weak interaction with newly formed GDY at the graphene platform resulting in the weakening of the diffusion barrier of the monomers, which diminishes the formation of the thin film of GDY. Hence tri-layered GDY films have been produced effectively.

In another report Hiyama coupling was employed for the formation of the GDY shown as figure 7.7. Zhou *et al* in their work they found that monomer not only becomes more stable than bare HB (deprived of the deprotection process), but also preventing oxidation as well as self-polymerization of the monomers over lengthy reaction times [44]. Further the group reported the formation of unbroken GDY thin film through surface templating method.

7.2.3 Gas phase method

On-surface covalent synthesis had already been utilized frequently to make ultra-thin 2D materials, and it is a good technique to regulate the formation of consistent new 2D nanostructures [45–47]. Precursors were coupled through a covalent connection over the metal surface throughout the reaction process, leading to the formation of a thin coating comprising 2D molecular structure. Either an scanning tunneling microscopy (STM) system (where ultra-high vacuum conditions are required) or even a chemical vapor deposition (CVD) system (working in inert

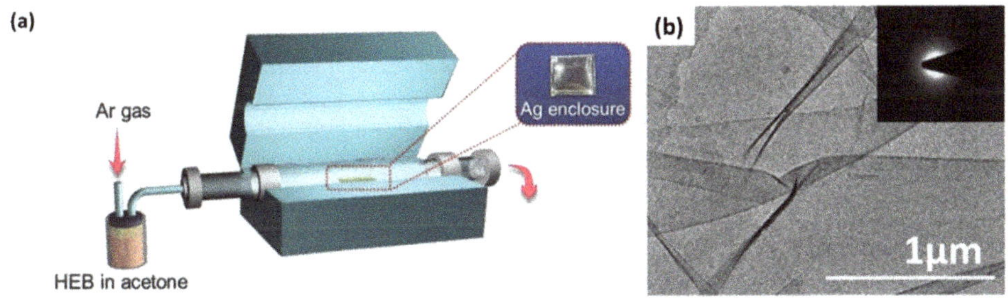

Figure 7.8. (a) A typical CVD system using HB as a precursor along with a small image of a carbon monolayer over a Ag surface. (b) TEM image. The inset shows the SAED pattern. (Reproduced with permission from [49]. Copyright 2017 Wiley.)

atmosphere conditions) can be used for this operation [48, 49]. The previous is thought to be a good technique for making and examining GDY-based substructures *in situ* that might help with real-time tracking of the coupling process, or intermediates, including by-products.

The typical set-up for the formation of a GDY layer over a Ag substrate is shown in figure 7.8(a), which was also confirmed by TEM, as depicted in figure 7.8(b), and no significant in-plane order was realized, indicating the formation of non-crystalline film, shown in the inset of figure 7.8(b). This may be due to the side reactions that took place between two adjacent terminal alkynes during the on-surface synthesis process [50].

Glaser–Hay type coupling is demonstrated, as well as a novel method of silver-acetylide chemistry. The alkyne homo-coupling occurs in ultra high vacuum (UHV) under mild conditions on the Ag(111) noble-metal surface, with volatile H_2 being the only by-product [51]. The group produced discrete molecules or polymeric networks with a conjugated backbone via a hierarchic reaction route that not only favors low thermal activation, but also results in outstanding yields with better selectivity.

On annealing of the coated surface at 300 K, the typical dimers reappear, interacting alongside unreacted precursor 1,3,5,-tris-(4-ethynylphenyl) benzene (Ext TEB) species. As a result, monomer desorption does not exceed the surface-assisted response, which is attributed to the greater surface contact. Ext-TEB dimers show spectacular scanning properties, in especially remarkable is the clarity of the interconnecting waistline, in addition to the total length, both of which were able to be characterized using the same concepts as their 1,3,5,-triethynyl-benzene (TEB) counterparts. As a result, the Ag(111) surface stimulates the Ext-TEB homo-coupling response in the same way in which the TEB species did, as a result, the same hierarchical strategy is provided in this challenge.

A great level of purity is demonstrated by a deeper examination of the structure. It also indicates that the covalently linked architecture has a number of characteristics that represent the paired connection of such terminal ethynyl groups as well as some molecular flexibility. These six-membered oligomeric cyclic subunits are the optimal outcome of a convergence homo-coupling process involving Ext-TEB

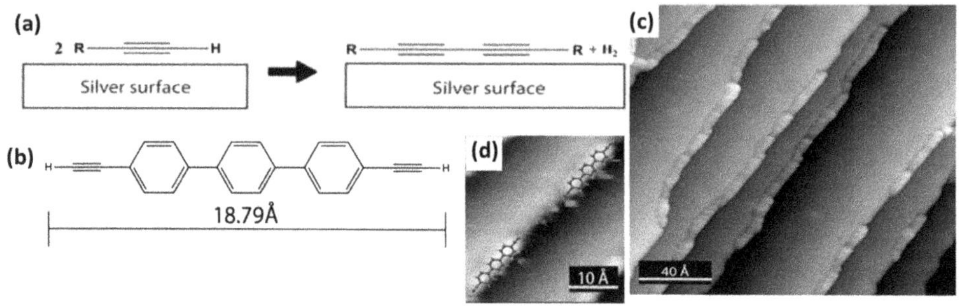

Figure 7.9. (a) The chemical structure of 4,4″-diethynyl-1,1′:4′,1″-terphenyl (**1**). (b) The distance between the terminal hydrogen atoms is 18.79 Å. (c) After annealing the sample at 450 K, extended connected structures are present at the (100) microfacets with resolved sub-molecular features. Panel (d) highlights two monomers with their models superimposed ($U_{Bias} = -1.0$ V, $I_t = 0.06$ nA). (Reproduced with permission from [52]. Copyright 2014 American Chemical Society.)

modules containing straight C–C connections. An enlarged view of the related honeycomb unit is shown in figure 7.9, along with a HYPERCHEM model which neatly matches the actual STM data. These extra five- as well as seven-fold cyclic configurations, on the other hand, suggest molecular module aberrations.

This indicates that the process is distinct from the well-known alkyne C–C coupling reactions on transition metal surfaces that target tiny aromatic molecules. Furthermore, a simple surface-assisted homo-coupling over Ag (111) is shown in figure 7.9(a). Encouraged by the templating conditions that Cirera *et al* adopted, the synthesis at the vicinal surface of the step edges of Ag(877) increases the chemo-selectivity of the linking process, allowing for the regulated creation of GDY wires. Figure 7.9(b) shows that the terminal distance between two hydrogen atoms is about 18.79 Å [52]. An STM image of the step edge decoration of Ag(877) is shown in figure 7.9(c). The molecules, which are seen as bright rods in figure 7.9(d), attach to the bottom step edges with an extended backbone parallel to that as well.

Moreover Sun *et al* employed a dehalogenative homo-coupling process to create the particular precursor shown in figure 7.10(a) containing terminal alkynyl bromide that might pair together over the Au (111) surface [48]. Because the C–Br bond has a lower bonding energy than that of the C–H bond, such an approach may result in greater selection as well as fewer by-products, resulting in an organized 2D porous network containing acetylenic connections, as shown in figure 7.10(b).

7.2.4 Doping oriented reformations

Single metal atom doping and heteroatom doping/co-doping have all been viewed as effective ways to improve the economic benefits or carbon allotropes in a variety of industries. The functionalization of the GDY framework with alkyne bonds and subnanoporous is expected to increase the reaction mechanisms due to the high number of reactive species and high yields, since this found to be a novel route towards tailoring GDY's electronic and physiochemical properties and the synthesis of novel carbon allotropes

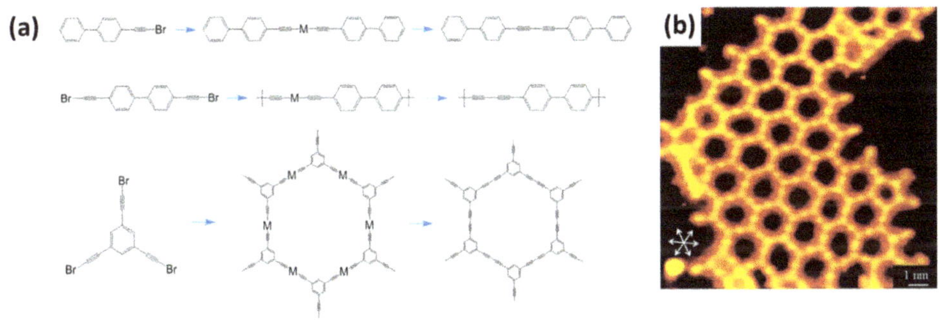

Figure 7.10. (a) Graphical representation of dehalogenative homocouplings involving BEBP, bBEBP, as well as tBEP molecules. (b) STM picture of the C–Au–C organometallic network formed following application of tBEP molecules on Au(111) at room temperature following mild annealing. (Reproduced with permission from [48]. Copyright 2016 American Chemical Society.)

7.2.4.1 Single atom doping

Doping with metal atoms (particularly single atoms) has also been suggested as a possible method for modifying GDY characteristics. Recently He *et al* demonstrated the adsorption behavior of single atoms using first-principles calculations [53]. The group found that sp-hybridized carbon atoms are important for the metal atom during the anchoring mechanism over GDY. Due to the electronic redistribution of the transition metal orbitals and also the charge transfer among transition metal atoms as well as GDY, metal atoms such as (V, Cr, Mn, Fe, Co, Ni) are favored to attach within the alkyne ring, and may easily regulate both the electrical and magnetic characteristics of GDY. Comparably, single atom noble metals have also been found theoretically to show good absorption ability. The metal atoms such as (Pd, Pt, Rh, and Ir) do not aggregate due to the characteristic high migration barrier over the GDY surface [54]. Xue *et al* have developed the first experimental single atom Fe(0) and Ni(0) doping over GDY [18]. The single atom doping over GDY was found to produce a better catalyst for hydrogen evolution.

7.2.4.2 Heteroatom doping

Various carbon allotropes can doped or co-doped, which changes their electrical, physical, as well as chemical characteristics [55]. Considering various defects that may be produced around the inserted heteroatoms with diverse inherent features, heteroatom doping with carbon materials is an effective way to create related carbon-based derivatives. Further, it has been found that the post-heating of GDY with a variety of precursors, which generally causes a domino effect over surface treatment with an ambiguous structure, and *in situ* doping are simple synthesis approaches that produce inhomogeneous heteroatom-doped GDY. Some essential basic qualities, such as shape and electrical structure, might be altered using this approach. Furthermore Mohajeri *et al* in their work, when compared to pure GDY, N-GDY pushed the Fermi level toward higher energies, inevitably resulting in n-type semiconducting performance, but for nitrogen–sulfur (N,S) co-doping the Fermi level was downshifted, producing a *p*-type doping [56]. In addition to altering

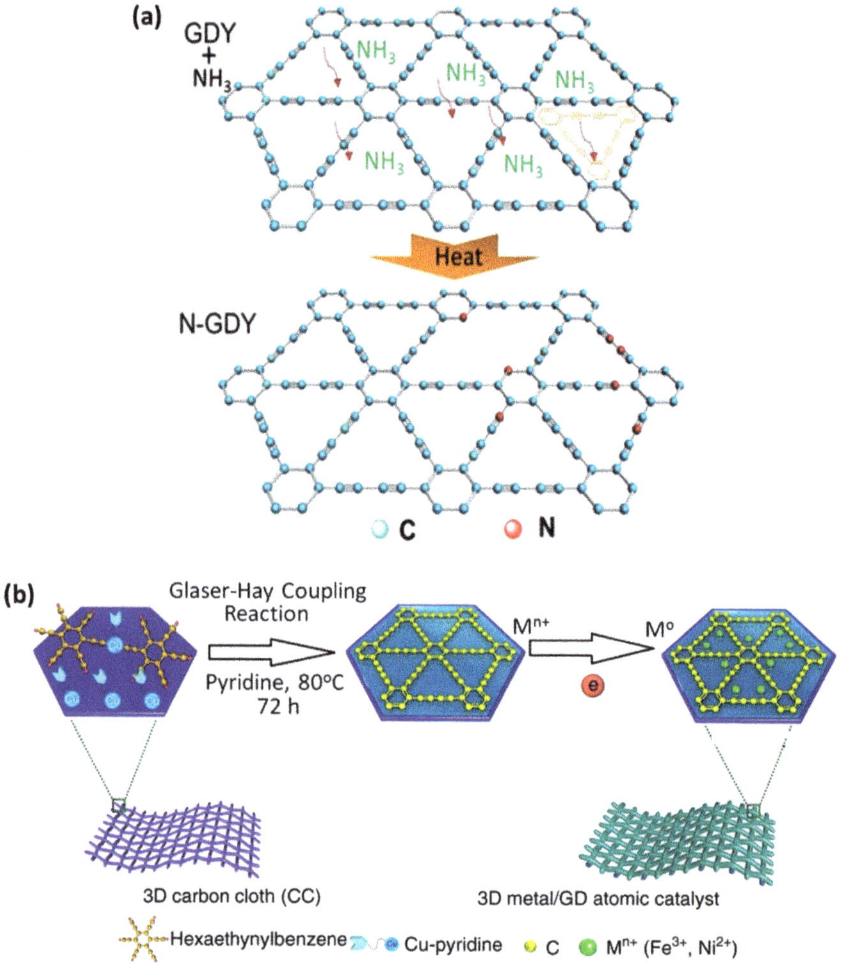

Figure 7.11. (a) Diagram exhibiting N-doped GDY made through annealing GDY in ammonia gas. (Reproduced with permission from [72]. Copyright 2016 American Chemical Society.) (b) Schematic view of the electrodeposition process for the synthesis of single atom Ni/GDY as well as Fe/GDY. (Reproduced with permission from [18]. CC BY 4.0.)

the concentrations of dopants, the energy gap may be controlled through a range of 1.0 eV. The band gap may be altered whenever GDY is co-doped using boron and nitrogen (B, N), however, the direct-band gap characteristics barely change, irrespective of the doping rate [57]. The presence of many acetylenic linkages, particularly in the context of GDY, might give additional doping sites for a range of heteroatoms, complicating the issue. To date, theoretical studies have already shown that non-metal heteroatoms [58–64], metal atoms [65, 66], certain transition metals [53], noble metals [67], and hybrid atoms [56, 68] could be used. Figure 7.11 shows the doping of nitrogen and metal atoms in GDY.

Zhang *et al* have analyzed the effect of the lower and the higher doping of the –OH groups over the optimized GDY structure [69]. The non-magnetic system is caused by the decreased doping concentration of –OH groups. When the quantity of –OH is sufficient, however, a chiral-like structure forms inside the GDY structure. Also, by raising the coverage rate for OH groups on GDY, the energy band gap of the hydroxylated GDY could change by 0.8 eV. Whenever the partial covering of OH groups reaches 18.2%, an evident semiconductor to metal transition occurs, which has also been proved by *ab initio* molecular dynamics computations. Remarkably, increasing the CO and COOH groups reduces the overall energy gap, but the rapid increase in energy gap is related to asymmetries throughout the functionalization on opposing edges [56]. A theoretical study found that with reduction in the coverage, butadiynic carbon atoms were the most likely sites of hydrogenation, followed by carbon atoms generating single bonds [70]. Again, depending on the concentration of hydrogen, the band gap of the system changes from 0.45 to 4.43 eV [58].

The above-mentioned doping procedures, as can be observed, nonetheless have associated weaknesses: (a) the location of the doping heteroatoms could perhaps take a variety of forms, making it more complicated to pinpoint the methodology of doping; (b) the percentage of doping heteroatoms cannot be monitored constantly; and (c) heteroatoms can be scattered widely on the GDY structure, potentially jeopardizing conjugation consistency [71].

Further, a new classical method was used for the formation of GDY, where the group replaced the monomer HB. Additional heteroatom based GDYs, such as chlorine- and fluorine-substituted GDYs, were also created by replacing other heteroatom GDYs [73, 74], as illustrated in figure 7.12. Furthermore the core benzene moiety within GDY was changed using pyridine, pyrimidine, as well as triazine moieties that were also joined with sp-hybridized butadiyne couplings.

Thus bottom-up techniques for including heteroatoms in GDY provide significant features: (a) the heteroatom largely distributed via doping sites is equally dispersed upon GDY, preserving the comprehensive conjugated skeleton inside the two-dimensional plane of the GDY; and (b) remarkable effects may be achieved by fine tunning of bond formation surroundings and heteroatoms resulting in unique physicochemical properties.

7.2.5 Fabrication of GDY-based composites

The presence of acetylene bonds within GDYs as well as their specific synthesis methods, not only expands the number of synthesis pathways for producing GDY-based composites, but moreover makes it easier to build highly controlled GDY-based composites. The fabrication of GDY-based composites can occur using the following procedures.

It is possible to create GDY adsorbed on top of specific nanomaterials using a variety of techniques for the production of GDY, mostly on the surface of any substrate. GDY was coated over FPCuSi paper (which is a composites of Si nanoparticles (SiNPs) and Cu nanowires (CuNWs)) *in situ* to create a composite

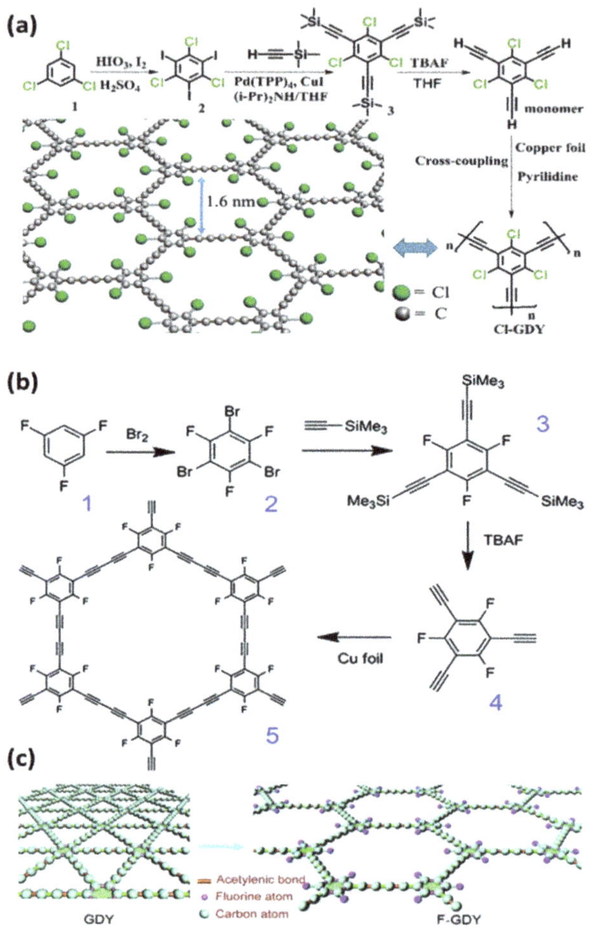

Figure 7.12. (a) Schematic showing a systematic approach for the formation of Cl-GDY. (Reproduced with permission from [73]. Copyright 2017 Wiley.) (b) Reaction mechanism with five major steps and (c) the formation of F-GDY. (Reproduced with permission from [74]. Copyright 2018 The Royal Society of Chemistry.)

structure [21] that successfully reduced the volume effect over silicon interface contact, which helps during charge and discharge in battery applications, as shown in figure 7.13(a), while figure 7.13(b) shows the FESEM image of the same. In addition to this work, another group has established a controlled *in situ* fabrication approach for ultra-thin GDY film placed over the surface of an aluminum negative electrode to create a composite structure. The well-formed GDY from this process was found to enhance the adjustable capacity as well as stability of the aluminum foil.

Moreover, the acetylene bond in GDY has a strong chemisorption with a single metal ion, which prevents metal ions from aggregating and improves the charge-transport behavior between metal atoms and GDY. In general, GDY as well as metal

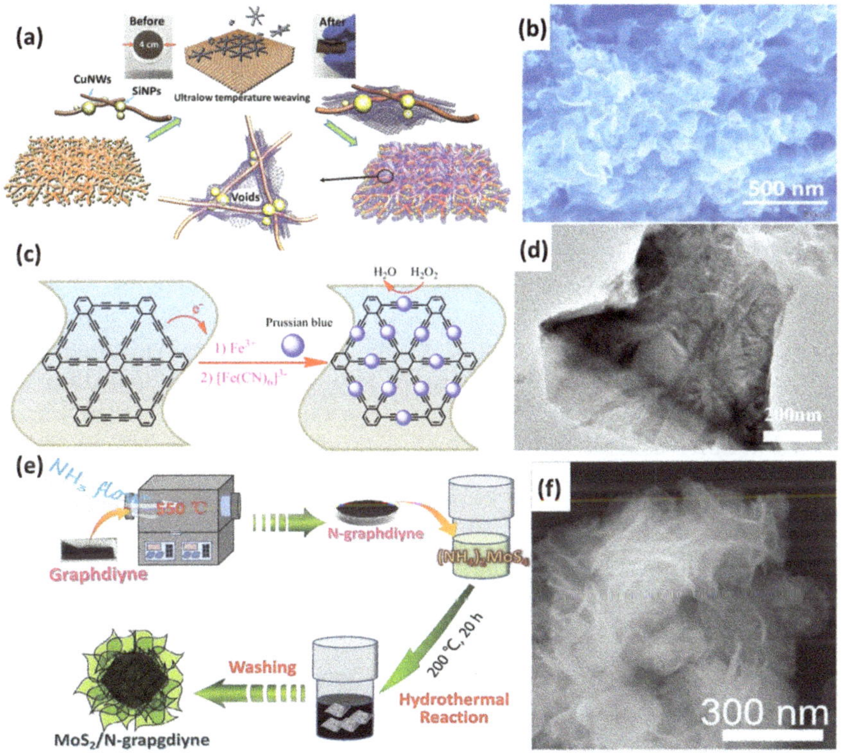

Figure 7.13. (a) Schematic examples of methods for *in situ* weaving a matrix of extremely thin GDY nanosheets over an Si anode (the insets show photographs of the electrode both with and without weaving GDY nanosheets). (b) Magnified SEM images of the GDY-coated FPCuSi cross section. (Reproduced with permission from [21]. Copyright 2018 Wiley.) (c) Synthesis process and (d) TEM of a Prussian blue/GDYO nanocomposite synthesized *in situ*. (Reproduced with permission from [75]. Copyright 2017 Elsevier.) (e) Schematic of the production of MoS_2/N-GDY by integrating 2D MoS_2 nanosheets using N-GDY nanolayers and (f) an SEM image. (Reproduced with permission from [89]. Copyright 2018 Wiley.)

ion precursors could be fully mixed together within the solution. Nevertheless, the nanoparticles were only spread throughout the nanoporous region of GDY once the corresponding chemical reactions of the ions around the acetyl bond had been taken place. Prussian blue (PB) nanoparticles attached to the top of GDYO were synthesized *in situ* to create a GDY-based composite [75]. This was made via dispersion of $FeCl_3$ in a GDYO solution as well as combining it with $Fe(CN)_6^{3-}$ at room temperature, shown in figure 7.13(c), while figure 7.13(d) displays a TEM image of the composite. The composite demonstrated the detection of hydrogen peroxide, with outstanding electrochemical catalytic performance and stability. Similar kinds of fabrication processes have been involved for different metal oxide/GDY [76, 77] and metal chalcogenide/GDY [78–80] composites. Apart from the formation of the heterojunction, the fabrication of a Z-scheme type composite was also attempted using graphene and graphitic carbon nitride (g-C_3N_4) [81, 82].

The fabrication of different GDY-based composites also involves the influence of temperature and pressure. In the literature it has been found that there are different compounds involving metal oxides, and the polymers in this category are TiO_2 [83] and ZnO [84] while some popular polymeric materials are phenyl-C61-butyric acid methyl ester (PCBM), poly(3-hexylthiophene) (P3HT) [85], P-*o*-FBDTP–C8DTBTff (PFC) [86], and g-C_3N_4 [87]. The methodology which is followed is similar to the following: (i) through the electrostatic interaction, GDY NPs are distributed over a surface with PRA-modified ZnO NPs, (ii) causing a ring-opening reaction between positively charged amine groups, mostly on the ZnO nanoparticles' surface as well as the negatively charged epoxy group of GDY NPs [84]. Furthermore, the fabrication of even doped polymeric potassium doped poly[3-(4-carboxybutyl) thiophene-2,5-diyl] (P3CT-K) [88] is quiet similar to the preparation of metal chalcogenides (i.e. N-doped GDY composite with 2D MoS_2), which was done in a controlled manner using a hydrothermal method [89]. Figures 7.13(e) and (f) show the stepwise formation of 2D-MoS_2/N-GDY composites along with an SEM image.

7.2.6 GDY analogues

New GDY analogues, such as gamma-graphyne (γ-GY) and graphtetrayne (GTY), have been developed. The amount of acetylenic linkages connecting two neighboring benzene rings is the key structural variation between these two GDY analogues. Some precursors containing calcium carbide as well as hexabromobenzene were fed inside a planetary ball mill, which can synthesis γ-GY powder while also carrying out a cross-coupling reaction utilizing the ball mill. The calcination technique has been shown to considerably minimize the structural flaws of γ-GY [90]. Gao *et al* used a cross-coupling process with di-iodobutadiene as a precursor to make GTY on the surface of copper foil [91], as shown in figure 7.14(a).

The benzene ring is replaced using some other moieties in the second type of GDY analogue, however, the quantity of acetylenic bonds between the center remains preserved. Two acetylene bonds involving ethylene as well as a neighboring ethylene sp^2 carbon atom, for example, make up β-GDY. In β-GDY the structure has a larger proportion of acetylenic bonds compared to γ-GDY (67% and 50%, respectively). Furthermore, theoretical simulations suggest that β-GDY is a material that includes a significant quantity of hexagonal pore spaces in addition to having a zero band gap. Again the work presented by Li *et al* validates the experimentally form β-GDY over a Cu surface through a Glaser coupling reaction, as depicted in figure 7.14(b) [92].

Using Cu salts, the identical Glaser–Hay coupling reaction of 1,3,5-tris-(4-ethynyl-phenyl)-benzene(TEPB) leads to the synthesis of 1,3-diyne-linked conjugated micro-porous polymer nanosheets (CMPNs), which are comprised of two acetylene bonds connecting four benzene ring groups as well as a benzene ring group [93].

According to Mutsuoka *et al*, a GDY analogue with a triphenylene core (TP-GDY) was produced using a liquid–liquid interfacial approach which appeared to have a maximal width of 220 nm [94]. In another work by Zhou *et al* employing 1,3,5-triethinyl-2,4,6-triphenyl-benzol as a precursor, a crystalline structure of GDY

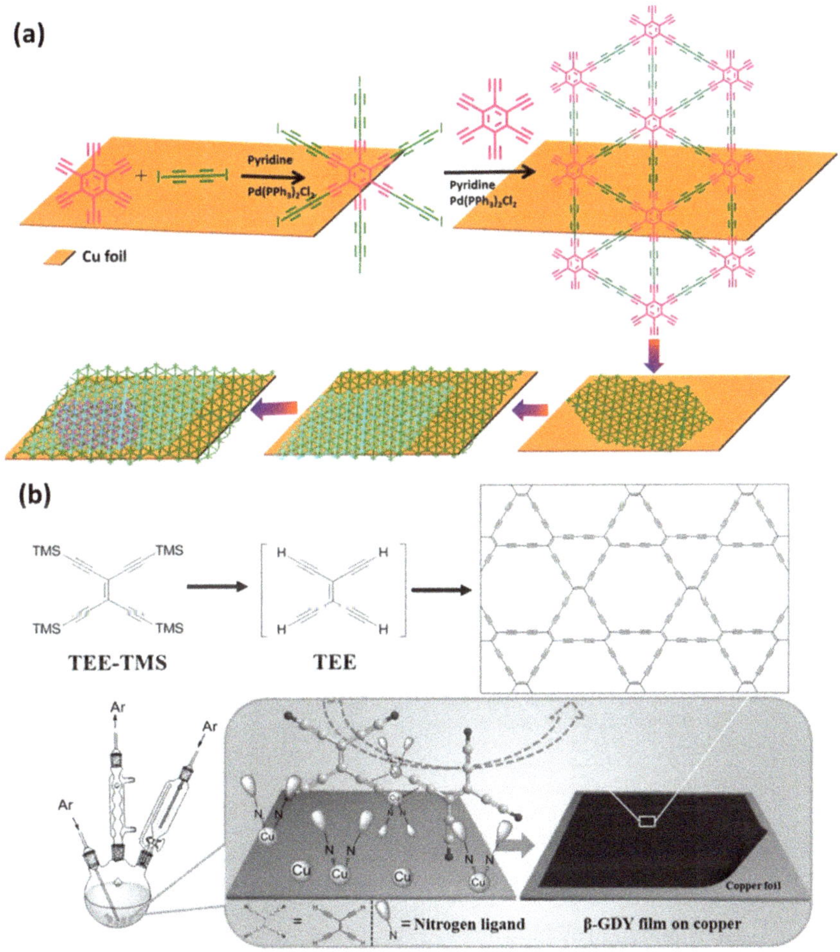

Figure 7.14. Methods for synthesizing GDY analogues. (a) The cross-coupling process is used to make graphtetrayne. (Reproduced with permission from [91]. Copyright 2017 Elsevier.) (b) The improved Glaser–Hay coupling reaction is used to make β-GDY. (Reproduce with permission from [92]. Copyright 2017 Wiley.)

analogue (i.e. Ben-GDY) was produced on top of copper foil [95]. The Ben-GDY appears to be made up of six 1,3,5-triphenylbenzene rings joined by butadiyne connections in a functional group.

Therefore, the above fabrication of the different GDY analogues results from the following basic steps that can occur in solution-phase reactions: (a) surface polymerization in the solution phase (i.e. Glaser coupling) [96]; (b) dual activity shown by the Cu substrates (i.e. they work both as a catalyst as well as a substrate); (c) the molecular structure of the precursor results in the reduction of the probability of the formation of side products; and (d) the precursor which is used to form the product is made through simple cross-coupling reactions. An overview for the preparation of GDY based upon the number of parameters have been enlisted in the Table form shown as Table 7.1.

Table 7.1. List of the different methods along with various parameters such as (precursors, catalysts, specific reactions over particular substrates and temperature) for the preparation of GDY.

Sl. No.	Methods	Precursor	Catalyst	Reaction	Substrate	Temperature (K)	Reference
1.	CVD	HB	—	TC	Ag foil	423	[49]
2.	EM	HB	—	TC	—	393	[28]
3.	EM	2,4,6-triethynyl-1,3,5-triazine 2,3,4,5,6-pentaethynyl pyridine	—	TC	—	393	[97]
4.	SM	HB	Cu	G–H	AAO	333	[40]
5.	SM	HB	Cu	G–H	Cu foil	333	[10]
6.	SM	HB	Cu/TMEDA	G–H	CF	323	[38]
7.	SM	Tetraethynylethene	CA	EG	GP	298	[98]
8.	SM	HB	CuCl	G–H	GP	338–343	[44]
9.	SM	HB	CA	G–H	CA/PVP	298	[42]
10.	SM	2,4,6-tristriethynylpyridine 6-triethynylpyrimidine	Cu	G–H	CF	343	[99]
11.	SM	HB	$Cu(OH)_2$	G–H	$Cu(OH)_2$ nanowires	323	[100]
12.	SM	HB	Cu/TMEDA	G–H	Cu foam	323	[16]
13.	SM	HB	Cu	G–H	CuNWs	298	[101]
14.	SM	1,3,6,8-tetraethylenepyrene	Cu/TMEDA	G–H	CF	323	[102]
15.	SM	HB	Cu envelope	G–H	Substrate	323	[41]
16.	SM	TET	Cu	G–H	CF	333	[103]
17.	SM	HB	Cu/TMEDA	G–H	Diatomite	333	[104]
18.	SM	1,3,5-trichloro-2,4,6-triethynylbenzene	Cu	G–H	CF	383	[73]
19.	SM	1,3,5-triethynylbenzene	Cu	G–H	CF	333	[105]
20	Two-phase	HB	CA	G–H	—	293	[43]
21.	Two-phase	TET	CA	G–H	—	293	[106]
	Two-phase	2,3,5,6-tetraethynylpyrazine	CA	G–H	—	293	
22.	VLS	HB	Zn	TC	ZnO nano arrays	813	[27]

Acronyms: HB = hexaethynylbenzene; EM = explosion method; SM = solution method; GP = graphene; TC = thermal coupling; G–H = Glaser–Hay method; TET = 2,4,6-triethynyl-1,3,5-triazine; EG = Eglinton method; CA = cupric acetate; CF = copper foil.

7.3 Applications of GDY

The basic characteristics of GDY (i.e. an sp^2- and sp-hybridized flat framework with π-conjugated structure) include a precise pore distribution, including customizable electrical characteristics. As a result of these porous distributions with above-mentioned features, GDY-based materials could be used in energy conversion, energy storage, electrical devices, magnets, detectors, biological applications, and environmental applications, such as gas separation and water purification, based on their interesting architectures and electronic characteristics. Many papers have previously demonstrated the practical implementation of GDY in the lab.

7.3.1 Energy conversion

Exceptional carbon-based catalysts must typically have a wide surface area, high porosity, and a stable structure. GDY features a two-dimensional interconnected and extremely porous plane that can boost reactive sites for a given reaction while simultaneously providing an atomic-level dispersed framework enabling additional effective catalysts and photoelectrocatalysts, as mentioned previously. Many GDY-based catalysts have already been created during the last decade, with promising applications in sectors including photocatalysis, electrochemical and photoelectrochemical water-splitting reactions, and so on.

7.3.1.1 Catalysts

GDY is very active in several organic and inorganic processes due to its fascinating electrical as well as structural features. In 2011 the effect of the GDY in the field of photocatalysts was witnessed when GDY was made into a nanocomposite with P25 [83]. Under UV irradiation, the composite outperformed not only the P25 and P25–carbon nanotube composite photocatalysts, but also the already popular P25–graphene composite photocatalysts in degrading methylene blue, which can be seen in figure 7.15(a). Yang *et al* also reported that the nanocomposite formed by TiO_2 with (001) phase and GDY has outstanding charge separation as well as oxidation properties, as well as the lengthiest lifetimes of the photoexcited carriers of all the 2D composites containing TiO_2 of various facets, both theoretically and experimentally [15]. Figure 7.15(b) shows an HRTEM image of well-formed TiO_2/GDY which was then followed by photodegradation of the TiO_2/GDY with respect to other composites, as shown in figure 7.15(c). Characteristically, the rate constant for this TiO_2(001)–GDY composite in photocatalytic removal of methylene blue was 1.63 ± 0.15 that of pristine TiO_2 (001).

Using a hydrothermal technique, researchers created a new GDY–ZnO nanohybrid and tested its photocatalytic activities on the breakdown of both methylene blue and rhodamine B [77]. The absorption spectra as well as total organic carbon studies revealed that the GDY–ZnO nanohybrids had better photocatalytic capabilities than the unmodified ZnO nanoparticles. On the photodegradation of both azo dyes, the rate constant of GDY–ZnO nanohybrids was around two-fold greater than that of bare ZnO nanoparticles.

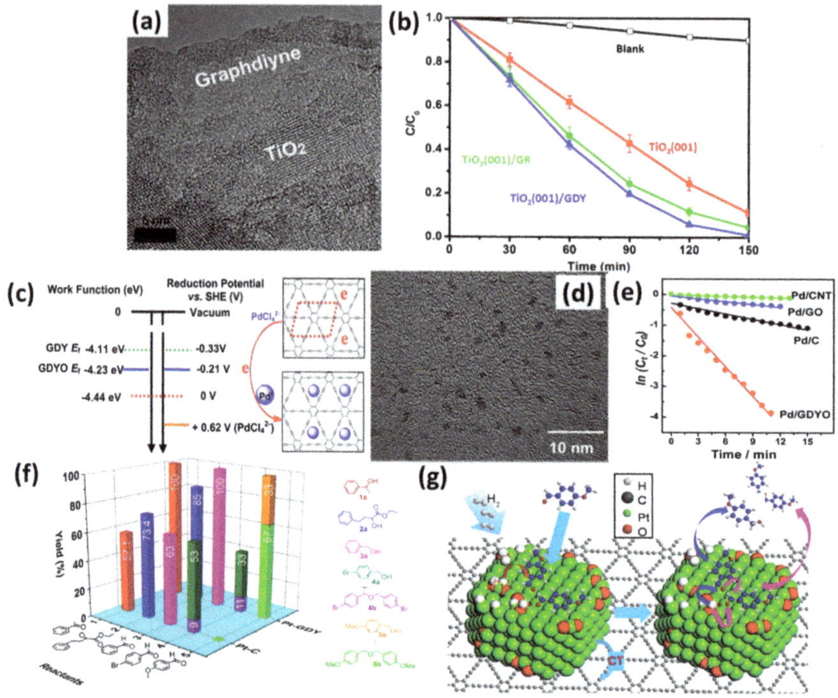

Figure 7.15. (a) HRTEM images of TiO_2(001)–GD. (b) Photocatalytic degrading of MB on TiO_2(001), TiO_2(001)–GD, and TiO_2(001)–graphene (GR) composites, as well as the control experiment. (Reproduced with permission from [15]. Copyright 2013 American Chemical Society.) (c) Diagrammatic representation of Pd/GDYO formation via electroless Pd deposition on GDYO, (d) HRTEM image of Pd/GDYO, (e) graph of $\ln(C_t/C_0)$ as a function of reaction time for four distinct catalysts catalyzing the reduction of 4-NP. (Reproduced with permission from [107]. Copyright 2015 American Chemical Society.) (f) Using Pt–GDY as a catalyst, hydrogenation of various ketones and aldehydes under the conditions employed by Shen *et al.* (g) Suggested catalytic mechanism for the hydrogenation of aldehydes and ketones to the corresponding dibenzylether. (Reproduced with permission from [108]. Copyright 2018 American Chemical Society.)

Because of its low reduction potential as well as highly conjugated electronic structure, GDY can be utilized as a reducing agent and stabilizer enabling electroless deposition, as shown in figure 7.15(d), which is not only a highly dispersed system but also has surfactant-free Pd clusters [107]. In addition, figure 7.15(e) shows a HRTEM image of Pd/GDYO whereas figure 7.15(f) shows the better degradation factor of the same. Furthermore, GDYO, the oxidation state of GDY, is shown to provide an even better substrate for depositing ultrafine Pd clusters, resulting in a Pd/GDYO nanocomposite with a high catalytic performance for 4-nitrophenol reduction. The synergetic phenomena shown in figure 7.15(f) that develop at the Pd/GDYO nanocomposite might be attributed to this excellent performance.

The use of GDY as a support can prevent Pt nanoparticles from aggregating and promote the interactions among Pt NPs as well as reactants [108]. When compared to the commercial Pt–C, this ultrastable PtNP catalyst performed well in the hydrogenation of aldehydes and ketones to their corresponding alcohols. Furthermore, as compared to commercial Pt–C, Pt NPs catalysts with a size of 2–3 nm demonstrated superior efficacy in the hydrogenation of aldehydes and along with ketones to the corresponding alcohols, as shown in figures 7.15(g) and (h).

Conclusively, it has been found that GDY is a potential substrate for fabricating metal nanoparticle-based heterogeneous catalysts, particularly for applications that need strong interactions involving metal nanoparticles and reactants.

7.3.1.2 Water splitting

We have discussed the catalytic behavior of GDY in the previous section and the characteristic behavior of GDY also makes it a good candidate for water-splitting applications. The separation of water molecules can be done in the presence or absence of sunlight. The process which involves sunlight is photoelectrochemical water splitting (PEC-WS) whereas electrochemical water splitting does not involve sunlight. GDY acts as a good catalyst to enhance the water-splitting behavior of different materials which can be seen in the following sections.

7.3.1.2.1 Photoelectrochemical water splitting

PEC-WS can provide clean, economical hydrogen fuel using high solar to chemical energy conversion, offering an effective alternative to increasingly depleted fossil fuels [109–114]. The formation of heterojunctions, doping, and even grafting of different co-catalysts over other wide band gap catalysts has been found to enhance the PEC performance of the system. As described earlier, the synthesis of $BiVO_4$/GDY was carried out through an Cu-envelope catalysis method [41]. In detail, $BiVO_4$ electrodes were enveloped inside a copper wrapped as well as submerged inside a TMEDA, pyridine, mixed with acetone. The solution was boiled at 50 °C for 2 h in an argon atmosphere. HB/acetone solution was then slowly added and it was treated for an extra 12 h at 50 °C within an argon environment enabling, another acetylenic coupling process, leading to the creation of GDY nanowalls on the $BiVO_4$ electrode. To eliminate monomers and oligomers, including catalysts, the GDY-grown electrodes were treated using a hot mixture of acetone, DMF, with pyridine, and subsequently dehydrated under a nitrogen flow. Figure 7.16(a) shows a schematic depiction of the incorporated photoanode as well as a suggested technique for improving the photoelectrochemical water-splitting efficacy. A well-formed $BiVO_4$/GDY nanocomposite was shown through FESEM and TEM observations, shown in figures 7.16(b) and (c), respectively. At the standard potential of 1.23 V versus reversible hydrogen electrode (RHE), a photocurrent density of around 1.32 mA cm^{-2} was observed, figure 7.16(d), which was about two times higher than that of a $BiVO_4$ photoanode. In addition, figure 7.16(e) shows that the hole injection yield at 1.23 V versus RHE can approach 60%, which is greater than that of the output for bare $BiVO_4$ (just under 30%), suggesting that GDY might speed up charge separation effectively.

Han *et al* developed a metal-free 2D/2D heterojunction comprising g-C_3N_4/GDY as an excellent photoelectrocatalyst for water splitting depicted in the FESEM and TEM images in figures 7.16(f) and (g), respectively [87]. Owing to the outstanding

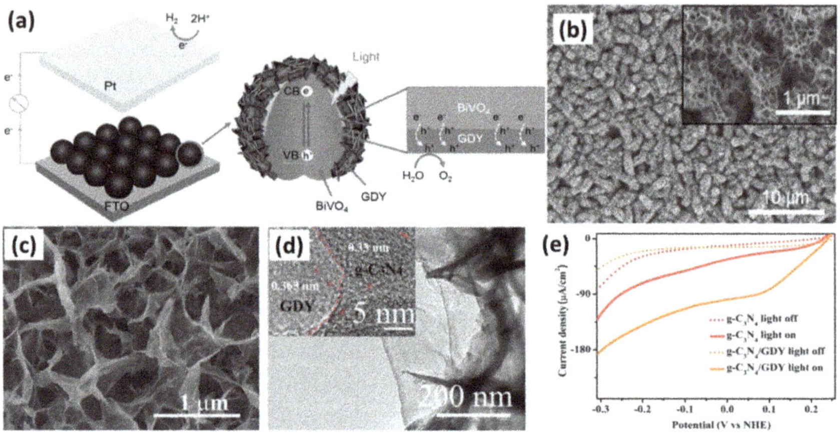

Figure 7.16. (a) Diagrammatic representation of GDY/$BiVO_4$ photoanodes inside a PEC arrangement showing photogenerated charges migrating only at the interface. (b) GDY/$BiVO_4$ FESEM images. (Reproduced with permission from [41]. Copyright 2016 Wiley.) (c) FESEM images of the 3D GDY nanosheet array, (d) TEM images of g-C_3N_4/GDY, i.e. 2D/2D heterojunction composite nanosheets (the side view shows an HRTEM image of g-C_3N_4/GDY). (e) Scanning linear sweep voltammetry across various photocathodes during dark as well as light conditions. (Reproduced with permission from [87]. Copyright 2018 Wiley.)

hole-transport characteristic of GDY and the structure of the g-C_3N_4/GDY photocathode with 2D/2D heterojunctions, it displays improved photocarrier separation. Figure 7.16(h) represents a tremendous increase in photocurrent density with g-C_3N_4/GDY against bare g-C_3N_4 photocathode using a neutral aqueous medium at a voltage of 0 V versus normal hydrogen electrode (NHE). The group have noted the following findings: (i) the extremely thin framework of g-C_3N_4/GDY could perhaps reduce photocarrier transport separation; (ii) it is advantageous for effective hole injection from g-C_3N_4 towards 2D GDY; (iii) the high hole mobility of GDY can quickly transfer holes by suppressing photocarrier recombination in g-C_3N_4; (iv) the GDY with a nanosheet array has a greater surface area, which is highly favorable for the photoelectrocatalytic reaction; and (v) deposition of co-catalysts onto g-C_3N_4 that is adorned with coordinated N atoms produces 'sub-nanopores' that enable dual-control over photo-induced electrons and holes, within g-C_3N_4.

7.3.1.2.2 Electrochemical water splitting

Electrochemical water splitting (EC-WS), which consists of either a hydrogen evolution reaction (HER) somewhere at the cathode or an oxygen evolution reaction (OER) at the anode, is thought to be another effective and reliable method for producing hydrogen, that might be a feasible solution to the impending energy catastrophe as well as environmental issues [115].

Zhang *et al* constructed a novel GDY-based heterostructure comprising GDY@$NiO_x(OH)_y$ based purely on the fact that the breakdown of H–OH with Ni-based hydroxides has been proven to be advantageous for boosting OER characteristics [116]. Te GDY@NiO$_x$(OH)$_y$||GDY@NiO$_x$(OH)$_y$ had 1.54 V at 10 mA cm^{-2} in alkaline over all WS electrolysis, exhibiting strong durability throughout 100 h at 20 mA cm^{-2}, far superior to the majority of the published standard WS electrocatalysts. Hui *et al* reported that the addition of spontaneously porous GDY nanolayers to an iron carbonate hydroxide (FeCH) surface gives the pure catalyst architectural benefits that help it to function better in catalysis [117]. The featured catalyst can drive from 10 mA cm^{-2} and 100 mA cm^{-2} at 1.49 and 1.53 V, respectively. The composite materials demonstrate remarkable long stability of 10 000 cycles in the OER as well as 9000 cycles during the HER under 1.0 M KOH, attributable to the preservation of resistant GDY nanolayers as well as close contact among GDY as well as FeCH.

For the case of composite materials with metal sulfides, such as GDY-WS_2, a 2D-nanohybrid (2D-NH) catalyst demonstrates better HER efficiency compared to a noble-metal-free catalyst [79], with a better onset potential as low as 140 mV, a Tafel slope decreased to 54 mV dec^{-1}, and a strong endurance of more than 2 h. The exceptional HER success is due to (i) the 2D-NH, which has a much higher conductivity as well as a stable hierarchical pattern; (ii) edge-rich, defect-abundant WS_2 nanosheets, which provide innumerable active surface areas; and (iii) a built-in electric field and efficient electron transfer between GDY and WS_2. Additionally, the researchers believe that this approach will be useful in the development of a variety of additional GDY–metal sulfide 2D-NH compounds. On the other hand, as discussed earlier, the innovative hybridization of MoS_2 plus N-GDY provides structural and physicochemical benefits for improving electro-catalytic performance, not only being better than commercial Pt but also better than practically all standardized electrocatalysts [89]. All of the findings show that N-GDY may be employed as an efficient catalytic promoter enabling hydrogen generation, as well as a new method for constructing excellent water-splitting electrodes.

Ternary sulfides such as $NiCo_2S_4$ NWs coated with GDY show outstanding bi-functional catalytic activity, as demonstrated in the work of Liu and Li [80]. Figure 7.17(a) shows the stepwise development of $NiCo_2S_4$ NWs/GDY. The OER activities of samples were estimated using linear sweep voltammetry (LSV) polarization curves in relation to RHE, shown in figure 7.17(b). Moreover, the transformation of Ni(II) into Ni(III) species, which is also widely thought to become the effective area for OER, caused the oxidation peak to form at 1.32 V versus RHE. The LSV curves for $NiCo_2S_4$ NW/GDY in figure 7.17(c) reveal an initial overpotential of just 112 mV. The team has demonstrated an elevated alkaline water electrolyzer exhibiting 10 and 20 mA cm^{-2} at very low cell voltages of 1.53 and 1.56 V. Furthermore, the composite shows impressive stability during 140 h of uninterrupted electrolysis at 20 mA cm^{-2}.

GDY may be employed as a new catalyst-support for maintaining metal NPs during OER generated using an *in situ* synthesis reduction approach, as described in [118]. Figure 7.17(d) shows an FESEM of the 3D Cu@GDY/Co and figure 7.17(e) is a HRTEM image of GDY/Co composites. In the alkaline solution, the produced electrode exhibits exceptional OER action including an overpotential of 0.3 V accompanied by substantial activity of 413 A g^{-1} around 1.60 V against RHE. The further smaller Tafel slope in the figure 7.17(f) suggests better conductivity,

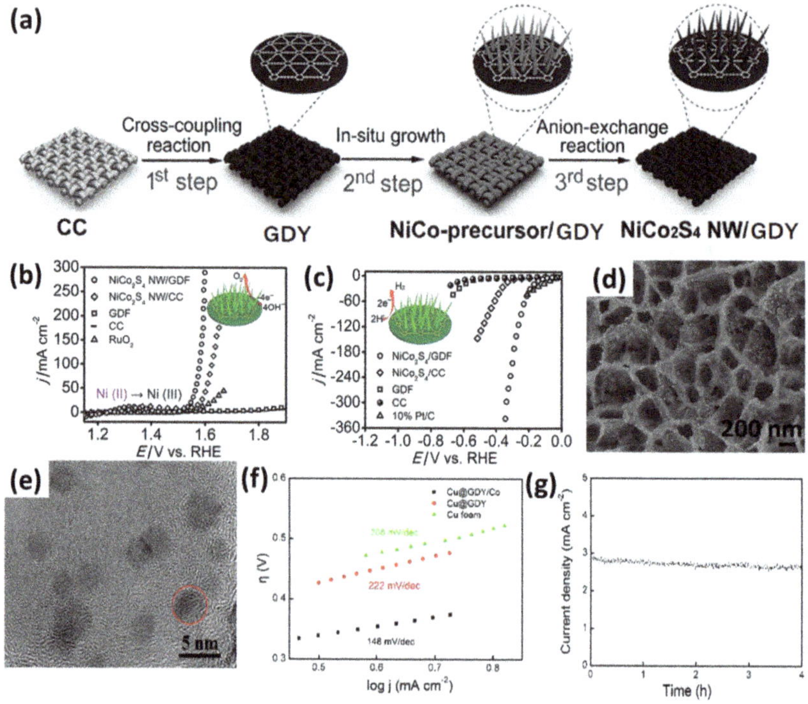

Figure 7.17. (a) Schematic depiction of the $NiCo_2S_4$ NW/GDY manufacturing method. Polarization curves for (b) OER and (c) HER for $NiCo_2S_4$ NW/GDF, $NiCo_2S_4$ NW/CC, GDF, CC, and RuO_2. (Reproduced with permission from [80]. Copyright 2017 Wiley.) (d) SEM images of Cu@GDY/Co electrodes. (e) HRTEM images of GDY/Co composites. (f) Tafel plots of the electrodes. (g) Cu@GDY/Co electrode stability test at 1.6 V versus RHE. (Reproduced with permission from [118]. Copyright 2017 American Chemical Society.)

supported by figure 7.17(g) which shows the stability of the composite electrode for 4 h. Despite the fact that the catalyst assistance is partially destroyed during electrolysis, the findings reported benefits from the addition of sufficient active catalytic locations and enhanced conductivity, which provides a somewhat simple and straightforward method for preparing the nanocomposite, but also expands the family of catalyst supports for NP stabilization. The best reported GDY-based electrocatalysts in the presence of an alkaline electrolyte are listed in table 7.2.

7.3.1.3 Electrocatalysts for ORR

A density functional theory (DFT) study of GDY showed the charge density distribution at every carbon atom in the GDY structure is not regular, implying the presence of positively charged carbon atoms that aid in the adsorption of O_2 as well as OOH^+ molecules [133]. Furthermore, Haley *et al* found that the inherent pores which are available inside the GDY plane are in the range of 2.5 μm in diameter, which aids in O_2 adsorption [134]. These are the fundamental benefits of GDY using metal-free ORR catalysts [135]. One of the classic approaches is heteroatom doping to fabricate GDY-based composites for ORR applications. For ORR, N-doped GD

Table 7.2. Performance of some popular GDY electrocatalysts for electrochemical water splitting.

Sl. No.	Catalyst	Current density ($mA\ cm^{-2}$)	Overpotential (mV)	Electrolyte	Reference
1.	CoP/NCNHP	10	1.64	1 M KOH	[119]
2.	Co_2P NC	10	1.56	1 M KOH	[120]
3.	Co_1Mn_1CH/NF	10	1.67	1 M KOH	[121]
4.	Fe–CoP	10	1.60	1 M KOH	[122]
5.	FeCH@GDY/NF	10	1.49	1 M KOH	[117]
		100	1.53		
6.	EG/$Co_{0.85}$SE/NiFe LDH	10	1.67	1 M KOH	[123]
		20	1.71		
7.	Hierarchical $NiCo_2O_4$ hollow microcuboids	10	1.65	1 M KOH	[124]
		20	1.74	1 M KOH	
8.	Ni_5P_4	10	1.70	1 M KOH	[125]
9.	NiFe–NiMo	10	1.51	1 M KOH	[126]
10.	NiFe LDH/Ni foam	10	1.70	1 M KOH	[127]
11.	NiFe LDH–NS@DG	20	1.50	1 M KOH	[128]
12.	$NiFeO_x$/CFP	10	1.51	1 M KOH	[129]
13.	np-$Co_{1.04}Fe_{0.96}P$	10	1.53	1 M KOH	[130]
14.	Ni–Co–P hollow nanobricks	10	1.62	1 M KOH	[131]
15.	Se-(NiCo)S/OH	10	1.60	1 M KOH	[132]

possesses exceptional electrocatalytic capabilities for the ORR, exhibiting electrocatalytic activity equivalent to commercial Pt/C catalysts. N-doping causes significant positive charges on the carbon atoms close to the N atom, which draw electrons from the anode more efficiently as well as promoting the ORR, according to detailed quantum mechanical calculations. As a result, developed a novel carbon-based metal-free ORR electrocatalyst was developed that was very productive. Other heteroatom-doped/co-doped GD catalysts may potentially be employed as point-of-saturation non-Pt as well as metal-free ORR electrocatalysts.

In half-cell and full-cell (primary Zn–air battery) experiments, nitrogen with fluorine co-doped, metal-free GDY demonstrates comparable electrocatalytic performance as commercial Pt/C in terms of the onset potential as well as limiting current density [136]. Furthermore, the novel catalyst outperforms conventional Pt/C in terms of stability, as well as resistance towards methanol crossover and CO poisoning effects. Lv *et al* have demonstrated nitrogen (N) doped in novel designs through exchanging sp-hybridized carbon atoms, that has been proved in this work utilizing combined theoretical and experimental technologies to be appropriate for ORR [71]. Further, three factors contribute to the high catalytic activity of N-GDY-900 °C and N′N-GDY: (i) to improve the conjugation effect as well as the electrical conductivity of both materials, tiny GDY fragments and oxygen-containing groups are always removed at high temperatures; (ii) following high-temperature treatment, the materials become porous, which is advantageous for increasing surface area and forming active sites; (iii) finally, additional imine N-2, imine N-2,2, and pyridinic-N have been doped into N-GDY-900 °C and N′N-GDY, which is advantageous for ORR active site creation.

Utilizing yet another process, iron and PANI were deposited on GDY nanocomposite at 900 °C [137]. This represents an excellent technique for the facile synthesis of inexpensive iron–nitrogen-doped carbon nanolayers folded with GDY core–shell electrocatalysts (Fe-PANI@GD-900) using ORR. The catalyst has a surprisingly high ORR activity (versus RHE) having onset as well as $E_{1/2}$ at 1.05 V and 0.82 V, respectively, although the mass activity of Fe-PANI@GD-900 is lower than for the Pt/C catalyst. Furthermore, the electrode also shows long durability, as it maintains a straight four-electron reduction route.

7.3.1.4 Solar cells

The fast development of solar cells in recent years has provided a significant answer to the global energy problem. After the work by Gratzel there has been a lot of work on dye-sensitized solar cells (DSSC) [138]. If the development of the solar cells is categorized in generations, then in this era lot of emphasis is placed on the third generation of solar cells, in which within the growth of the DSSC [139, 140] there is the emergence of other crucial types of solar cells, such as organic [141], inorganic [142–146], hybrid [147, 148], quantum-dot solar cells (QDSCs) [149], and perovskite solar cells [150]. Researchers from all across the globe are working to build a low-cost photovoltaic system that is both efficient and cost-effective. GDY has exceptional physical capabilities due to its highly conjugated two-dimensional structure, broad interlayer distance, and variable electronic band gaps. Photovoltaic technologies based on GDY, such as perovskite, dye-sensitized, and quantum dots, have demonstrated exceptional performance.

7.3.1.4.1 Organic solar cells

For the first time, through the approach of Ren *et al* [151], the interactions between Pt atoms and '–C≡C–' in GDY were carried out theoretically on composites of the Pt_2 dimer with GDY. Further, the electrostatic potential surfaces with negative polarization of electrostatic potential (ESP) of '–C≡C–' were shown as red whereas blue displayed positive polarization of same. Furthermore the group also reported concentrate on the left hand (I_3/I^-) redox pair though it is the something that affects the performance of counter electrode. Because of their peculiar p–n intersection structure featuring increased catalytic activity and high electron transfer capability, the overall energy conversion efficiency of DSSCs using PtNP–graphdiyne nanosheets (GDNS) counter electrodes has been enhanced vastly. The implementation seems equivalent to expensive Pt foil and superior to Pt nanoparticles as well as rGO/Pt nanoparticle composites, according to experimental data. This research might pave the way to enable GD nanocomposites to be used as elevated photoelectric counter electrodes.

In another report GDY has been integrated into polymer solar cells by Du *et al* [152] and, moreover, their report demonstrated that introducing GDY into the active layer improves the power efficiency of polymer photovoltaic systems.

The device with 2.5 wt% GDY has a 2.4 mA cm^{-2} increase in J_{sc} and the greatest power conversion efficiency (PCE) of 3.53%. The increased charge transfer capabilities of GDY and the creation of effective percolation channels inside the active layer are responsible for the enhanced quality.

7.3.1.4.2 Quantum-dot solar cells

GDY may also be used directly as a hole transfer material since its theoretical computed hole mobility is 10^4 cm^2 V^{-1} s^{-1} [14]. CdSe QDs containing 4-mercapto-pyridine surface functionalization were subsequently stuck on the GDY surface of film by the π–π interaction [153], as shown in the conceptual image in figure 7.18(a).

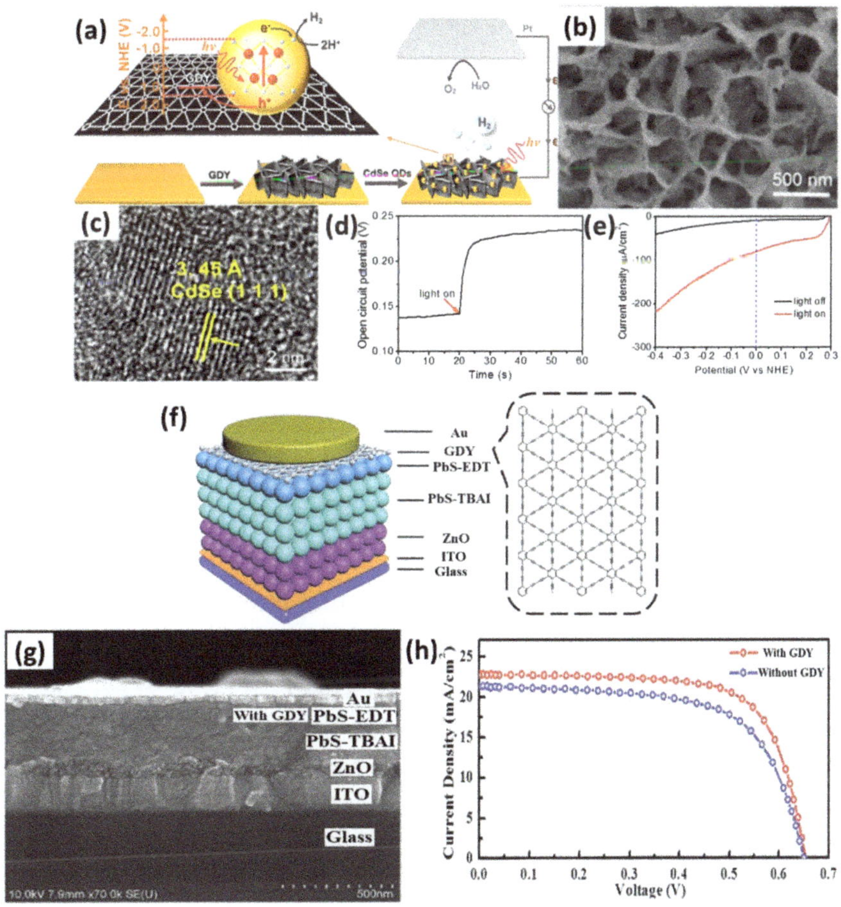

Figure 7.18. (a) Integrated CdSe QD/GDY photocathode and Pt wire for the counter electrode with the associated interfacial mechanism of the photogenerated excitons in the designed circuit. (b) FESEM and (c) HRTEM images of the assembled CdSe QD/GDY. (d) CdSe QD/GDY photocathode open-circuit potential response. (e) LSV scanning from 0.3 to 0.4 V at 2 mV s^{-1} in light as well as dark conditions for the CdSe QD/GDY photocathode. (Reproduced with permission from [153]. Copyright 2016 American Chemical Society.) (f) Diagrammatic representation of both the PbS CQD solar cells with and without a GDY anode buffer layer. (g) Device cross-sectional SEM image. (h) *J*–*V* characteristics. (i) EQE characteristics with and without GDY. (Reproduced with permission from [154]. Copyright 2016 Wiley.)

Furthermore, FESEM and HRTEM images of the quantum-dot/GDY composite are shown in figures 7.18(b) and (c), respectively. The photo-induced holes might be effortlessly transferred to GDY from CdSe QDs, since the potential value of the valence band (VB) of GDY has been more negative than that of CdSe QDs, as revealed by the open-circuit potential measurement shown in figure 7.18(d). GDY is used as a potential hole transport material (HTM) again for photocathode since it has properties that reduce charge recombination, improve conductivity, and lower photocathode resistance. Figure 7.18(e) shows the system working as QDSC, which also demonstrated good Faradaic efficiency and considerable photoactivity enabling hydrogen generation for 12 h.

Another work presented by the Sargent and Li group has noted evidence of the insertion of GDY to form a composite in terms of QDSCs [154]. Figure 7.18(f) shows a colloidal quantum-dot solar cell which can also be understood through the cross-sectional FESEM image of figure 7.18(g). Power conversion efficiency is notably improved to 10.64% from 9.49%, associated with the QDSC without the incorporation of GDY as shown in figure 7.18(h). The results show that only by lowering the work function of the CQDSC with GDY can the hole transport through the quantum-dot solid active layer towards the anode be significantly improved. Moreover, the all-carbon buffer layer is found to extend the carrier lifetime while lowering surface recombination only on the photovoltaic device's hitherto ignored rear side.

7.3.1.4.3 Perovskite solar cells

GDY can significantly enhance two important parameters, namely PCE and short-circuit current density (J_{sc}), of polymer solar cells. GDY has a strong charge-transport capability and therefore can generate a highly effective percolation route inside the active layer, and thus this enhancement may be attributed to it [85, 152, 155, 156].

P3HT as well as the GDY nanoparticle coating have a significant π-π stacking relationship (as shown in figure 7.19(a)), which encouraged hole transport and hence increased cell performance [85]. When compared to bare P3HT, hole removal occurred which increased within P3HT/GDY HTM, as shown as the band alignment of the composite in figure 7.19(b). In addition, the combination of GDY in P3HT films produced excellent scattering. Enhanced absorption of long-wavelength illumination within equivalent perovskite solar cell (PSC) was due to the decreased transmittance caused by scattering. P3HT/GDY HTM achieved a better photoelectric conversion efficiency of 14.58% compared to PSCs constructed over pure P3HT.

Li *et al* reported that in PSCs using an $MAPbI_3$-like dynamic coating, including some other hole-transport polyelectrolyte by means of a comparable arrangement, called P3CT-K, continues to provide an extra function as a dopant for GDY, as shown in figure 7.19(e) [88]. Figure 7.19(f) shows an about 19.5% increase in conversion efficiency, which may be attributed to a better J_{sc} as well as fill factor (FF).

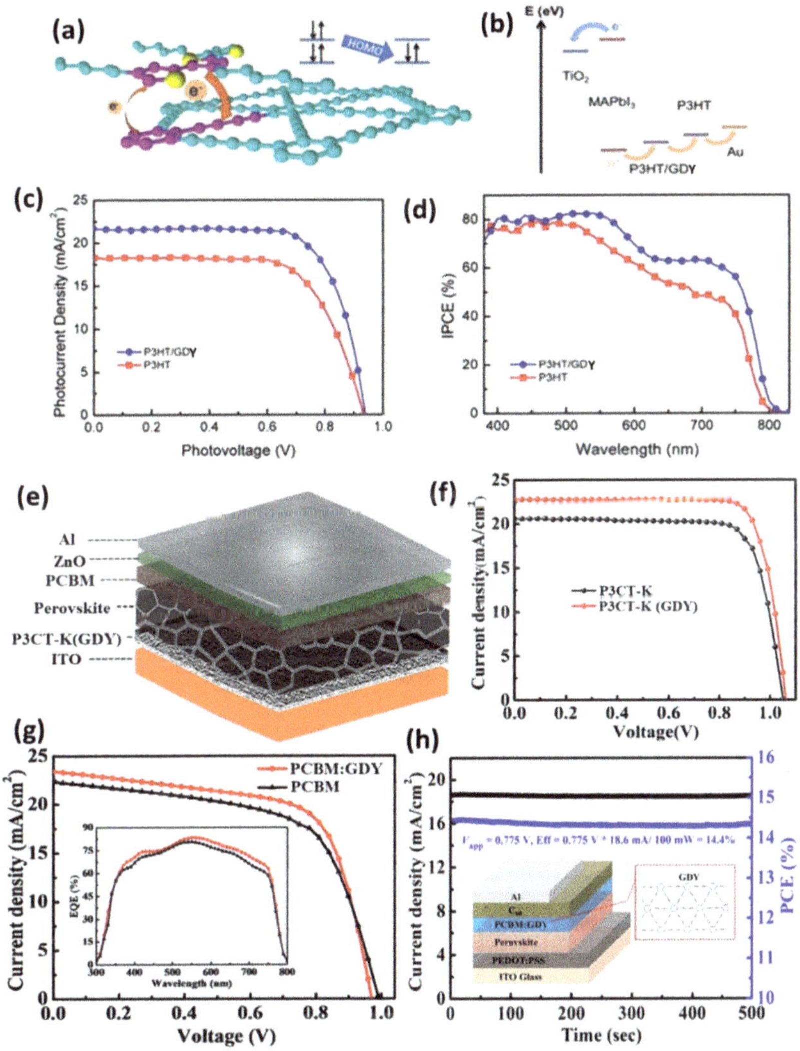

Figure 7.19. (a) Schematic of the interaction of P3HT and GDY. (b) Energy levels of PSCs with P3HT/GDY. (c) *J–V* curves and (d) IPCE spectra of P3HT/GDY- and P3HT-based PSCs. (Reproduced with permission from [85]. Copyright 2015 Wiley.) (e) Schematic showing the device arrangement of the P3CT-K (GDY) based PSC. (f) *J–V* curves with and without GDY over P3CT-K based perovskite structure. (Reproduced with permission from [88]. Copyright 2018 American Chemical Society.) (g) *J–V* characteristic results of bare PCBM and the composite of PCBM with GDY. The inset graph shows the (EQE) spectra. (h) Photocurrent density (black) and steady-state efficiency (blue) of the PCBM:GDY-based PSC device where the inset shows the device structure of PSC with GDY as the interfacial layer. (Reproduced with permission from [157]. Copyright 2015 American Chemical Society.)

GDY may improve the charge-extraction capability, conductivity, including mobility mostly within electron transport layer (ETL) as a dopant, as well as the contact condition between a certain ETL layer as well as the perovskite layer, which is critical for data reproducibility [157]. In PSCs including an inverted mechanism

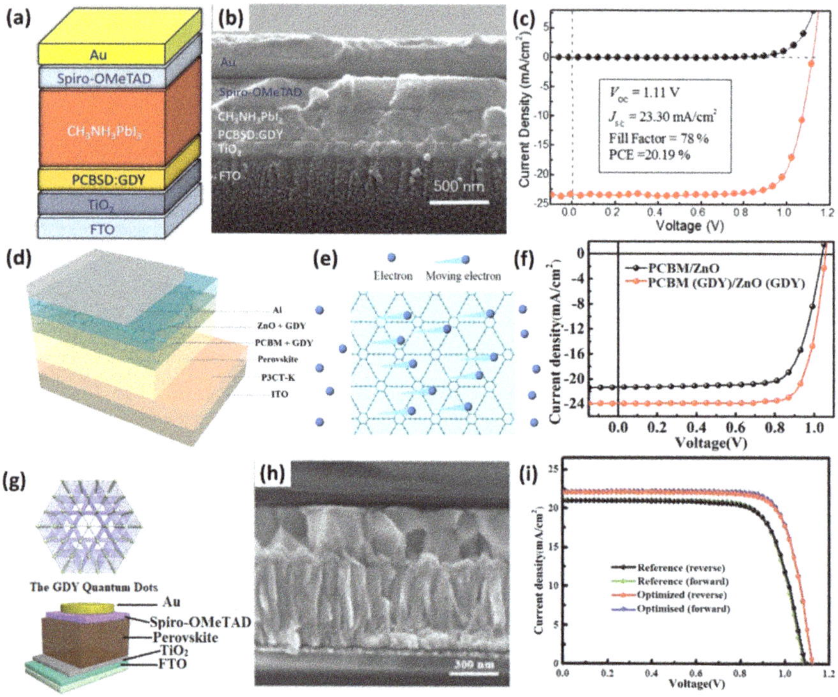

Figure 7.20. (a) Configuration of a planar PSC. (b) Cross-sectional FESEM view of a usual PSC including C-PCBSD:GDY established ETL, (c) *J–V* curves of the champion device. (Reproduced with permission from [158]. Copyright 2017 Elsevier.) (d) Device displayed as p–i–n architecture. (e) Construction of GDY used for device assembly. (f) *J–V* characteristic curves of PCBM/ZnO and PCBM(GDY)/ZnO(GDY) based perovskite solar cells. (Reproduced with permission from [159]. Copyright 2018 Elsevier.) (g) Schematic illustration of the GDY QDs based perovskite solar cells. (h) Cross-sectional SEM image of the same structure. (i) *J–V* characteristics under both backward and front scan orientations. (Reproduced with permission from [160]. Copyright 2017 Wiley.)

architecture, a PCBM layer doped using GDY might increase electron transport. GDY could potentially remain useful for passivating grain edges and effectively preventing recombination by lowering interface trap states. Insertion of the GDY to form a composite with PCBM shows a change in the PCE of 14.8% optimal power conversion with negligible hysteresis. Also, the external quantum efficiency (EQE), the inset of figure 7.19(g), shows the same trend as that of the *J–V* characteristics. The inset of figure 7.19(h) shows the PCBM:GDY ETL-based PSCs featuring planar heterojunction structures, and the photocurrent density as well as the steady-state efficiency are shown in figure 7.19(h).

According to previous studies, the interfacial layer is a critical factor that influences the performance and reliability of PSCs. Researchers have developed novel ETLs combining GDY and fullerene derivatives (i.e. PCBSD), as shown in the schematic of figure 7.20(a) [158]. The intermolecular interactions between both the cross-bridge PCBSD as well as the conjugation of GDY were the reasons for this (figure 7.20(b)). Again, from cross-sectional FESEM images, the width of the ETL including combinations of GDY with PCBSD for conventional devices can be seen in figure 7.20(c). A PCE of 20.19% was found in the best photovoltaic device, shown in figure 7.20(c). Furthermore, compared to other comparable devices using TiO_2 as the ETL, GDY-containing devices showed improved stability that is attributed to the solvent repellent feature of the 2D GDY with fullerene derivatives. The addition of GDY in the ETL enhanced electron transport from the active layer toward the electrode successfully, and was credited with the exceptional system efficiency.

The novel dual doping approach is used for fabricating a cascade structure GDY-containing an electron transport layer, as shown in figure 7.20(d) [159]. The excellent dispersal of GDY in organic solvents such as chlorobenzene ensures that PCBM or ZnO will be doped to GDY in the future, which is helpful for electron transport through each layer of the device, as shown in figure 7.20(e). As a consequence, photovoltaic devices employing $MAPbI_3$ as the active material as well as GDY/ZnO along with GDY/PCBM as two-fold ETLs had a superior PCE of 20.0% and less *JV* hysteresis, which can be seen in figure 7.20(f). The introduction of GDY to the dual ETL prevents electron build-up effectively and lowers transport resistance, improving the reliability of GDY-containing devices.

GDY quantum dots (QDs), as shown in figure 7.20(g), were employed by Zhang *et al* [160] as holistic dopants for the active layer, and the efficiency of PSCs using $CH_3NH_3PbI_3$ as an effective material with TiO_2 as well as Spiro-OMeTAD is indicated in figure 7.20(h). Figure 7.20(i) shows that the PCEs of GDY-doped cells would be in the range of 17.17%–19.89%, which is significantly better, demonstrating that the doping of GDY QDs improves perovskite photovoltaic device performance.

7.3.1.5 Electrochemical actuators

Electrochemical actuators are instruments which use electrochemical processes to transform electrical energy into mechanical energy [161]. These actuators are used in soft robotics, artificial muscles, micropumps, sensors, and other industries. The development of bendable and robust electrode materials is still a difficult task. Existing actuators have a 1.0% energy transduction efficiency. The available electrode materials are limited due to their microstructure missing active units and inherent features that could not be achieved by assembly processes.

Lu *et al* have developed a GDY-based electrochemical actuator which shows 6.03% electromechanical transduction efficiency with a 11.5 kJ m^{-3} energy density (figure 7.21) [19]. Following 100 000 repetitions at 1 Hz under 2.5 V, the actuator shows no signs of wear. The group discovered that when the surrounding ions are deposited on GDY under electric stimuli, the bond length of GDY changes due to bending of the actuator. This may be because of surrounding ions that interact through alkyne bonds, causing alkyne bonds to change into alkene bonds despite differences in bond length as per their empty orbitals. The dimensional shift inside the GDY structure is caused by the bond length modification of alkene–alkyne complex transitions. The transformation remains reversible and may be influenced by external electric fields.

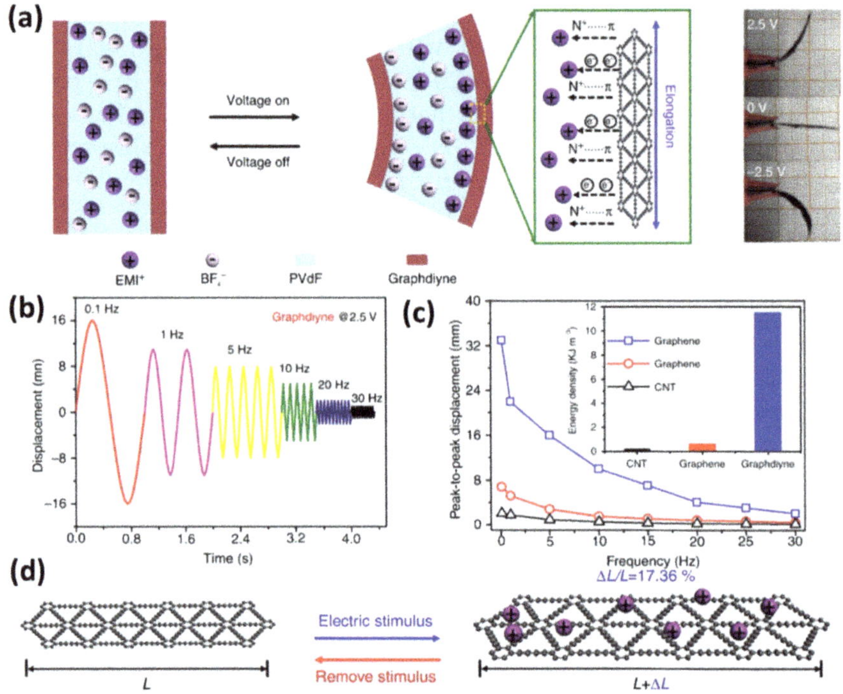

Figure 7.21. (a) Schematic showing the working mechanism of a GDY-based electrochemical actuator. (b) Graphical representation of the strain of the GDY actuator with increasing frequency. (c) Frequency-dependent displacement and related power densities. (d) GDY is exposed to an electric stimulus. (Reproduced from [19]. CC BY 4.0.)

7.3.2 Energy storage applications

GDY is not only used in energy conversion, it is also applied in energy storage due to its value added characteristics. In this section, we will cover the different energy storage applications such as supercapacitors and batteries.

7.3.2.1 Supercapacitors

GDY nanostructures were produced by a cross-coupling process utilizing HB as the precursor [162]. The GDY nanostructured thin films possessed both the behavior of an electric double layer capacitor (EDLC) and pseudocapacitance. A charge–discharge study of the GDY electrodes revealed a specific capacitance of 71.4 F g^{-1} once under a discharge current density of 3.5 A g^{-1}. Furthermore, GDY electrodes have a capacitance retention of roughly 97% after 1000 cycles, implying that they might be used as supercapacitor electrodes. The excellent GDY adhesion contributes to the supercapacitor's remarkable performance as shown by Wang *et al*, which is demonstrated in the schematic of figure 7.22(a) and the HRTEM image in figure 7.22(b) [29]. Figure 7.22(c) shows the current–voltage (CV) curves at different scan rates. The discharge curves of figure 7.22(d) reveal a substantial specific area capacitance of 53.66 mF cm^{-2} at 0.2 mA cm^{-2}, which was sustained at 49.42 mF cm^{-2}

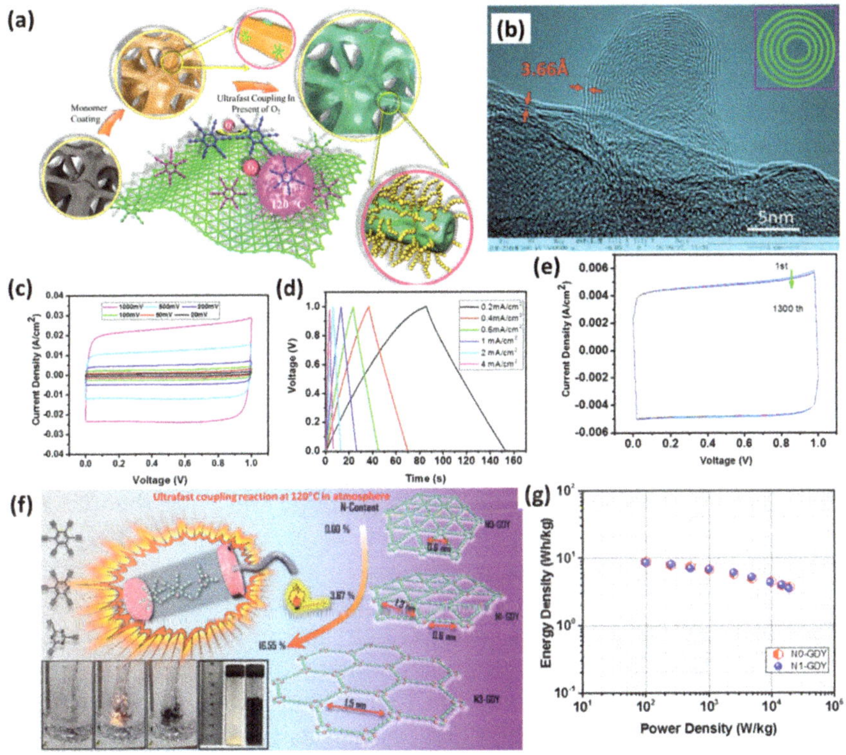

Figure 7.22. (a) Illustration of the design of the weaving process of a GDY nanostructure on 3D foam-like substrates. (b) Photograph of modified GDY with typical structures of the GDY nanochains sprouting from the GDY nanofilm. (c) CV curves obtained at different scanning rates. (d) GCD profiles under different current densities. (e) Long-term CV at a scan rate of 200 mV s $^{-1}$. (Reproduced with permission from [29]. Copyright 2019 American Chemical Society.) (f) Schematic depiction of preparation procedures with varied architectures of N0-GDY, N1-GDY, and N3-GDY. (g) Ragone plot. (Reproduced with permission from [97]. Copyright 2017 Elsevier.)

at 4 mA cm^{-2}, showing a stable rate capability. Furthermore, following 1300 cycles of CV testing on the as-fabricated device, the GDY-based electrode demonstrated outstanding long-term endurance exhibiting nearly no current density deterioration, as shown in figure 7.22(e).

Shang *et al* have created a new approach for fabricating N-doped GDY nanostructures, shown in figure 7.22(f) [97]. These N-patterns with spongy structures for GDYs may all be efficiently tuned using this approach, demonstrating high cornering grip. The GDY for supercapacitors has high specific capacitance, energy density, and power density as shown in figure 7.22(g).

For broadening the applicability of GDY-centered products, these products can be used as substrates enabling *in situ* MnO_2 formation, yielding MnO_2@GDYO nanoheterostructures that can be used for pseudocapacitor electrode materials [163]. This MnO_2@GDYO electrode had a specific capacitance of 301 F g^{-1} at 0.2 A g^{-1} and showed 98% retention after 3000 cycles, outperforming other transition metal pseudocapacitor electrodes.

7.3.2.2 *Batteries*

The presence of sp-hybridized carbon inside a coupled carbon background, as well as the enlarged pores in the GDY framework, offer adequate storage sites as well as diffusion passages for metal atoms such as lithium, sodium, magnesium, and others, enhancing the volume and effectiveness of rechargeable batteries when GDY acts as an electrode. The unevenly arranged electrons, semiconducting characteristics, and porous 2D structure of GDY, in particular, could create additional paths for technical achievements in many types of energy storage batteries [164–169]. In this chapter we have focused on the two different commonly used batteries, namely Li-ion batteries (LIBs) and Na-ion batteries (SIBs), in which researchers have found some experimental results.

7.3.2.2.1 Li-ion batteries

Theoretical investigations by Srinivasu *et al* show that GDY has much higher specific capacities as well as lithiation potentials than graphite-based electrode materials [170]. GDY has lithiation potentials ranging from 2.7 to 2.1 V. LiC_3 might represent the in GDY, than LiC_6 dipped in graphite since the theoretical specific capacity is nearby 744 mAh g^{-1}, double that of graphite [167, 171]. An energy barrier of about 0.52 eV was achieved for the Li atoms to permeate over a single layer of GDY, as shown in figures 7.23(a)–(d). Because the energy barrier to breach a single layer of GDY is 0.35 eV, Li atoms might spread widely on any edges of the GDY layer. Li atoms can occupy all three symmetrical corners of the large triangle perforations in the GDY plane, giving Li a higher binding energy (BE) to GDY. The cohesive energy of Li next to GDY is greater than its BE GDY has the potential to

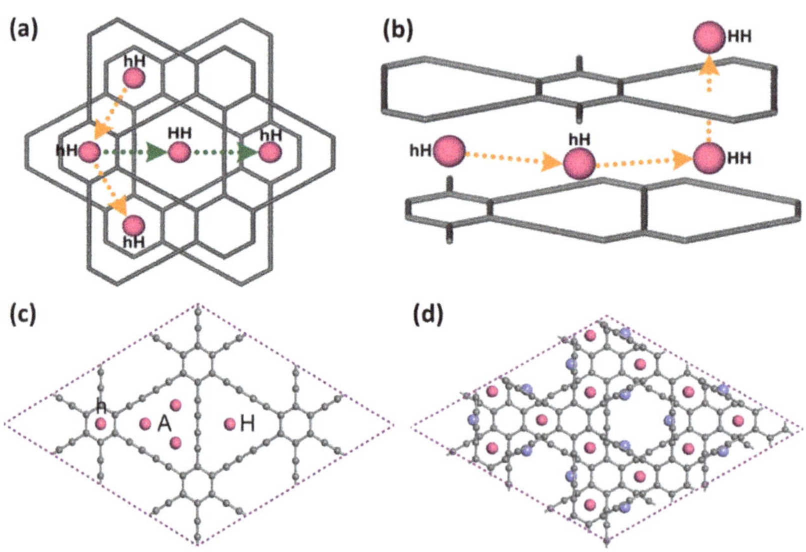

Figure 7.23. (a) Schematic showing probable in-plane diffusion in GDY and (b) diffusion through the lithium GDY layer. (Reproduced with permission from [167]. Copyright 2011 American Chemical Society.) (c) In single-layer GDY there are three potential Li storage locations. (d) Multidimensional GDY intercalated by Li in an effective arrangement. (Reproduced with permission from [171]. Copyright 2013 AIP Publishing.)

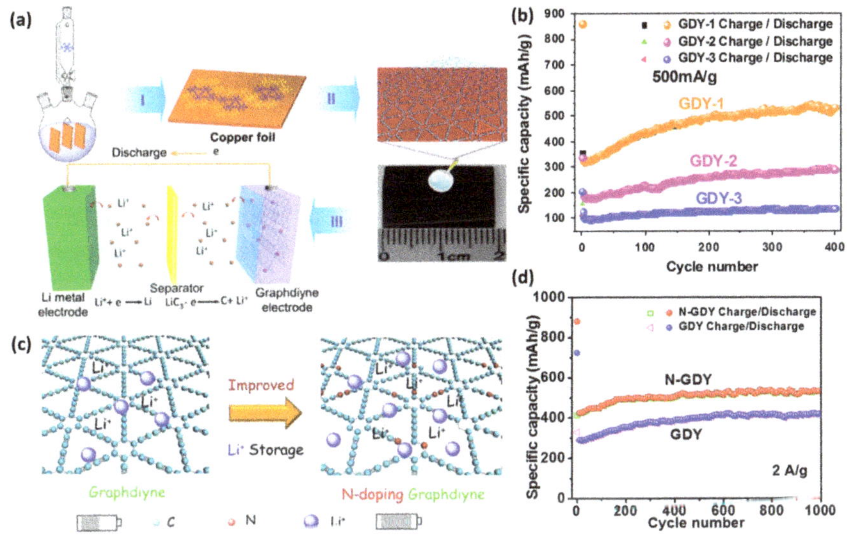

Figure 7.24. (a) Diagrammatic representation of GDY synthesis as well as the LIB creation process. (b) Progression efficiency of bare GDY electrodes at a constant current density of 500 mA g^{-1}. (Reproduced with permission from [172]. Copyright 2014 Elsevier.) (c) Model consisting of N-doping, which is a new effective technique for the electrochemical behavior of GDY materials. (d) The cycle capability of as-prepared GDY and N-GDY electrodes at a constant current density of 2 A g^{-1}. (Reproduced with permission from [72]. Copyright 2016 American Chemical Society.)

be used as a high-capacity lithium storage material because of its practical accessibility and strong Li mobility. As it can be diffused quickly in three dimensions, it is employed as an anode material for LIBs with high rate capacity.

It was Huang *et al* who demonstrated that GDY can be grown on Cu, which may be used as an electrode designed for the production of superior functioning LIBs, as shown in figure 7.24(a) [172]. As shown in figure 7.24(b) at about 400 cycles with a current density of 500 mA g^{-1}, the pristine GDY-created battery demonstrated an adjustable capacity of 520 mAh g^{-1}. Over 1000 cycles, the GDY-based battery with a greater current density of 2 A g^{-1} maintained its outstanding specific capacity of 420 mAh g^{-1}. The results demonstrate that GDY-based LIBs showed superior electrochemical performance, including large specific capacities, fast throughput performance, as well as strong cycle ability.

Through the process of analyzing as well as examining mass density for GDY against actual data, researchers were able to determine the question of whether the mechanism of intercalation and adsorption was primary component of lithium storage. This was found to be entirely consistent with the theoretical findings. For LIBs, GDY powder has been used as a negative electrode [173]. Such LIBs have a high capacity, strong cycle stability, as well as a remarkable coulombic efficiency. Reversible capacity of 552 $mAhg^{-1}$ was maintained for 200 cycles at 50 mAg^{-1} and the specific area increased with the chemical resistance of GDYThe porous structure of GDY, together with its low lithium-ion diffusion resistance, allows for fast passage of ions and electrons over its surface, making it suitable for Li storage, in particular under high charge–discharge rates.

There have been numerous attempts to develop GDY-based LIBs with efficient power density in addition to specific capacity. The use of N-doping to improve the efficiency of GDY-based electrode materials was found to be a successful strategy. Zhang *et al* made N-GDY films through annealing GDY in an ammonia environment as well as using them as electrodes to test the characteristics of LIBs, as shown in figure 7.24(c) [72]. The authors observed that perhaps the presence of N atoms in N-GDY resulted in increased heteroatom defects as well as effective spots. As a result, the electrochemical performance was enhanced. Figure 7.24(d) shows the stability following 1000 cycles, and these N-GDY-based electrodes displayed a specific capacitance of around 510 mAh g^{-1} at a current density of 2 A g^{-1}.

7.3.2.2.2 Na-ion batteries

Different metal ion batteries, aside from LIBs, are gaining some traction and becoming viable options for upcoming energy storage technologies. Considering that sodium is chemically the closest to lithium among the metal elements found on our planet, SIBs are an excellent replacement for LIBs. Due to its intriguing features, GDY can play an important role in the production of SIBs [28, 174–179]. The use of GDY-based nanomaterials as anode in SIBs has been verified theoretically and empirically. A typical SIB is shown in figure 7.25(a).

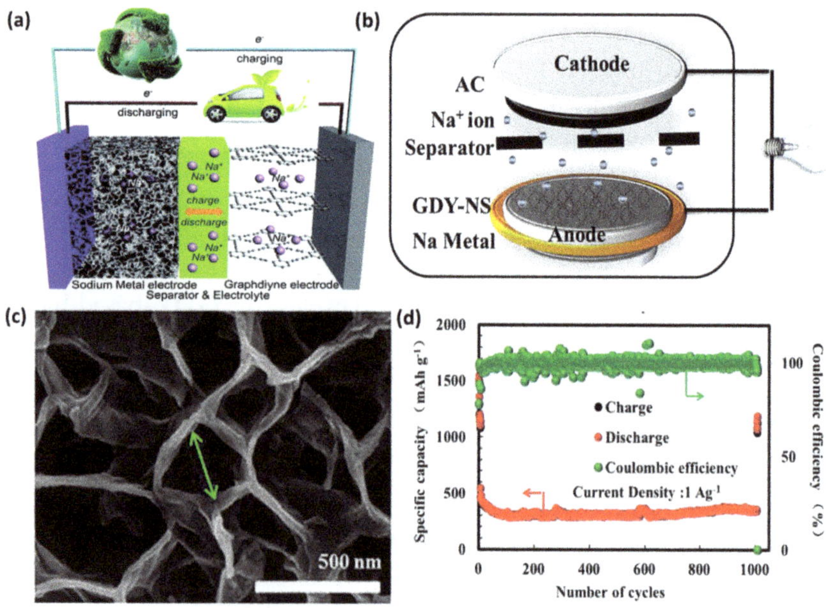

Figure 7.25. Schematic representation of (a) a Na-ion battery including GDY. (Reproduced with permission from [175]. Copyright 2017 The Royal Society of Chemistry.) (b) Schematic showing the use of a GDY Na-ion battery. (c) FESEM image of well grown GDY nanosheets over Cu. (d) Performance of a Na-ion battery at a current density of 1 A g^{-1}. (Reproduced with permission from [174]. Copyright 2017 American Chemical Society.)

The Lv group performed DFT simulations that explored the interaction of Na with single-layered as well as bulk GDY, and they found that inserting Na into GDY results in a stable structure of NaC_3, allowing for the highest efficient Na storage [177]. The group found fast transportation of Na by a reduced energy barrier to almost 0.4 eV. Farokh Niaei *et al* calculated the maximum capacity not only of a single GDY layer ($NaC_{2.57}$, corresponding 497 mAh g^{-1}) but also bulk GDY ($NaC_{5.14}$, equivalent to 316 mAh g^{-1}) for Na atoms employing an analogous approach [178].

Lan and co-workers [175] tested the use of bulk GDY powder as well as a 3D design of GDY nanosheets as SIB anodes, as shown in figure 7.25(b), and that fabricated by Wang *et al* is depicted in figure 7.25(c) [174]. Upon merging the results from the above groups the researchers had reasonable capacities of 211 mAh g^{-1} at 100 mA g^{-1} following 1000 cycles in addition 405 mAh g^{-1} at 1 A g^{-1} for 1000 cycles. Further, GDY powder provides 98.2% of capacity preservation, as shown in figure 7.25(d) [175].

The heteroatom-doping process has been used to manufacture new GDY-based SIBs with the goal of improving efficiency. HsGDY, for example, may be used as a primary anode in SIBs [105] and had a specific capacitance of 650 mAh g^{-1} at 100 mA g^{-1}, identical to Cl-GDY-based LIBs [176]. Again, the work presented by Li's group shows the production of a boron-doped GDY (BGDY), which has good electrochemical performances for SIBs, both theoretically and practically, because of its peculiar bonding, superior conductivity, and comparatively low band gap. In addition to LIBs and SIBs, recent research has shown that sulfide-doped GDY (SGDY) based cathodes which were utilized as Li–S as well as Mg–S batteries exhibited impressive electrochemical performance [179].

7.3.3 Other electronic device applications

At 300 K, the hole as well as electron mobilities for GDYs containing a single layer are anticipated to be about 2×10^4 and 2×10^5 cm^2 V^{-1} s^{-1}, respectively [180]. At ambient temperature, the overall electron mobility for GDY nanoribbons (NRs) appears to be greater than the hole mobility, which is around 10^5 cm^2 V^{-1} s^{-1}. The group further revealed that increasing the width of the GDY-NR can boost its charge mobility. Furthermore, GDY-NRs featuring an armchair edge have more mobility than GDY-NRs with a zigzag edge. Deformation potential theory has produced analogous conclusions concerning the mobility of GDY-NRs [181]. The application of GDY in different electronic devices is discussed in the following sections.

7.3.3.1 Transistors

GDY has advantages in direct use in various photoelectric devices since it is an inherent semiconductor with no extra band gap modulation. GDY has displayed an effective role in field effect transistor (FETs), according to both theoretical and experimental evidence [27, 86, 182]. GDY–metal (i.e. aluminum, silver, copper, gold, iridium, platinum, nickel, and palladium) connections were explored theoretically by the Yang and Lu group [182], with the creation of ohmic or quasi-ohmic contacts in the middle of GDY along with aluminum, silver, copper, and even Schottky contact involving GDY, gold, iridium, platinum, nickel, and palladium. A GDY-centered FET using Al electrodes, a gated two-probe model, resulted in a high on–off ratio of 10^4 coupled with a massive on-state current (1.3×10^4 mA mm^{-1}). Subsequently, FETs have been made using a substantial patterned GDY film which had an efficient field effect mobility around 100 cm^2 V^{-1} s^{-1} and left room for additional device improving efficiency by tweaking the manufacturing and testing processes [27]. GDY was combined with conjugated *p*-*o*-fluoro-*p*-alkoxyphenyl-substituted polymers to create a new hybrid semiconductor, GDY/PFC, where an elevated FET was built and examined [86]. The GDY/PFC-based devices had a significantly better on/off ratio as well as threshold voltage than the pristine device. Most crucially, the overall mobility of GDY/PFC films was two orders of magnitude greater than for non-GDY films. All of this indicates that using GDY might be a successful technique for improving organic field effect transistors (OFET) effectiveness.

Li's group synthesized heterostructure nanowires comprising GDY/CuS core/shell formation, demonstrating flawless diode behavior with a high rectification ratio of (238.08 at 0.5 V through heterojunction area ~0.0162 μm^2) [183]. The interfacial dimensions of the GDY/CuS heterostructure NWs were modified effectively by varying the deposition environments. In addition, an increasing heterojunction area led to a decrease in the rectification ratio, representing a high dependence on heterojunction interfacial area for diode efficiency. Moreover, this kind of report suggests a new approach enabling size-controlled performance characteristic modulation of diverse 1D nanomaterials.

7.3.3.2 Sensors

Again, GDY with its different characteristics was found to show promise as one of the best candidates to work as both humidity and biosensors. The current understanding of the function of GDY in such sensors is discussed in the following subsections.

7.3.3.2.1 Humidity sensors

Humidity sensors measure and track the quantity of water in the air. Such sensors are employed extensively in a variety of sectors, including semiconductors, biomedical devices, textiles, food processing, pharmaceuticals, meteorology, microelectronics, agriculture, structural health monitoring, and environmental monitoring [184]. In the area of GDYO, Yan *et al* achieved significant advances [185]. They observed that even when the alkyne bonds within GDY partly corroded generating O-holding moieties, the GDYO had a reaction time of just ~ 7 ms, which is unparalleled. This corresponds to a three-fold increase in speed over GO of identical thickness as well as O/C ratio. GDYO has more acetylenic bonds than GO, so it can bind water molecules more quickly. The research revealed GDYO's remarkable low volume sensing capabilities and promising application possibilities, which also provided further information regarding the structural relationship with 2D carbon materials.

7.3.3.2.2 Biosensors

Using the van der Waals force as well as stacking connections involving GDY/GDYO along with nucleobases, GDY combined with GDYO was first used as a nanoquencher in the fluorescence detection of biomolecules in 2016, as shown in figure 7.26(a) [186].

Pravin *et al* discovered that very few GDY nanosheets had a strong affinity as well as quenching ability for single and double-stranded DNA, as shown in figure 7.26(b) [17], as well as an impressive detection capability of 25×10^{-12} m, an elevated signal ratio, and satisfactory selectivity, demonstrating the biosensor potential of GDY. In a subsequent theoretical investigation, however, the amino acids recognized GDY with high sensitivity.

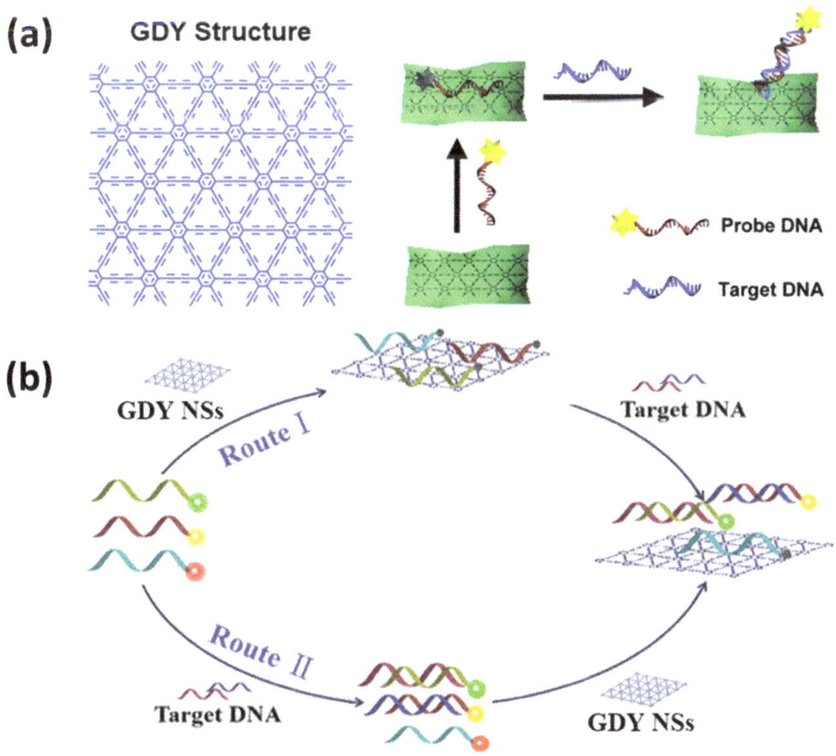

Figure 7.26. (a) A diagram of GD structure with a GD based fluorometric DNA test. (Reproduced with permission from [186]. Copyright 2016 The Royal Society of Chemistry.) (b) Plan intended for multiplexed GD-based DNA detection via several channels. Route I: Initially GDY, NSs coupled dye labeled ssDNA with variable fluorescence to produce dyes. Formation of dsDNA in the presence of target DNA (T1: red; T2: blue) triggers the removal of dye-labeled DNA off surfaces of GD nanosheets (NSs), resulting in fluorescence retrieval and completing recognition. (Reproduced with permission from [17]. Copyright 2017 Wiley.)

7.3.3.3 Electronic detectors

Exceptional performance in a variety of optoelectronic devices has been demonstrated by GDY. ZnO nanoparticles were modified using *n*-propylamine in order to operate as a platform created for profitable innovation of GDY. This modification resulted in GDY:ZnO heterostructure that remained synthesized along with integration through an UV photodetector [84]. As well as a rapid rising decay time of 6.1–2.1 s, a significant photoresponse (1260 A W^{-1}) was detected, which was related to the enhancement of carrier exchange capacity. Similarly, another infrared photodetector erected on the top of heterojunction was examined for minimizing the BE of excitons by combining high standard semiconducting s-single-walled nanotubes (s-SWNTs) using GDY [187]. As an infrared detector it consistently achieved an ideal responsivity pf about 0.4 mA W^{-1} plus a detectivity of 5×10^6 with a reaction time of less than 1 ms in comparison to the s-SWNTs based device. There was a high on/off ratio (above 10^5) and strong carrier mobility (approximately 25 cm^2 V^{-1} s^{-1}).

7.3.4 Magnetism applications

Carbon materials exhibiting magnetic properties have gained a lot of research attention because of their usefulness in implementations as light nonmetallic magnets [53, 188–192]. The inherent magnetism in freshly manufactured GDY powder has been investigated in practice. The impact of N-doping on the magnetic properties of powdered GDY has been studied [193].

Figures 7.27(a) and (b) show the typical temperature dependent susceptibility for GDY and N-GDY. At a temperature of 2 K, GDY has a noticeable saturating ΔM value of 0.51 B g^{-1}, but following N-doping, the saturation moment increases to 0.96 emu g^{-1}, as shown in figure 7.27(c). The overall doping of N atoms, including its significance to localized magnetic moment, has been studied extensively using the peak-differentiation version of the spin-polarized DFT calculation (figure 7.27(d)). An important local ΔM value of 0.98 μ_B when asymmetric pyridinic-N was substituted is shown in figure 7.27(e). Studies have revealed that GDY replaced with a N atom on an alkyne unit with two proportional pyridinic-N atoms has a better chance of improving magnetism than other N-doping locations.

Further, Du and Tang with others employed a cross-coupling process to produced GDY, which has a spin-half paramagnetism as shown in figures 7.27(g) and (i) [194]. During annealing, GDY displays a significant increase of spin density, with antiferromagnetism developing after annealing at 600 °C, as displayed in figure 7.27(j). This same hydroxyl group only on the chain of the GDY sheet has been interpreted as (i) a significant magnetic sources and (ii) a high barrier energy of 1.73 eV for trying to migrate from the ring location to the chain location, that can also protect the hydroxyl groups from agglomerating and may tend to support antiferromagnetism inside the heat treated GDY sheet.

7.3.5 Biological applications

Li *et al* presented, for the first time, GDY nanosheets containing PEGylating for use in a photothermo-acoustic wave nanotransducer for *in vivo* photoacoustic imaging

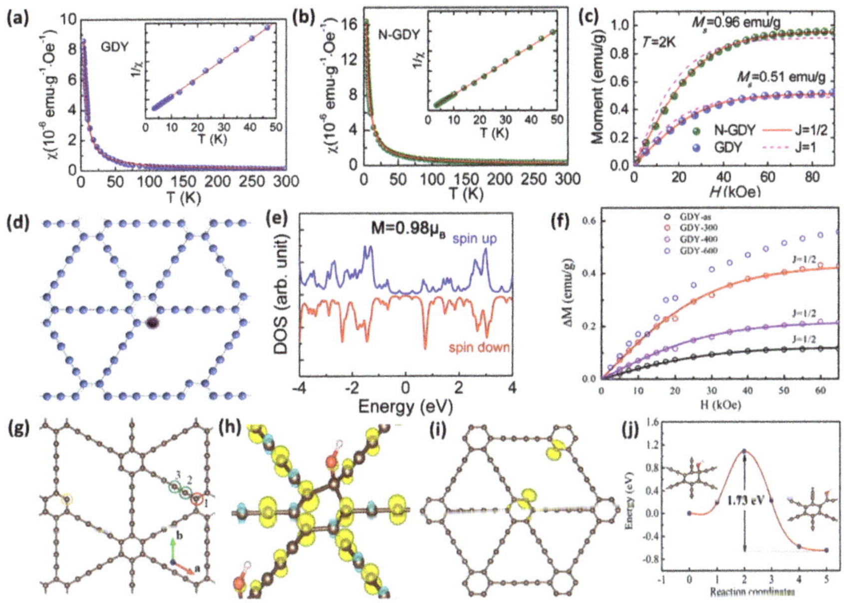

Figure 7.27. χ–T was evaluated for (a) GDY and (b) N-GDY at temperatures ranging from 2 to 300 K. (c) At T = 2 K, magnetization plots comprising N-GDY and GDY were produced. (d) N-doped GDY graphical representations with (e) spin-resolved DOS calculation. (Reproduced from [193]. CC BY 4.0.) (f) At 2 K, ΔM is subtracted from the linear diamagnetic background of GDY. (g) GDY single-vacancy model that engages hydroxyl and carbonyl sitting on various locations using red shaded circle along with N-substituting C on colored circle sites. GDY spin-charge density division in (h) vacancy and (i) hydroxyl absorption. (j) Hydroxyl traveling through the GDY toward the attachment portion faces an energy barrier. (Reproduced with permission from [194]. Copyright 2017 AIP Publishing.)

(PAI) as well as photothermal therapy during cancer treatment [195] (figures 7.28(a)–(c)) to investigate this highly improved theranostic platform in cancer treatment. The GDY-PEG had an excellent photoacoustic response along with photothermal performance, including a significant photothermal conversion efficiency of 42%, which can be attributed to GDY- PEG's extinction coefficient in the near infrared regions. Even after laser irradiation, the GDY-PEG material remained stable. Furthermore, PAI combined with photothermal ablation for cancer cells in live mice utilizing GDY-PEG produced a beneficial outcome.

The model drug doxorubicin (DOX) was used to create an additional GDY-based drug delivery system. DOX was discovered to be capable of loading effectively on GDY [22]. In live mice, mixed photothermal/chemotherapy testing was performed with such GDY/DOX using 808 nm irradiation and strong photothermal capacity, which exhibited a significantly improved cancer reduction rate (figure 7.28(d)).

Zhao and colleagues created GDY/bovine serum albumin nanoparticles (GDY-BSA NPs) targeting free radical removal in addition to radio shielding [196]. Through animal research cell models, the GDY structure, shown in figure 7.28(e), changed to GDY-BSA NPs and shielded bone marrow DNA satisfactorily from

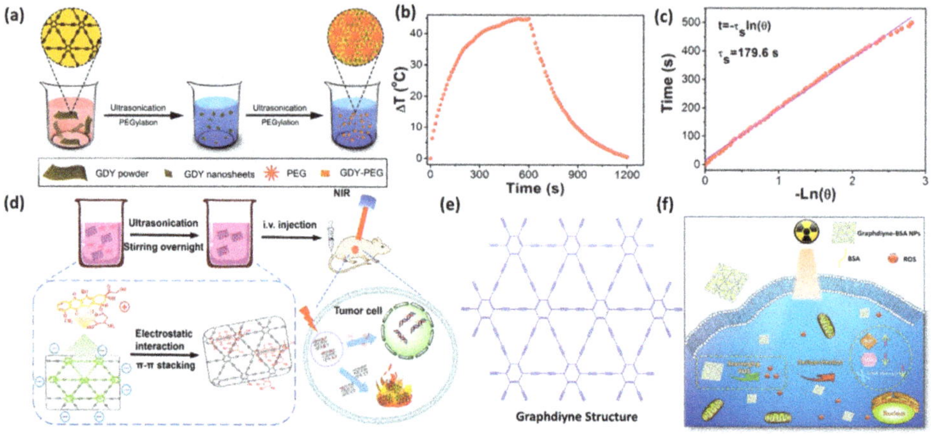

Figure 7.28. (a) A fabrication procedure producing GDY nanosheets with pegylation (GDY-PEG) is depicted in the schematic diagram. (b) Photothermal reaction of GDY-PEG solution exposed to an 808 laser aimed for 5 min before being turned off. (c) Time shown against the negative natural logarithm of temperature which is the driving force obtained during the cooling phase. (Reproduced with permission from [195]. Copyright 2017 American Chemical Society.) (d) Depiction of GDY/DOX and cancer combination therapy. (Reproduced with permission from [22]. Copyright 2018 American Chemical Society.) (e) and (f) Diagrammatic representation of GDY nanoparticles used for radiation shielding through treatment of cancer. (Reproduced with permission from [196]. Copyright 2018 American Chemical Society.)

radiation-induced injury and supported the healing process of superoxide dismutase as well as malondialdehyde (two types of critical determinants of radiation-induced injury) to conventional standards, shown in the schematic of figure 7.28(f). The GDY-BSA NPs also had good biocompatibility and no toxicity, in addition to their exceptional efficiency in free radical elimination.

7.3.6 Environmental applications

Issues involving environmental remediation, which comprises the removal of contaminants from different environmental media, have found to be amenable to the use of GDY. In the following sections, the use of GDY in addressing two major environmental issues is discussed.

7.3.6.1 Gas separation

The potential for the isolation of mixed gases lies in the two-dimensional porous structure of GDY. The overall pore size of the triangle pores is determined by the amount of acetylenic connections, enabling preferential penetration of molecules of various sizes. Heteroatom-substituted GDY has varied pore diameters, and the origin of selection is more diversified since the electronic structure varies. Also there is a degree of selection for molecules of comparable size. Researchers have addressed a myriad of factors based on this concept, as can be seen in figure 7.29(a).

From the standpoint of simulated results, Jiao *et al* investigate possibility of GDY for extracting H_2 from methane (CH_4) and carbon monoxide (CO) [197]. They

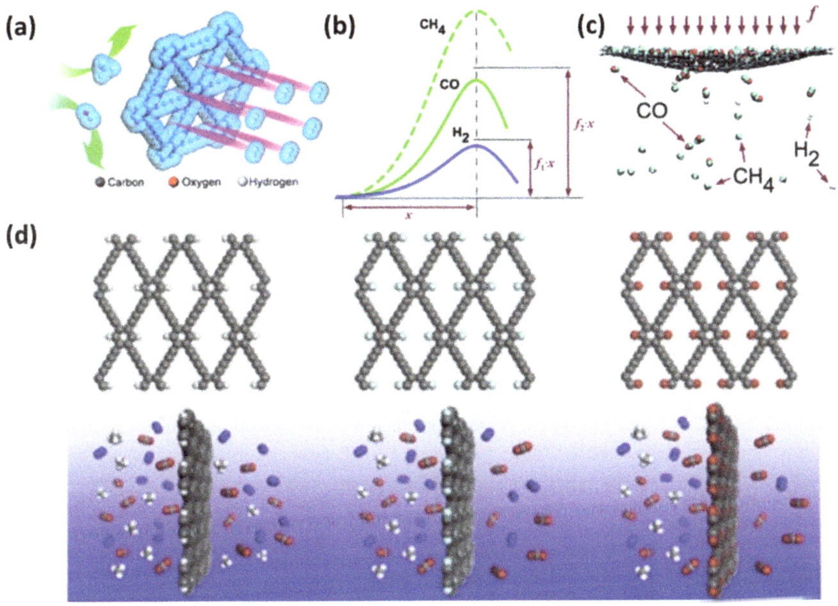

Figure 7.29. (a) Dissociation of H_2 via GDY depicted schematically. (b) Force applied variations at specific temperature (300 K). Using more force to cross the GDY membrane can lower the amount of energy required, enabling selected filtering at key force levels. (Reproduced with permission from [197]. Copyright 2011 The Royal Society of Chemistry.) (c) Simulation snapshot at $T = 300$ K and $f \approx 70$ pN molecular force, with H_2, CO, and CH_4 passing through the membrane efficiently. The hydrogen travels across the membrane. (Reproduced with permission from [198]. Copyright 2012 The Royal Society of Chemistry.) (d) GDY membranes treated with H, F, and O could distinguish $CO_2/N_2/CH_4$ combinations. (Reproduced with permission from [23]. Copyright 2017 Elsevier.)

computed the efficiency of the relatively close adsorption–diffusion mechanism of H_2 and discovered that it had the minimum energy but also the best permeability. Moreover, the researchers assessed that GDY has a 10^4-fold quicker H_2 penetration rate than a porous graphene (GR) structure, on applying the Arrhenius equation.

Cranford *et al* employed molecular dynamics to study this situation [198]. The authors investigated the preferential diffusion characteristics of GDY to hydrogen, methane, and carbon monoxide gas combinations using complete molecular dynamics simulations, as shown in figures 7.29(b) and (c). Furthermore, the group observed that the gas must pass through the filter at a specified pressure. Hydrogen required the least amount of air pressure of the three gases; only H_2 could permeate the GDY coating at atmospheric pressure. Guo *et al* used DFT and molecular dynamics simulations to investigate selective dissociation of $CO_2/N_2/CH_4$ via replacing GDY (H-GDY, O-GDY, and F-GDY) as shown in figure 7.29(d) [23]. Over a wide range of temperatures, researchers demonstrated that F-GDYF as well as O-GDY films can successfully distinguish N_2 and CH_4. Furthermore, at temperatures below 300 K, O-GDY may successfully separate CO_2/N_2 gas mixtures. Theoretical investigations show the potential applications of GR-like acetylene films in gas separation, and this is extremely useful in support of future research.

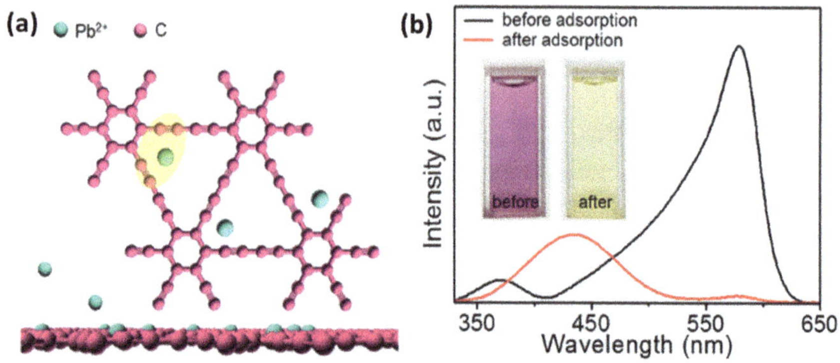

Figure 7.30. (a) During adsorption, an aqueous solution often includes lead ions, resulting towards the occasional appearance of an orange colored Xylenol molecule. (b) Absorption spectra of Xylenol orange onto a lead ion containing aqueous phase. Inset: before and after adsorption images of water solution which contains lead ions with additional xylenol orange. (Reproduced with permission from [206]. Copyright 2017 Wiley.)

7.3.6.2 Water purification

Self-cleaning, anti-fog, oil–water separation, and energy-related areas all benefit from superhydrophobic substances [199–202]. Both the investigation as well as design of such elements has received more attention in recent years. Established superhydrophobic materials have already been developed and produced by combining very rough surface features with low-surface-energy coatings [203, 204]. GDY, a novel candidate within the 2D carbon group, has been employed extensively in a variety of industries due to its exceptional physicochemical characteristics [82, 84, 100, 137, 186, 195, 205]. GDY-containing regular porous structures, for example, may be made on Cu substrates using a wet chemical technique [10].

Zhang's group was intrigued by the above facts and developed a strong superhydrophobic foam material via synthesizing 3D honeycomb-like GDY on stiff Cu foam *in situ* [16]. As-prepared GDY-based foam displays superhydrophobicity (static contact angle ~160.11°) as well as performing excellent separation of diverse oil–water combinations after being coated with a limited surface energy PDMS layer.

Furthermore, as illustrated in figures 7.30(a) and (b), GDY-based foam has been used for cleaning lead-ion-polluted water. The pristine GDY filter may well capture lead ions in water effectively, exhibiting elimination an capacity of 99.6%, as a result of impending difficult affiliation of acetylenic linkages in GDY using metal ions [206].

7.4 Conclusion

Following the efficient fabrication of GDY for the first time, many experts have been concentrating considerable efforts toward studying this novel planar layered material. Multiple operational models had anticipated a direct intrinsic band gap in GDYs as well as a Dirac cone structure, which would be usually derived from overlapping carbon $2p_z$ orbitals which would result from the inhomogeneous bonding of carbon atoms with various hybridizations. The predicted optical band gap is around 0.44–1.47 eV. GDYs can be employed as promising materials in a variety of applications due to their unusual structure and electrical characteristics. BN-doping, inducing strain, hydrogenation, and the creation of NRs are approaches that have been used to change the GDY band gap. A variety of GDY morphologies have been generated successfully, comprising NWs, NTs, nanowalls, and NSs, including structured striped arrays. GDY research has not been confined to speculation; the aforementioned practical production has comprehensive potential applications, such as energy harvesting and storage, electrical devices, optoelectronic devices, magnetic materials, and biological and other environmental applications. GDY, the novel carbon allotrope, has appealing structural as well as functional properties, as well as better promise towards environmental as well as energy sustainability beyond conventional carbon compounds.

7.5 Future prospects

For novel materials such as GDY, the creation of cheaper and much more effective methods is critical. Significantly, the interlayer stacking for 2D materials has a significant impact on its characteristics. As a consequence, a single crystalline sample is needed to analyze the basic fundamental properties of a novel material such as GDY. Moreover, it is critical to establish a reliable process for producing high-quality single-layer GDY. Designing appropriate procedures to exfoliate as-prepared GDY powders is yet another good option to obtain one- or few-layer GDY materials.

A second critical approach should be toward establishing clean transfer techniques for GDY. GDY film production, including amorphous impurities, may be guaranteed using the PMMA-assisted transfer approach. However, it is unknown whether hydrofluoric acid has any effect on GDY's structure. As a result, developing a nondestructive as well as clean transformation technique for GDY manufacturing is critical to the field's advancement.

Finally, finding a simple and adaptable approach to rapidly assess the quality in addition to arrangement of GDY is critical. Currently, the most common methods for characterizing GDY are typical 2D material characterization methods, that do not have its particular unique properties. Furthermore, the diverse growing procedures may produce samples of varying quality. As a result, it is critical to devise a method for properly determining the polymerization degree of GDY samples.

References

[1] Georgakilas V, Perman J A, Tucek J and Zboril R 2015 Broad family of carbon nanoallotropes: classification, chemistry, and applications of fullerenes, carbon dots, nanotubes, graphene, nanodiamonds, and combined superstructures *Chem. Rev.* **115** 4744–822

[2] Kroto H W, Heath J R, O'Brien S C, Curl R F and Smalley R E 1985 C_{60}: buckminsterfullerene *Nature* **318** 162–3

[3] Iijima S 1991 Helical microtubules of graphitic carbon *Nature* **354** 56–8

[4] Novoselov K S, Geim A K, Morozov S V, Jiang D, Zhang Y, Dubonos S V, Grigorieva I V and Firsov A A 2004 Electric field effect in atomically thin carbon films *Science* **306** 666–9
[5] Dai H 2002 Carbon nanotubes: synthesis, integration, and properties *Acc. Chem. Res.* **35** 1035–44
[6] Zhu Y, Murali S, Cai W, Li X, Suk J W, Potts J R and Ruoff R S 2010 Graphene and graphene oxide: synthesis, properties, and applications *Adv. Mater.* **22** 3906–24
[7] Zhang J *et al* 2016 Carbon science in 2016: status, challenges and perspectives *Carbon* **98** 708–32
[8] Ivanovskii A L 2013 Graphynes and graphdyines *Prog. Solid State Chem.* **41** 1–19
[9] Wang N, He J, Wang K, Zhao Y, Jiu T, Huang C and Li Y 2019 GDY-based materials: preparation and application for electrochemical energy storage *Adv. Mater.* **31** 1803202
[10] Li G, Li Y, Liu H, Guo Y, Li Y and Zhu D 2010 Architecture of GDY nanoscale films *Chem. Commun.* **46** 3256–8
[11] Li Y, Xu L, Liu H and Li Y 2014 GDY and graphyne: from theoretical predictions to practical construction *Chem. Soc. Rev.* **43** 2572–86
[12] Jia Z, Li Y, Zuo Z, Liu H, Huang C and Li Y 2017 Synthesis and properties of 2D carbon —GDY *Acc. Chem. Res.* **50** 2470–8
[13] Haley M M, Brand S C and Pak J J 1997 Carbon networks based on dehydrobenzoannulenes: synthesis of GDY substructures *Angew. Chem. Int. Ed. Engl.* **36** 836–8
[14] Long M, Tang L, Wang D, Li Y and Shuai Z 2011 Electronic structure and carrier mobility in GDY sheet and nanoribbons: theoretical predictions *ACS Nano* **5** 2593–600
[15] Yang N, Liu Y, Wen H, Tang Z, Zhao H, Li Y and Wang D 2013 Photocatalytic properties of GDY and graphene modified TiO_2: from theory to experiment *ACS Nano* **7** 1504–12
[16] Gao X, Zhou J, Du R, Xie Z, Deng S, Liu R, Liu Z and Zhang J 2016 Robust superhydrophobic foam: a GDY-based hierarchical architecture for oil/water separation *Adv. Mater.* **28** 168–73
[17] Parvin N *et al* 2017 Few-layer GDY nanosheets applied for multiplexed real-time DNA detection *Adv. Mater.* **29** 1606755
[18] Xue Y, Huang B, Yi Y, Guo Y, Zuo Z, Li Y, Jia Z, Liu H and Li Y 2018 Anchoring zero valence single atoms of nickel and iron on GDY for hydrogen evolution *Nat. Commun.* **9** 1460
[19] Lu C *et al* 2018 High-performance GDY-based electrochemical actuators *Nat. Commun.* **9** 752
[20] Li L, Zuo Z, Shang H, Wang F and Li Y 2018 *In situ* constructing 3D GDY as all-carbon binder for high-performance silicon anode *Nano Energy* **53** 135–43
[21] Shang H, Zuo Z, Yu L, Wang F, He F and Li Y 2018 Low-temperature growth of all-carbon GDY on a silicon anode for high-performance lithium-ion batteries *Adv. Mater.* **30** 1801459
[22] Jin J *et al* 2018 Graphdiyne nanosheet-based drug delivery platform for photothermal/chemotherapy combination treatment of cancer *ACS Appl. Mater. Interfaces.* **10** 8436–42
[23] Zhao L, Sang P, Guo S, Liu X, Li J, Zhu H and Guo W 2017 Promising monolayer membranes for $CO_2/N_2/CH_4$ separation: GDYs modified respectively with hydrogen, fluorine, and oxygen atoms *Appl. Surf. Sci.* **405** 455–64
[24] Ge C, Chen J, Tang S, Du Y and Tang N 2019 Review of the electronic, optical, and magnetic properties of GDY: from theories to experiments *ACS Appl. Mater. Interfaces* **11** 2707–16

[25] Ciesielski A and Samorì P 2014 Graphene via sonication assisted liquid-phase exfoliation *Chem. Soc. Rev.* **43** 381–98

[26] Hernandez Y *et al* 2008 High-yield production of graphene by liquid-phase exfoliation of graphite *Nat. Nanotechnol.* 3 563–8

[27] Qian X, Liu H, Huang C, Chen S, Zhang L, Li Y, Wang J and Li Y 2015 Self-catalyzed growth of large-area nanofilms of two-dimensional carbon *Sci. Rep.* **5** 7756

[28] Zuo Z, Shang H, Chen Y, Li J, Liu H, Li Y and Li Y 2017 A facile approach for GDY preparation under atmosphere for an advanced battery anode *Chem. Commun.* **53** 8074–7

[29] Wang F, Zuo Z, Shang H, Zhao Y and Li Y 2019 Ultrafastly interweaving GDY nanochain on arbitrary substrates and its performance as a supercapacitor electrode *ACS Appl. Mater. Interfaces* **11** 2599–607

[30] Diederich F and Rubin Y 1992 Synthetic approaches toward molecular and polymeric carbon allotropes *Angew. Chem. Int. Ed. Engl.* **31** 1101–23

[31] Huynh C and Linstrumelle G 1988 A short route to dehydro[12]annulenes *Tetrahedron* **44** 6337–44

[32] Siemsen P, Livingston R C and Diederich F 2000 Acetylenic coupling: a powerful tool in molecular construction *Angew. Chem. Int. Ed.* **39** 2632–57

[33] Glaser C 1869 Beiträge zur Kenntniss des Acetenylbenzols *Ber. Dtsch. Chem. Ges.* **2** 422–4

[34] Fomina L, Vazquez B, Tkatchouk E and Fomine S 2002 The Glaser reaction mechanism. A DFT study *Tetrahedron* **58** 6741–7

[35] Eglinton G and Galbraith A R 1959 182. Macrocyclic acetylenic compounds. Part I. Cyclotetradeca-1 :3-diyne and related compounds *J. Chem. Soc.* 889–96

[36] Nishihara Y, Ikegashira K, Hirabayashi K, Ando J, Mori A and Hiyama T 2000 Coupling reactions of alkynylsilanes mediated by a Cu(I) salt: novel syntheses of conjugate diynes and disubstituted ethynes *J. Org. Chem.* **65** 1780–7

[37] Xin G *et al* 2022 Ultrathin GDY film on graphene through solution-phase van der Waals epitaxy *Sci. Adv.* **4** eaat6378

[38] Zhou J *et al* 2015 Synthesis of GDY nanowalls using acetylenic coupling reaction *J. Am. Chem. Soc.* **137** 7596–9

[39] Wang S-S, Liu H-B, Kan X-N, Wang L, Chen Y-H, Su B, Li Y-L and Jiang L 2017 Superlyophilicity-facilitated synthesis reaction at the microscale: ordered GDY stripe arrays *Small* **13** 1602265

[40] Li G, Li Y, Qian X, Liu H, Lin H, Chen N and Li Y 2011 Construction of tubular molecule aggregations of GDY for highly efficient field emission *J. Phys. Chem.* C **115** 2611–5

[41] Gao X *et al* 2017 Direct synthesis of GDY nanowalls on arbitrary substrates and its application for photoelectrochemical water splitting cell *Adv. Mater.* **29** 1605308

[42] Zhao F, Wang N, Zhang M, Sápi A, Yu J, Li X, Cui W, Yang Z and Huang C 2018 *In situ* growth of GDY on arbitrary substrates with a controlled-release method *Chem. Commun.* **54** 6004–7

[43] Matsuoka R, Sakamoto R, Hoshiko K, Sasaki S, Masunaga H, Nagashio K and Nishihara H 2017 Crystalline GDY nanosheets produced at a gas/liquid or liquid/liquid interface *J. Am. Chem. Soc.* **139** 3145–52

[44] Zhou J, Xie Z, Liu R, Gao X, Li J, Xiong Y, Tong L, Zhang J and Liu Z 2019 Synthesis of ultrathin GDY film using a surface template *ACS Appl. Mater. Interfaces* **11** 2632–7

[45] Grill L, Dyer M, Lafferentz L, Persson M, Peters M V and Hecht S 2007 Nano-architectures by covalent assembly of molecular building blocks *Nat. Nanotechnol.* **2** 687–91

[46] Federico P D F R 2009 Extending polymer conjugation into the second dimension *Science* **323** 216–7

[47] Cai J *et al* 2010 Atomically precise bottom-up fabrication of graphene nanoribbons *Nature* **466** 470–3

[48] Sun Q, Cai L, Ma H, Yuan C and Xu W 2016 Dehalogenative homocoupling of terminal alkynyl bromides on Au(111): incorporation of acetylenic scaffolding into surface nanostructures *ACS Nano* **10** 7023–30

[49] Liu R, Gao X, Zhou J, Xu H, Li Z, Zhang S, Xie Z, Zhang J and Liu Z 2017 Chemical vapor deposition growth of linked carbon monolayers with acetylenic scaffoldings on silver foil *Adv. Mater.* **29** 1604665

[50] Cirera B, Zhang Y-Q, Klyatskaya S, Ruben M, Klappenberger F and Barth J V 2013 2 D self-assembly and catalytic homo-coupling of the terminal alkyne 1,4-bis(3,5-diethynylphenyl)butadiyne-1,3 on Ag(111) *Chem. Cat. Chem.* **5** 3281–8

[51] Zhang Y-Q *et al* 2012 Homo-coupling of terminal alkynes on a noble metal surface *Nat. Commun.* **3** 1286

[52] Cirera B, Zhang Y-Q, Björk J, Klyatskaya S, Chen Z, Ruben M, Barth J V and Klappenberger F 2014 Synthesis of extended GDY wires by vicinal surface templating *Nano Lett.* **14** 1891–7

[53] He J, Ma S Y, Zhou P, Zhang C X, He C and Sun L Z 2012 Magnetic properties of single transition-metal atom absorbed GDY and graphyne sheet from DFT+U calculations *J. Phys. Chem.* C **116** 26313–21

[54] Lu Z, Li S, Lv P, He C, Ma D and Yang Z 2016 First principles study on the interfacial properties of NM/GDY (NM=Pd, Pt, Rh and Ir): the implications for NM growing *Appl. Surf. Sci.* **360** 1–7

[55] Dai L, Xue Y, Qu L, Choi H-J and Baek J-B 2015 Metal-free catalysts for oxygen reduction reaction *Chem. Rev.* **115** 4823–92

[56] Mohajeri A and Shahsavar A 2017 Tailoring the optoelectronic properties of graphyne and GDY: nitrogen/sulfur dual doping versus oxygen containing functional groups *J. Mater. Sci.* **52** 5366–79

[57] Bu H, Zhao M, Zhang H, Wang X, Xi Y and Wang Z 2012 Isoelectronic doping of GDY with boron and nitrogen: stable configurations and band gap modification *J. Phys. Chem.* A **116** 3934–9

[58] Tan J, He X and Zhao M 2012 First-principles study of hydrogenated graphyne and its family: stable configurations and electronic structures *Diam. Relat. Mater.* **29** 42–7

[59] Lu R, Rao D, Meng Z, Zhang X, Xu G, Liu Y, Kan E, Xiao C and Deng K 2013 Boron-substituted graphyne as a versatile material with high storage capacities of Li and H_2: a multiscale theoretical study *Phys. Chem. Chem. Phys.* **15** 16120–6

[60] Beheshtian J, Peyghan A A, Bagheri Z and Tabar M B 2014 Density-functional calculations of HCN adsorption on the pristine and Si-doped graphynes *Struct. Chem.* **25** 1–7

[61] Kang B, Shi H, Wu S, Zhao W, Ai H and Lee J Y 2017 Revealing the importance of nitrogen doping site in enhancing the oxygen reduction reaction on β-graphyne *Carbon* **123** 415–20

[62] Guo S, Yan H, Wu F, Zhao L, Yu P, Liu H, Li Y and Mao L 2017 GDY as electrode material: tuning electronic state and surface chemistry for improved electrode reactivity *Anal. Chem.* **89** 13008–15

[63] Enyashin A N and Ivanovskii A L 2013 Fluorographynes: stability, structural and electronic properties *Superlattices Microstruct.* **55** 75–82
[64] Koo J, Huang B, Lee H, Kim G, Nam J, Kwon Y and Lee H 2014 Tailoring the electronic band gap of graphyne *J. Phys. Chem.* C **118** 2463–8
[65] Yun J, Zhang Z, Yan J, Zhao W and Xu M 2015 First-principles study of B or Al-doping effect on the structural, electronic structure and magnetic properties of γ-graphyne *Comput. Mater. Sci.* **108** 147–52
[66] Liu J, Xin Z, Yan X, Li H and Yu M 2017 Structural, phononic and electronic properties of Ge-doped γ-graphynes: a first-principles study *Solid State Commun.* **258** 38–44
[67] Ma D W, Li T, Wang Q, Yang G, He C, Ma B and Lu Z 2015 Graphyne as a promising substrate for the noble-metal single-atom catalysts *Carbon* **95** 756–65
[68] Yun J, Zhang Y, Xu M, Yan J, Zhao W and Zhang Z 2017 DFT study of the effect of BN pair doping on the electronic and optical properties of graphyne nanosheets *J. Mater. Sci.* **52** 10294–307
[69] Zhang P, Ma S and Sun L Z 2016 Hydroxylated graphyne and GDY: first-principles study *Appl. Surf. Sci.* **361** 206–12
[70] Autreto P A S, de Sousa J M and Galvao D S 2014 Site-dependent hydrogenation on GDY *Carbon* **77** 829–34
[71] Lv Q *et al* 2017 Nitrogen-doped porous GDY: a highly efficient metal-free electrocatalyst for oxygen reduction reaction *ACS Appl. Mater. Interfaces* **9** 29744–52
[72] Zhang S, Du H, He J, Huang C, Liu H, Cui G and Li Y 2016 Nitrogen-doped GDY applied for lithium-ion storage *ACS Appl. Mater. Interfaces* **8** 8467–73
[73] Wang N *et al* 2017 Synthesis of chlorine-substituted GDY and applications for lithium-ion storage *Angew. Chem. Int. Ed.* **56** 10740–5
[74] He J, Wang N, Yang Z, Shen X, Wang K, Huang C, Yi Y, Tu Z and Li Y 2018 Fluoride GDY as a free-standing electrode displaying ultra-stable and extraordinary high Li storage performance *Energy Environ. Sci.* **11** 2893–903
[75] Zhuang X, Mao L and Li Y 2017 *In situ* synthesis of a Prussian blue nanoparticles/GDY oxide nanocomposite with high stability and electrocatalytic activity *Electrochem. Commun.* **83** 96–101
[76] Lin Z, Liu G, Zheng Y, Lin Y and Huang Z 2018 Three-dimensional hierarchical mesoporous flower-like TiO_2@GDY with superior electrochemical performances for lithium-ion batteries *J. Mater. Chem.* A **6** 22655–61
[77] Thangavel S, Krishnamoorthy K, Krishnaswamy V, Raju N, Kim S J and Venugopal G 2015 GDY–ZnO nanohybrids as an advanced photocatalytic material *J. Phys. Chem.* C **119** 22057–65
[78] Lv J-X, Zhang Z-M, Wang J, Lu X-L, Zhang W and Lu T-B 2019 *In situ* synthesis of CdS/GDY heterojunction for enhanced photocatalytic activity of hydrogen production *ACS Appl. Mater. Interfaces* **11** 2655–61
[79] Yao Y, Jin Z, Chen Y, Gao Z, Yan J, Liu H, Wang J, Li Y and Liu S 2018 GDY-WS2 2D-nanohybrid electrocatalysts for high-performance hydrogen evolution reaction *Carbon* **129** 228–35
[80] Xue Y, Zuo Z, Li Y, Liu H and Li Y 2017 GDY-supported $NiCo_2S_4$ nanowires: a highly active and stable 3D bifunctional electrode material *Small* **13** 1700936
[81] Zhang X, Zhu M, Chen P, Li Y, Liu H, Li Y and Liu M 2015 Pristine GDY-hybridized photocatalysts using graphene oxide as a dual-functional coupling reagent *Phys. Chem. Chem. Phys.* **17** 1217–25

[82] Si H-Y *et al* 2018 Z-scheme Ag_3PO_4/GDY/g-C_3N_4 composites: enhanced photocatalytic O_2 generation benefiting from dual roles of GDY *Carbon* **132** 598–605

[83] Wang S, Yi L, Halpert J E, Lai X, Liu Y, Cao H, Yu R, Wang D and Li Y 2012 A novel and highly efficient photocatalyst based on P25–GDY nanocomposite *Small* **8** 265–71

[84] Jin Z, Zhou Q, Chen Y, Mao P, Li H, Liu H, Wang J and Li Y 2016 Graphdiyne:ZnO nanocomposites for high-performance UV photodetectors *Adv. Mater.* **28** 3697–702

[85] Xiao J, Shi J, Liu H, Xu Y, Lv S, Luo Y, Li D, Meng Q and Li Y 2015 Efficient $CH_3NH_3PbI_3$ perovskite solar cells based on GDY (GD)-modified P3HT hole-transporting material *Adv. Energy Mater.* **5** 1401943

[86] Cui W, Zhang M, Wang N, He J, Yu J, Long Y, Yan S and Huang C 2017 High-performance field-effect transistor based on novel conjugated p-o-fluoro-p-alkoxyphenyl-substituted polymers by GDY doping *J. Phys. Chem.* C **121** 23300–6

[87] Han Y-Y, Lu X-L, Tang S-F, Yin X-P, Wei Z-W and Lu T-B 2018 Metal-free 2D/2D heterojunction of graphitic carbon nitride/GDY for improving the hole mobility of graphitic carbon nitride *Adv. Energy Mater.* **8** 1702992

[88] Li J, Zhao M, Zhao C, Jian H, Wang N, Yao L, Huang C, Zhao Y and Jiu T 2019 GDY-doped P3CT-K as an efficient hole-transport layer for $MAPbI_3$ perovskite solar cells *ACS Appl. Mater. Interfaces* **11** 2626–31

[89] Yu H, Xue Y, Hui L, Zhang C, Zhao Y, Li Z and Li Y 2018 Controlled growth of MoS_2 nanosheets on 2D N-doped GDY nanolayers for highly associated effects on water reduction *Adv. Funct. Mater.* **28** 1707564

[90] Li Q, Li Y, Chen Y, Wu L, Yang C and Cui X 2018 Synthesis of γ-graphyne by mechanochemistry and its electronic structure *Carbon* **136** 248–54

[91] Gao J, Li J, Chen Y, Zuo Z, Li Y, Liu H and Li Y 2018 Architecture and properties of a novel two-dimensional carbon material-graphtetrayne *Nano Energy* **43** 192–9

[92] Li J *et al* 2017 Architecture of β-GDY-containing thin film using modified Glaser–Hay coupling reaction for enhanced photocatalytic property of TiO_2 *Adv. Mater.* **29** 1700421

[93] Wang L *et al* 2017 Conjugated microporous polymer nanosheets for overall water splitting using visible light *Adv. Mater.* **29** 1702428

[94] Matsuoka R *et al* 2019 Expansion of the GDY family: a triphenylene-cored analogue *ACS Appl. Mater. Interfaces* **11** 2730–3

[95] Zhou W *et al* 2019 Direct synthesis of crystalline GDY analogue based on supramolecular interactions *J. Am. Chem. Soc.* **141** 48–52

[96] Gao H-Y, Wagner H, Zhong D, Franke J-H, Studer A and Fuchs H 2013 Glaser coupling at metal surfaces *Angew. Chem. Int. Ed* **52** 4024–8

[97] Shang H, Zuo Z, Zheng H, Li K, Tu Z, Yi Y, Liu H, Li Y and Li Y 2018 N-doped GDY for high-performance electrochemical electrodes *Nano Energy* **44** 144–54

[98] Li J, Xiong Y, Xie Z, Gao X, Zhou J, Yin C, Tong L, Chen C, Liu Z and Zhang J 2019 Template synthesis of an ultrathin β-GDY-like film using the Eglinton coupling reaction *ACS Appl. Mater. Interfaces* **11** 2734–9

[99] Yang Z *et al* 2019 GDY containing atomically precise N atoms for efficient anchoring of lithium ion *ACS Appl. Mater. Interfaces* **11** 2608–17

[100] Gao X, Ren H, Zhou J, Du R, Yin C, Liu R, Peng H, Tong L, Liu Z and Zhang J 2017 Synthesis of hierarchical GDY-based architecture for efficient solar steam generation *Chem. Mater.* **29** 5777–81

[101] Shang H, Zuo Z, Li L, Wang F, Liu H, Li Y and Li Y 2018 Ultrathin GDY nanosheets grown *in situ* on copper nanowires and their performance as lithium-ion battery anodes *Angew. Chem. Int. Ed.* **57** 774–8

[102] Prabakaran P, Satapathy S, Prasad E and Sankararaman S 2018 Architecting pyrediyne nanowalls with improved inter-molecular interactions, electronic features and transport characteristics *J. Mater. Chem.* C **6** 380–7

[103] Yang Z *et al* 2018 Triazine-GDY: a new nitrogen-carbonous material and its application as an advanced rechargeable battery anode *Carbon* **137** 442–50

[104] Li J, Xu J, Xie Z, Gao X, Zhou J, Xiong Y, Chen C, Zhang J and Liu Z 2018 Diatomite-templated synthesis of freestanding 3D GDY for energy storage and catalysis application *Adv. Mater.* **30** 1800548

[105] He J *et al* 2017 Hydrogen substituted GDY as carbon-rich flexible electrode for lithium and sodium ion batteries *Nat. Commun.* **8** 1172

[106] Kan X, Ban Y, Wu C, Pan Q, Liu H, Song J, Zuo Z, Li Z and Zhao Y 2018 Interfacial synthesis of conjugated two-dimensional N-GDY *ACS Appl. Mater. Interfaces* **10** 53–8

[107] Qi H, Yu P, Wang Y, Han G, Liu H, Yi Y, Li Y and Mao L 2015 GDY oxides as excellent substrate for electroless deposition of Pd clusters with high catalytic activity *J. Am. Chem. Soc.* **137** 5260–3

[108] Shen H, Li Y and Shi Z 2019 A novel GDY-based catalyst for effective hydrogenation reaction *ACS Appl. Mater. Interfaces* **11** 2563–70

[109] Majumder S, Quang N D, Hung N M, Chinh N D, Kim C and Kim D 2021 Deposition of zinc cobaltite nanoparticles onto bismuth vanadate for enhanced photoelectrochemical water splitting *J. Colloid Interface Sci.* **599** 453–66

[110] Majumder S, Gu M and Hyeon Kim K 2022 Facile fabrication of $BiVO_4/Bi_2S_3/NiCoO_2$ for significant photoelectrochemical water splitting *Appl. Surf. Sci.* **574** 151562

[111] Majumder S, Quang N D, Hien T T, Chinh N D, Hung N M, Yang H, Kim C and Kim D 2021 Effect of SILAR-anchored $ZnFe_2O_4$ on the $BiVO_4$ nanostructure: an attempt towards enhancing photoelectrochemical water splitting *Appl. Surf. Sci.* **546** 149033

[112] Majumder S, Quang N D, Kim C and Kim D 2021 Anion exchange and successive ionic layer adsorption and reaction-assisted coating of $BiVO_4$ with Bi_2S_3 to produce nano-structured photoanode for enhanced photoelectrochemical water splitting *J. Colloid Interface Sci.* **585** 72–84

[113] Majumder S, Gu M and Kim K H 2022 Effect of annealing of β-Bi_2O_3 over enhanced photoelectrochemical performance *Mater. Sci. Semicond. Process.* **141** 106439

[114] Majumder S, Quang N D, Thi Hien T, Chinh N D, Yang H, Kim C and Kim D 2021 Nanostructured β-Bi_2O_3/PbS heterojunction as np-junction photoanode for enhanced photoelectrochemical performance *J. Alloys Compd.* **870** 159545

[115] Fang Y, Xue Y, Hui L, Yu H, Liu Y, Xing C, Lu F, He F, Liu H and Li Y 2019 *In situ* growth of GDY based heterostructure: toward efficient overall water splitting *Nano Energy* **59** 591–7

[116] Zhang C, Xue Y, Hui L, Fang Y, Liu Y and Li Y 2021 GDY@$NiO_x(OH)_y$ heterostructure for efficient overall water splitting *Mater. Chem. Front.* **5** 5305–11

[117] Hui L, Jia D, Yu H, Xue Y and Li Y 2019 Ultrathin GDY-wrapped iron carbonate hydroxide nanosheets toward efficient water splitting *ACS Appl. Mater. Interfaces* **11** 2618–25

[118] Li J, Gao X, Jiang X, Li X-B, Liu Z, Zhang J, Tung C-H and Wu L-Z 2017 GDY: a promising catalyst–support to stabilize cobalt nanoparticles for oxygen evolution *ACS Catal.* **7** 5209–13

[119] Pan Y *et al* 2018 Core–shell ZIF-8@ZIF-67-derived CoP nanoparticle-embedded N-doped carbon nanotube hollow polyhedron for efficient overall water splitting *J. Am. Chem. Soc.* **140** 2610–8

[120] Li H *et al* 2018 Colloidal cobalt phosphide nanocrystals as trifunctional electrocatalysts for overall water splitting powered by a zinc–air battery *Adv. Mater.* **30** 1705796

[121] Tang T, Jiang W-J, Niu S, Liu N, Luo H, Chen Y-Y, Jin S-F, Gao F, Wan L-J and Hu J-S 2017 Electronic and morphological dual modulation of cobalt carbonate hydroxides by Mn doping toward highly efficient and stable bifunctional electrocatalysts for overall water splitting *J. Am. Chem. Soc.* **139** 8320–8

[122] Tang C, Zhang R, Lu W, He L, Jiang X, Asiri A M and Sun X 2017 Fe-doped CoP nanoarray: a monolithic multifunctional catalyst for highly efficient hydrogen generation *Adv. Mater.* **29** 1602441

[123] Hou Y, Lohe M R, Zhang J, Liu S, Zhuang X and Feng X 2016 Vertically oriented cobalt selenide/NiFe layered-double-hydroxide nanosheets supported on exfoliated graphene foil: an efficient 3D electrode for overall water splitting *Energy Environ. Sci.* **9** 478–83

[124] Gao X, Zhang H, Li Q, Yu X, Hong Z, Zhang X, Liang C and Lin Z 2016 Hierarchical $NiCo_2O_4$ hollow microcuboids as bifunctional electrocatalysts for overall water-splitting *Angew. Chem. Int. Ed.* **55** 6290–4

[125] Ledendecker M, Krick Calderón S, Papp C, Steinrück H-P, Antonietti M and Shalom M 2015 The synthesis of nanostructured Ni_5P_4 films and their use as a non-noble bifunctional electrocatalyst for full water splitting *Angew. Chem. Int. Ed.* **54** 12361–5

[126] Sahasrabudhe A, Dixit H, Majee R and Bhattacharyya S 2018 Value added transformation of ubiquitous substrates into highly efficient and flexible electrodes for water splitting *Nat. Commun.* **9** 2014

[127] Luo J, Im J-H, Mayer M T, Schreier M, Nazeeruddin M K, Park N-G, Tilley S D, Fan H J and Grätzel M 2014 Water photolysis at 12.3% efficiency via perovskite photovoltaics and earth-abundant catalysts *Science* **345** 1593–6

[128] Jia Y *et al* 2017 A heterostructure coupling of exfoliated Ni–Fe hydroxide nanosheet and defective graphene as a bifunctional electrocatalyst for overall water splitting *Adv. Mater.* **29** 1700017

[129] Wang H, Lee H-W, Deng Y, Lu Z, Hsu P-C, Liu Y, Lin D and Cui Y 2015 Bifunctional non-noble metal oxide nanoparticle electrocatalysts through lithium-induced conversion for overall water splitting *Nat. Commun.* **6** 7261

[130] Tan Y, Wang H, Liu P, Shen Y, Cheng C, Hirata A, Fujita T, Tang Z and Chen M 2016 Versatile nanoporous bimetallic phosphides towards electrochemical water splitting *Energy Environ. Sci.* **9** 2257–61

[131] Hu E, Feng Y, Nai J, Zhao D, Hu Y and Lou X W 2018 Construction of hierarchical Ni–Co–P hollow nanobricks with oriented nanosheets for efficient overall water splitting *Energy Environ. Sci.* **11** 872–80

[132] Hu C, Zhang L, Zhao Z-J, Li A, Chang X and Gong J 2018 Synergism of geometric construction and electronic regulation: 3D Se-(NiCo)S_x/$(OH)_x$ nanosheets for highly efficient overall water splitting *Adv. Mater.* **30** 1705538

[133] Wu P, Du P, Zhang H and Cai C 2012 Graphyne as a promising metal-free electrocatalyst for oxygen reduction reactions in acidic fuel cells: a DFT study *J. Phys. Chem.* C **116** 20472–9

[134] Haley M M 2008 Synthesis and properties of annulenic subunits of graphyne and GDY nanoarchitectures *Pure Appl. Chem.* **80** 519–32

[135] Liu R, Liu H, Li Y, Yi Y, Shang X, Zhang S, Yu X, Zhang S, Cao H and Zhang G 2014 Nitrogen-doped GDY as a metal-free catalyst for high-performance oxygen reduction reactions *Nanoscale* **6** 11336–43

[136] Zhang S *et al* 2016 Heteroatom doped GDY as efficient metal-free electrocatalyst for oxygen reduction reaction in alkaline medium *J. Mater. Chem.* A **4** 4738–44

[137] Li Y, Guo C, Li J, Liao W, Li Z, Zhang J and Chen C 2017 Pyrolysis-induced synthesis of iron and nitrogen-containing carbon nanolayers modified GDY nanostructure as a promising core–shell electrocatalyst for oxygen reduction reaction *Carbon* **119** 201–10

[138] O'Regan B and Grätzel M 1991 A low-cost, high-efficiency solar cell based on dye-sensitized colloidal TiO_2 films *Nature* **353** 737–40

[139] Baviskar P K, Dubal D P, Majumder S, Ennaoui A and Sankapal B R 2016 'Basic idea, advance approach': efficiency boost by sensitization of blended dye on chemically deposited ZnO films *J. Photochem. Photobiol.* A **318** 135–41

[140] Sankapal B, Tirpude A, Majumder S and Baviskar P 2015 1-D electron path of 3-D architecture consisting of dye loaded CdS nanowires: dye sensitized solar cell *J. Alloys Compd.* **651** 399–404

[141] Almuntaser F M A, Majumder S, Sali J V and Sankapal B R 2017 Inverted organic solar cell with ultrasonic spray deposited active layer *Optik (Stuttg)* **131** 1079–84

[142] Majumder S, Baviskar P K and Sankapal B R 2016 Light-induced electrochemical performance of 3D-CdS nanonetwork: effect of annealing *Electrochim. Acta* **222** 100–7

[143] Majumder S, Mendhe A C and Sankapal B R 2019 Nanoheterojunction through PbS nanoparticles anchored CdS nanowires towards solar cell application *Int. J. Hydrog. Energy* **44** 7095–107

[144] Majumder S and Sankapal B R 2017 Facile fabrication of CdS/CdSe core–shell nanowire heterostructure for solar cell applications *New J. Chem.* **41** 5808–17

[145] Majumder S, Mendhe A C, Kim D and Sankapal B R 2019 CdO nanonecklace: effect of air annealing on performance of photo electrochemical cell *J. Alloys Compd.* **788** 75-82

[146] Majumder S, Baviskar P K and Sankapal B R 2016 Straightening of chemically deposited CdS nanowires through annealing towards improved PV device performance *Ceram. Int.* **42** 6682–91

[147] Almuntaser F M A, Baviskar P K, Majumder S, Tarkas H S, Sali J V and Sankapal B R 2019 Role of polyaniline thickness in polymer-zinc oxide based solid state solar cell *Mater. Sci. Eng.* B **244** 23–8

[148] Almuntaser F M A, Majumder S, Baviskar P K, Sali J V and Sankapal B R 2017 Synthesis and characterization of polypyrrole and its application for solar cell *Appl. Phys.* A **123** 555

[149] Sankapal B R, Salunkhe D B, Majumder S and Dubal D P 2016 Solution-processed CdS quantum dots on TiO_2: light-induced electrochemical properties *RSC Adv.* **6** 83175–84

[150] Asghar M I, Zhang J, Wang H and Lund P D 2017 Device stability of perovskite solar cells—a review *Renew. Sustain. Energy Rev.* **77** 131–46

[151] Ren H *et al* 2015 A new GDY nanosheet/Pt nanoparticle-based counter electrode material with enhanced catalytic activity for dye-sensitized solar cells *Adv. Energy Mater.* **5** 1500296

[152] Du H *et al* 2011 The effect of GDY doping on the performance of polymer solar cells *Synth. Met.* **161** 2055–7

[153] Li J, Gao X, Liu B, Feng Q, Li X-B, Huang M-Y, Liu Z, Zhang J, Tung C-H and Wu L-Z 2016 GDY: a metal-free material as hole transfer layer to fabricate quantum dot-sensitized photocathodes for hydrogen production *J. Am. Chem. Soc.* **138** 3954–7

[154] Jin Z *et al* 2016 Efficient hole transporter for stable high-performance colloidal quantum dot solar cells *Adv. Funct. Mater.* **26** 5284–9

[155] Pan L D, Zhang L Z, Song B Q, Du S X and Gao H-J 2011 Graphyne- and GDY-based nanoribbons: density functional theory calculations of electronic structures *Appl. Phys. Lett.* **98** 173102

[156] Chakraborti H 2016 Giant first hyperpolarizabilities of donor–acceptor substituted graphyne: an *ab initio* study *Spectrochim. Acta* A **153** 226–30

[157] Kuang C *et al* 2015 Highly efficient electron transport obtained by doping PCBM with GDY in planar-heterojunction perovskite solar cells *Nano Lett.* **15** 2756–62

[158] Li M, Wang Z-K, Kang T, Yang Y, Gao X, Hsu C-S, Li Y and Liao L-S 2018 GDY-modified cross-linkable fullerene as an efficient electron-transporting layer in organometal halide perovskite solar cells *Nano Energy* **43** 47–54

[159] Li J, Jiu T, Duan C, Wang Y, Zhang H, Jian H, Zhao Y, Wang N, Huang C and Li Y 2018 Improved electron transport in $MAPbI_3$ perovskite solar cells based on dual doping GDY *Nano Energy* **46** 331–7

[160] Zhang X, Wang Q, Jin Z, Chen Y, Liu H, Wang J, Li Y and Liu S 2018 GDY quantum dots for much improved stability and efficiency of perovskite solar cells *Adv. Mater. Interfaces* **5** 1701117

[161] Pang D, Alhabeb M, Mu X, Dall'Agnese Y, Gogotsi Y and Gao Y 2019 Electrochemical actuators based on two-dimensional $Ti_3C_2T_x$ (MXene) *Nano Lett.* **19** 7443–8

[162] Krishnamoorthy K, Thangavel S, Chelora Veetil J, Raju N, Venugopal G and Kim S J 2016 GDY nanostructures as a new electrode material for electrochemical supercapacitors *Int. J. Hydrog. Energy* **41** 1672–8

[163] Xu J, Li J, Yang Q, Xiong Y and Chen C 2017 *In situ* synthesis of MnO_2@GDY oxides nanocomposite with enhanced performance of supercapacitors *Electrochim. Acta* **251** 672–80

[164] Chandra Shekar S and Swathi R S 2013 Rattling motion of alkali metal ions through the cavities of model compounds of graphyne and GDY *J. Phys. Chem.* A **117** 8632–41

[165] Guo Y, Jiang K, Xu B, Xia Y, Yin J and Liu Z 2012 Remarkable hydrogen storage capacity in Li-decorated graphyne: theoretical predication *J. Phys. Chem.* C **116** 13837–41

[166] Guo Y, Lan X, Cao J, Xu B, Xia Y, Yin J and Liu Z 2013 A comparative study of the reversible hydrogen storage behavior in several metal decorated graphyne *Int. J. Hydrog. Energy* **38** 3987–93

[167] Zhang H, Zhao M, He X, Wang Z, Zhang X and Liu X 2011 High mobility and high storage capacity of lithium in sp–sp^2 hybridized carbon network: the case of graphyne *J. Phys. Chem.* C **115** 8845–50

[168] Hwang H J, Koo J, Park M, Park N, Kwon Y and Lee H 2013 Multilayer graphynes for lithium ion battery anode *J. Phys. Chem.* C **117** 6919–23

[169] Mohajeri A and Shahsavar A 2016 Li-decoration on the edge oxidized graphyne and GDY: a first principles study *Comput. Mater. Sci.* **115** 51–9

[170] Srinivasu K and Ghosh S K 2012 Graphyne and GDY: promising materials for nanoelectronics and energy storage applications *J. Phys. Chem.* C **116** 5951–6
[171] Zhang H, Xia Y, Bu H, Wang X, Zhang M, Luo Y and Zhao M 2013 GDY: a promising anode material for lithium ion batteries with high capacity and rate capability *J. Appl. Phys.* **113** 44309
[172] Huang C, Zhang S, Liu H, Li Y, Cui G and Li Y 2015 GDY for high capacity and long-life lithium storage *Nano Energy* **11** 481–9
[173] Zhang S, Liu H, Huang C, Cui G and Li Y 2015 Bulk GDY powder applied for highly efficient lithium storage *Chem. Commun.* **51** 1834–7
[174] Wang K, Wang N, He J, Yang Z, Shen X and Huang C 2017 Preparation of 3D architecture GDY nanosheets for high-performance sodium-ion batteries and capacitors *ACS Appl. Mater. Interfaces* **9** 40604–13
[175] Zhang S, He J, Zheng J, Huang C, Lv Q, Wang K, Wang N and Lan Z 2017 Porous GDY applied for sodium ion storage *J. Mater. Chem.* A **5** 2045–51
[176] Wang N, Li X, Tu Z, Zhao F, He J, Guan Z, Huang C, Yi Y and Li Y 2018 Synthesis and electronic structure of boron-GDY with an sp-hybridized carbon skeleton and its application in sodium storage *Angew. Chem. Int. Ed.* **57** 3968–73
[177] Xu Z, Lv X, Li J, Chen J and Liu Q 2016 A promising anode material for sodium-ion battery with high capacity and high diffusion ability: graphyne and GDY *RSC Adv.* **6** 25594–600
[178] Farokh Niaei A H, Hussain T, Hankel M and Searles D J 2017 Sodium-intercalated bulk GDY as an anode material for rechargeable batteries *J. Power Sources* **343** 354–63
[179] Du H, Zhang Z, He J, Cui Z, Chai J, Ma J, Yang Z, Huang C and Cui G 2017 A delicately designed sulfide GDY compatible cathode for high-performance lithium/magnesium–sulfur batteries *Small* **13** 1702277
[180] Chen J, Xi J, Wang D and Shuai Z 2013 Carrier mobility in graphyne should be even larger than that in graphene: a theoretical prediction *J. Phys. Chem. Lett.* **4** 1443–8
[181] Bai H, Zhu Y, Qiao W and Huang Y 2011 Structures, stabilities and electronic properties of GDY nanoribbons *RSC Adv.* **1** 768–75
[182] Pan Y *et al* 2015 GDY–metal contacts and GDY transistors *Nanoscale* **7** 2116–27
[183] Xue Z *et al* 2016 Controlling the interface areas of organic/inorganic semiconductor heterojunction nanowires for high-performance diodes *ACS Appl. Mater. Interfaces* **8** 21563–9
[184] Shivananju B N, Hoh H Y, Yu W and Bao Q 2019 Optical biochemical sensors based on 2D materials *Fundamentals and Sensing Applications of 2D Materials Woodhead Publishing Series in Electronic and Optical Materials* ed M Hywel, C S Rout and D J Late (Cambridge: Woodhead), ch 10 pp 379–406
[185] Yan H, Guo S, Wu F, Yu P, Liu H, Li Y and Mao L 2018 Carbon atom hybridization matters: ultrafast humidity response of GDY oxides *Angew. Chem. Int. Ed.* **57** 3922–6
[186] Wang C, Yu P, Guo S, Mao L, Liu H and Li Y 2016 GDY oxide as a platform for fluorescence sensing *Chem. Commun.* **52** 5629–32
[187] Zheng Z, Fang H, Liu D, Tan Z, Gao X, Hu W, Peng H, Tong L, Hu W and Zhang J 2017 Nonlocal response in infrared detector with semiconducting carbon nanotubes and GDY *Adv. Sci.* **4** 1700472
[188] Pan J, Du S, Zhang Y, Pan L, Zhang Y, Gao H-J and Pantelides S T 2015 Ferromagnetism and perfect spin filtering in transition-metal-doped graphyne nanoribbons *Phys. Rev.* B **92** 205429

[189] He J, Zhou P, Jiao N, Ma S Y, Zhang K W, Wang R Z and Sun L Z 2014 Magnetic exchange coupling and anisotropy of 3D transition metal nanowires on graphyne *Sci. Rep.* **4** 4014
[190] Bhattacharya B, Singh N B and Sarkar U 2015 Tuning the magnetic property of vacancy-defected graphyne by transition metal absorption *AIP Conf. Proc.* **1665** 50066
[191] Chen X, Gao P, Guo L, Wen Y, Zhang Y and Zhang S 2017 Two-dimensional ferromagnetism and spin filtering in Cr and Mn-doped GDY *J. Phys. Chem. Solids* **105** 61–5
[192] Zhang M, Wang X, Sun H, Yu J, Wang N, Long Y and Huang C 2018 Preparation of room-temperature ferromagnetic semiconductor based on GDY-transition metal hybrid *2D Mater.* **5** 35039
[193] Zhang M, Wang X, Sun H, Wang N, Lv Q, Cui W, Long Y and Huang C 2017 Enhanced paramagnetism of mesoscopic GDY by doping with nitrogen *Sci. Rep.* **7** 11535
[194] Zheng Y, Chen Y, Lin L, Sun Y, Liu H, Li Y, Du Y and Tang N 2017 Intrinsic magnetism of GDY *Appl. Phys. Lett.* **111** 33101
[195] Li S, Chen Y, Liu H, Wang Y, Liu L, Lv F, Li Y and Wang S 2017 GDY materials as nanotransducer for *in vivo* photoacoustic imaging and photothermal therapy of tumor *Chem. Mater.* **29** 6087–94
[196] Xie J *et al* 2019 GDY nanoparticles with high free radical scavenging activity for radiation protection *ACS Appl. Mater. Interfaces* **11** 2579–90
[197] Jiao Y, Du A, Hankel M, Zhu Z, Rudolph V and Smith S C 2011 GDY: a versatile nanomaterial for electronics and hydrogen purification *Chem. Commun.* **47** 11843–5
[198] Cranford S W and Buehler M J 2012 Selective hydrogen purification through GDY under ambient temperature and pressure *Nanoscale* **4** 4587–93
[199] Xu D, Lena M, Hans-Jürgen B and Doris V 2012 Candle soot as a template for a transparent robust superamphiphobic coating *Science* **335** 67–70
[200] Gao X, Yan X, Yao X, Xu L, Zhang K, Zhang J, Yang B and Jiang L 2007 The dry-style antifogging properties of mosquito compound eyes and artificial analogues prepared by soft lithography *Adv. Mater.* **19** 2213–7
[201] Zhang F, Bin Zhang W, Shi Z, Wang D, Jin J and Jiang L 2013 Nanowire-haired inorganic membranes with superhydrophilicity and underwater ultralow adhesive superoleophobicity for high-efficiency oil/water separation *Adv. Mater.* **25** 4192–8
[202] Zhu J, Hsu C-M, Yu Z, Fan S and Cui Y 2010 Nanodome solar cells with efficient light management and self-cleaning *Nano Lett.* **10** 1979–84
[203] Feng L, Li S, Li Y, Li H, Zhang L, Zhai J, Song Y, Liu B, Jiang L and Zhu D 2002 Super-hydrophobic surfaces: from natural to artificial *Adv. Mater.* **14** 1857–60
[204] Wang S, Liu K, Yao X and Jiang L 2015 Bioinspired surfaces with superwettability: new insight on theory, design, and applications *Chem. Rev.* **115** 8230–93
[205] Liu J, Shen X, Baimanov D, Wang L, Xiao Y, Liu H, Li Y, Gao X, Zhao Y and Chen C 2019 Immobilized ferrous ion and glucose oxidase on GDY and its application on one-step glucose detection *ACS Appl. Mater. Interfaces* **11** 2647–54
[206] Liu R, Zhou J, Gao X, Li J, Xie Z, Li Z, Zhang S, Tong L, Zhang J and Liu Z 2017 GDY filter for decontaminating lead-ion-polluted water *Adv. Electron. Mater.* **3** 1700122

IOP Publishing

Nanocarbon Allotropes Beyond Graphene
Synthesis, properties and applications
Arpan Kumar Nayak and Santosh K Tiwari

Chapter 8

Synthesis and application of onion-like carbons

A Thennarasi, Pamula Siva, C Sreelakshmi, Lakshmi Sajeev and Kuraganti Vasu

Carbon nano-onions (CNOs) are zero-dimensional (0D) quasi-spherical nanoparticles consisting of concentric graphitic carbon shells which are defective and disordered in nature. Due to their ultra-small size and closed cage structure, CNOs area a most fascinating nanocarbon form with exotic physical and chemical properties. Their features, such as high surface area, electrical conductivity, and thermal stability, make CNOs attractive electrode materials for electrocatalytic energy conversion and storage applications. To date, many methods have been used for the synthesis of CNOs and their derivatives structures. Here, we discuss the various physical and chemical routes of synthesis for CNOs and their structures in detail. Further, the recent advances in the use of CNOs and their derivatives, such as nitrogen-doped CNO (N-CNO), metal nanoparticle core encapsulated CNO shells (Pt@N-CNO, CoFe@N-CNO), and metal oxide–CNO nanocomposite (Pd–MnO_2@CNO) structures, for electrochemical water splitting, fuel cells, supercapacitors, and battery applications, are discussed.

8.1 Introduction

Over the past two decades, carbon nanostructures, including zero-dimensional carbon quantum dots (0D-CQDs), one-dimensional carbon nanotubes (1D-CNTs), and two-dimensional graphene (2D-G), have received widespread research interest due to their fascinating properties and potential applications in energy, electronic, and optoelectronic devices [1–6]. The new allotropic carbon nano-onions (CNOs) or onion-like carbons (OLCs) are another nanocarbon structure, belonging to the fullerene family, with spherical and polyhedral shapes consisting of concentric defective graphitic layers close to one another [7, 8]. The CNO structures are composed of stacked sp^2 hybridized graphene layers containing hexagonal and pentagonal carbon rings. The carbon atoms present at the vertices of the ring form two single (—) bonds and one double (=) bond with the neighboring carbon atoms

doi:10.1088/978-0-7503-5177-5ch8

with a delocalized pi (π) electron in the molecule. The elimination of the highly energetic dangling bonds present at the graphitic sheet edge results in the formation of a curl and closure structure with onion-like symmetry [9]. The interlayer separation between two adjacent layers in a closed cage CNO is 0.34 nm, which is approximately equal to the distance between two planes in a graphitic structure. The CNOs have a self-compression property when they are subjected to electron or ion beam irradiation at high-temperatures [10]. As a result, the interlayer separation in self-compressed CNOs is decreased to 0.22 nm, a value close to the layer separation in nanodiamond [11].

Several physical and chemical methods have been used to synthesize CNO structures (a detailed discussion is provided in the next section). Sizes in the range of 5–100 nm and spherical and polyhedral shape (with strong faceted) morphologies are frequently obtained in any method of preparation. The optical and electronic properties of CNOs are dependent on the size, shape, doping, and surface functional groups [12]. Due to the high degree of curvature, the surfaces of small CNOs (2–10 nm) are chemically active and allow covalent functionalization of various organic and inorganic molecules [13–16]. CNOs have already been used in a variety of applications including chemical and bio-sensors [17], lubricant additives [18, 19], and field emitters [20–22].

The highly graphitized closed and spherical carbon shells can provide interesting catalytic and magnetic properties. CNOs modified with doping, surface functionalization, composite, and core–shell structures improve the surface area, thermal stability, and electrical conductivity, and these properties make them promising materials for use in energy conversion and storage applications [8, 23–26]. A variety of types of CNOs and their derivatives structures, including N-doped CNOs (N-CNOs), metal nanoparticle encapsulated core/shell CNOs (Co–C/CNOs, CoFe/N-CNOs, and Pt/N-CNOs), and metal oxide–CNO composites (Pd–MnO_2/CNO and Pd–CeO_2/CNO) were prepared by employing physical and chemical methods. In this chapter, we first discuss in detail the various preparation methods used for the synthesis of CNOs. Following this, the recent advances in CNOs for electrocatalytic water splitting, fuel cell, supercapacitor, battery, and spintronic applications are discussed in section 8.2.

8.2 Synthesis of carbon nano-onions

8.2.1 Overview

Historically, CNOs were first observed in 1980 by Sumio Iijima in vacuum deposited carbon films using transition electron microscopy [27]. However, the structures become popular after Ugarte explain the formation mechanism under electron beam irradiation of amorphous carbon particles in 1992 [9]. To date, many physical and chemical methods have been used for the synthesis of CNOs from various forms of carbon sources. The widely used methods are shown in figure 8.1 [7, 23]. The methods of thermal annealing, submerged arc discharge, ion implantation, electron beam irradiation, and laser ablation fall under the category of physical methods, whereas the chemical vapor deposition, hydrothermal, and flame pyrolysis synthesis

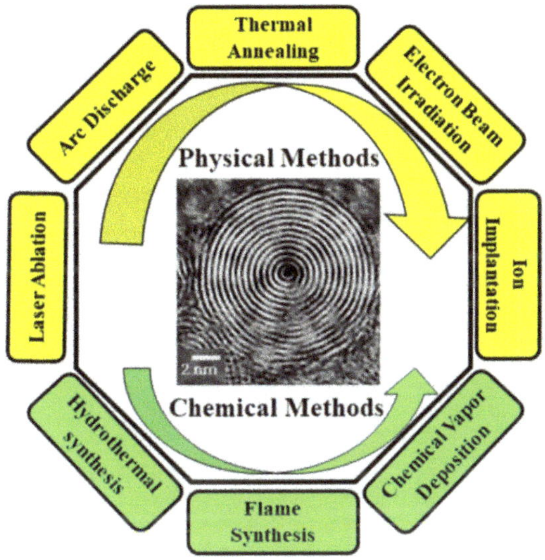

Figure 8.1. Schematic representing different physical and chemical methods used to synthesize CNO structures. (TEM image reproduced with permission from [41]. Copyright 2004 AIP Publishing.)

methods fall under the category of chemical methods. Depending on the preparation method and process parameters used, CNOs of various sizes (small or big), shapes (spherical or polyhedral), and core structures (dense/hallow) can be obtained [7].

8.2.2 Physical methods

8.2.2.1 Thermal annealing of nanodiamond

Thermal annealing of detonated nanodiamond (ND) powder under an inert atmosphere is a popular method used for gram-scale production of CNOs [7, 23, 28–30]. Due to its high purity, non-toxic, low-cost preparation, and ultra-fine particle size (average particle diameter 5 nm) ND is used widely as a precursor for the preparation of CNOs by the thermal annealing method [31, 32]. Figure 8.2(a) shows the schematic transformation of sp^3 hybridized ND to sp^2 hybridized CNO in the thermal annealing method. The structural transformation is a multi-stage process. It starts with the removal of oxygen and hydrogen-containing surface functional groups at a temperature close to 200 °C, followed by the emission of CO and CO_2 gases when the temperature of the ND particles reaches 700 °C–800 °C. The onset of graphitization begins when the temperature is in the range of 900 °C–1100 °C, leading to surface reconstruction with a high degree of sp^2 carbon atom networks forming core–shell structures (a small diamond particle surrounded by surface reconstructed graphitic shells). Further increasing the temperature to 1500 °C, the diamond completely transforms into a graphitic structure with 6–8 graphene layers. An annealing temperature of more than 1700 °C leads to the formation of CNOs with closed cage structures with an average size in the range of 5–6 nm. Transmission electron microscopy (TEM) is a powerful tool to study the structural

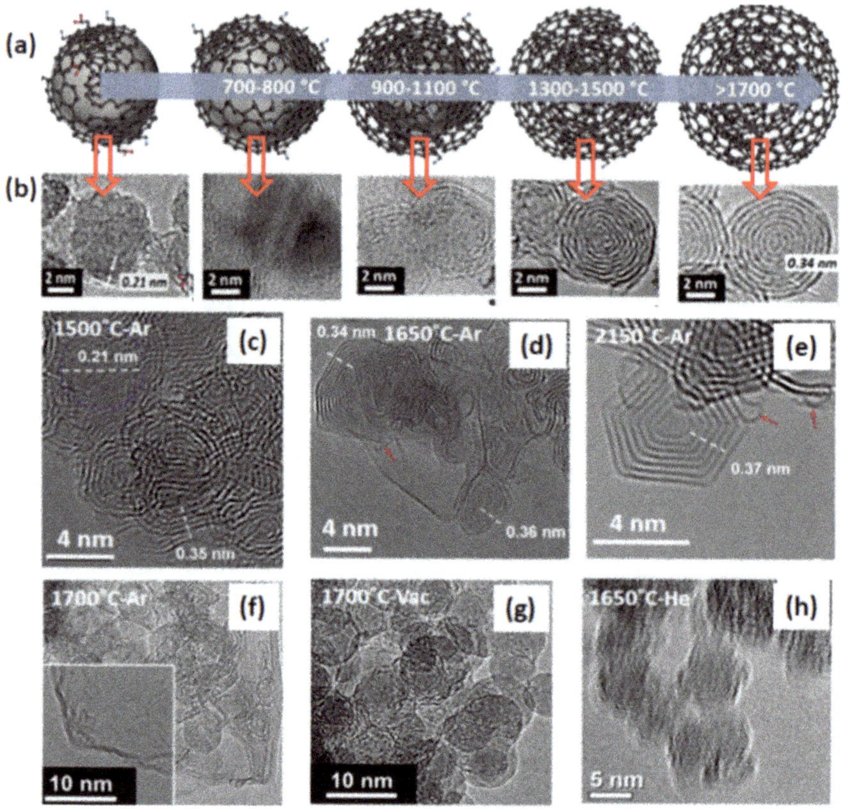

Figure 8.2. (a) Schematic of the transformation of diamond nanoparticles into CNO with (b) corresponding TEM images at different temperatures in the thermal annealing method. (Data reproduced from [23]. CC BY 3.0.) HRTEM images of CNOs formed by thermal annealing of NDs in Ar atmosphere at different temperatures: (c) 1500 °C, (d) 1650 °C, and (e) 2150 °C, respectively. (HRTEM images reproduced with permission from [34]. Copyright 2014 Elsevier.) HRTEM images of a CNO were obtained at a constant temperature in different annealing environments: (f) Ar and (g) vacuum. (HRTEM images reproduced with permission from [35]. Copyright 2015 Elsevier.) (h) HRTEM images of a CNO were obtained in a He environment annealing at 1650 °C. (HRTEM images reproduced with permission from [36]. Copyright 2014 Elsevier.)

and microstructural properties of carbon-based materials [33]. Figure 8.2(b) shows high-resolution TEM images acquired at each stage during the transformation of ND particles into CNOs. The pristine ND particle has a regular lattice arrangement with an interlayer spacing of 0.21 nm. The diamond core, surrounded by a few outer graphitic layers, is visible in the image when the annealing temperature is in the range of 900 °C–1100 °C. The complete transformation of CNO structures is evident at a temperature of 1300 °C–1500 °C and a further increase in the temperature (>1700 °C) results in a concentric closed shell with interlayer separation of 0.34 nm corresponding to the (002) plane *d*-spacing of graphite [23].

The properties, sizes, and shapes of CNOs derived from the thermal annealing of NDs are dependent on the process parameters, such as heating rate, annealing temperature, duration, and heating environment (vacuum, hydrogen, argon, nitrogen, or helium). Figures 8.2(c)–(e) show HRTEM images of CNOs obtained by thermal annealing of ND in an argon atmosphere at different temperatures. The diamond core surrounded by few-layer graphitic shells is seen at the temperature $T = 1500$ °C (figure 8.2(c)) and a complete transformation to a CNO occurred at 1600 °C (figure 8.2(d)). Further, by increasing the temperature to 2150 °C, pentagon shaped hollow CNOs are formed (figure 8.2(e)) [34]. This suggests that the annealing temperature has a significant effect on the CNO structure in the same medium. In another study, Zeiger *et al* investigated the effect of the annealing environment on the structure and properties of CNOs obtained at a temperature $T = 1700$ °C (figures 8.2(f) and (g)). CNOs interconnected with few-layer graphene nanoribbons were observed in Ar flow (figure 8.2(f)), whereas spherical shape CNOs were observed in a vacuum medium (figure 8.2(g)) [35]. Figure 8.2(h) shows the CNOs obtained by annealing nanodiamond particles at a temperature $T = 1600$ °C in He atmosphere [36]. Heteroatom (N/S and B/N) doped CNOs have been synthesized by the thermal annealing method in recent years [37, 38].

8.2.2.2 Arc discharge method

Non-vacuum-based arc discharge is another popular and inexpensive method for the synthesis of CNOs. Sano *et al* synthesized the first CNO structures by arc discharge between two graphite electrodes submerged in a water medium with an applied voltage of 16 V and a current of 30 A [39, 40]. This method also yields other nanocarbon structures (graphene, CNT, nanohorns), when the liquid environment (organic or inorganic solvents) and applied voltage (/current) parameters are changed [41–43]. CNOs can be obtained at optimized conditions of arc discharge in a specific liquid medium. Figure 8.3(a) shows a schematic of an arc discharge experimental set-up we employed for the synthesis of CNOs in formamide medium. Two high purity (99.999%) graphite rods 5 mm in diameter are fixed at the cathode and a movable anode site inside a SS Dewar filled with formamide. The arc discharge is generated between submerged graphite electrodes after touching each other at an applied voltage and current of 38 V and 40 A, respectively. In our work, the CNOs are obtained in a formamide medium and the corresponding HRTEM image is presented in figure 8.3(b) [42]. Figures 8.3(c) and (d) show HRTEM images of spherically shaped CNOs obtained in arc discharge in water and liquid N_2 (LN_2) medium at a constant arc current of 30 A [41]. Kim *et al* synthesized nitrogen-doped defective CNOs (N-CNOs) by arc discharge in NH_3 containing water medium at an applied DC voltage of 1000 V and a current of 30 A. The TEM image of N-CNO is presented in figure 8.3(e) with defect sites indicated by red arrow marks [44]. The high-voltage pulsed (4 kV amplitude, 2 μs pulse width, and 70 Hz repetition rate) spark discharge between the two Cu electrodes without graphite rods in hydrocarbon (heptane, cyclohexane, or toluene) medium results in the formation of a Cu core encapsulated by a CNO shell. Figure 8.3(f) shows a TEM image of a submicronic Cu particle surrounded by a large CNO ring obtained in toluene. The HRTEM image (figure 8.3(g)) clearly shows the lattice planes of the Cu core and the graphitic shells of the CNO [45]. Mo–Ni core nanoparticles encapsulated

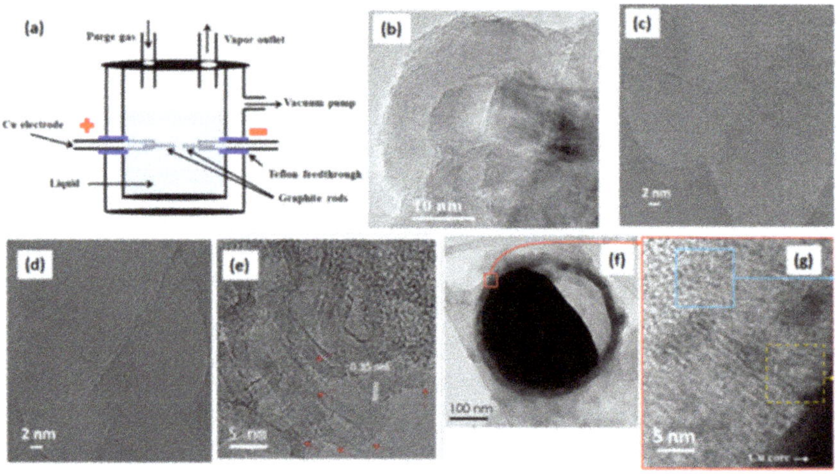

Figure 8.3. (a) Schematic of the arc discharge method for CNO synthesis. HRTEM images of CNOs obtained in (b) formamide, (c) water, (d) liquid nitrogen, and (e) nitrogen-doped CNO in NH_3 contain water medium, respectively. (f) and (g) TEM and HRTEM images of Cu submicronic particle shell encapsulated CNO obtained by modified graphite rod electrode arc discharge in ethanol medium. ((a) and (b) Reproduced with permission from [42]. Copyright 2013 Institute of Physics Publishing. (c) Reproduced with permission from [41]. Copyright 2004 AIP Publishing. (d,e) Reproduced with permission from [44]. Copyright 2016 Elsevier. (f) and (g) Reproduced with permission from [45]. CC BY 4.0.)

with CNO shell structure were prepared by arc discharge using a modified graphite rod electrode with Ni and Mo content in an ethanol medium [46]. Despite the different liquid media, the arc discharge method results in producing CNOs with diameters ranging between 15 and 25 nm with a hollow-cage structure [7].

8.2.2.3 Electron beam irradiation

Electron beam irradiation (from the source of an electron microscope (TEM or FESEM) unit) on the surface induces defects and structural transformation in carbon nanostructures [47–50]. The electron beam can be considered as a tool for *in situ* chemical reactions on the surfaces, and thereby the method is also used to synthesize various types of nanostructures [51]. In the electron beam irradiation method, the CNOs are obtained by the transformation of amorphous carbon, graphite, and nanodiamond particles [9, 52–55]. In a recent work, Lin *et al* reported the *in situ* transformation of diamond nanoparticles (figure 8.4(a)) into defective CNO (figure 8.4(b)) under 200 keV electron beam irradiation [55]. For CNO synthesis, the electron beam irradiation method has not gained much attention due to the low yield of production and the maintenance of electron microscope units (TEM or FESEM).

8.2.2.4 Ion implantation

The carbon ion implantation method of preparing CNOs has the potential to obtain a large area of solid-state CNO film on a catalyst and realize the use of CNOs in

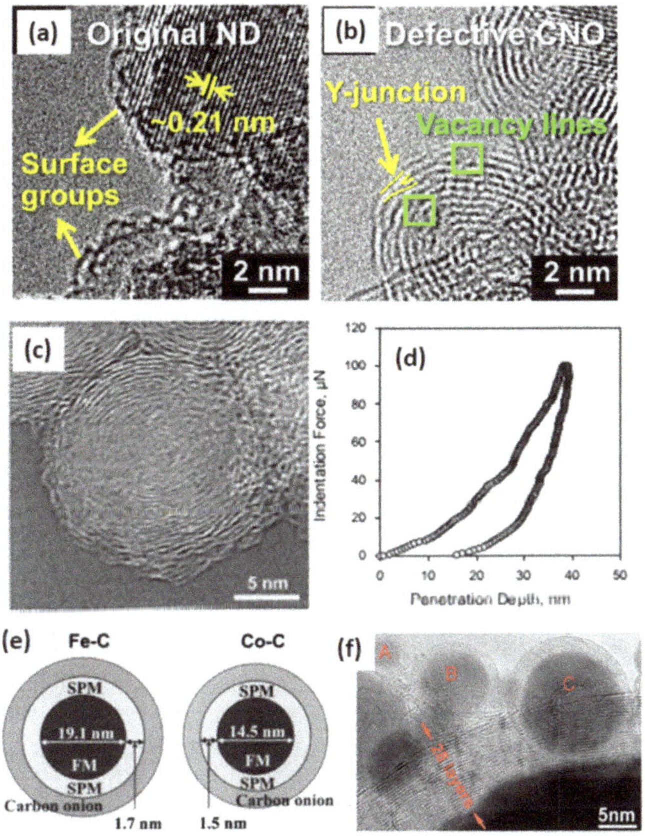

Figure 8.4. HRTEM images of the transformation of (a) ND particles into (b) defective CNO under 200 kV electron beam irradiation. (Reproduced with permission from [55]. Copyright 2022 Elsevier.) (c) HRTEM image and (d) load–displacement curve of CNOs obtained by methane plasma ion implantation on Ag substrate. (Reproduced with permission from [60]. CC BY 4.0.) (e) The model explains the formation of an FM core surrounded by an amorphous SPM (Fe–C and Co–C) shell encapsulated by CNOs obtained during Nd-YAG laser (λ = 532 nm) ablation in Ar medium. (Reproduced with permission from [70]. Copyright 2006 AIP Publishing.) (f) HRTEM images of different sized amorphous Co_3C/CNOs core/shell nanoparticles obtained by Nd-YAG laser (λ = 532 nm) ablation. (Reproduced with permission from [71]. Copyright 2013 Elsevier.)

particular applications such as batteries, photovoltaic cells, fuel cells, hard coatings, and solid lubricants. Generally, $^{12}C^+$ and $^{13}C^+$ ions with a fluence range of 10^{16}–10^{17} ions/cm^2 and an acceleration energy of more than 100 keV are implanted on a metal (Ag or Cu) catalyst surface. As carbon is immiscible with Ag or Cu, it is expected that carbon precipitates on the surface of the metal catalyst during implantation. Cabioc'h *et al* first reported film state CNO synthesis by ion implantation onto Ag and Cu catalyst with an ion implantation energy of 120 keV and fluence greater than 10^{16} ions/cm^2 in the temperature range 500 °C–800 °C [56, 57]. Spherically shaped CNOs with sizes in the range of 20–50 nm were obtained using this method. The same group also conducted a study to understand the growth mechanism of CNOs by an ion implantation method [58, 59]. In a recent study, Matsumoto *et al* used a plasma-based ion implantation method to synthesize CNOs using methane gas as a C source at an applied voltage of 20 kV pulsed power supply on Ag film [60]. Figure 8.4(c) shows a HRTEM image of spherical CNOs of average size of 18 nm obtained by methane plasma ion implantation on a Ag substrate. From the nanoindentation test, it is found that the CNO films show good mechanical strength as determined by the load–displacement curve shown in figure 8.4(d). The hardness and elastic modulus of CNO films at a penetration depth of 75 nm were measured as 6.5 and 152 GPa, respectively [60].

8.2.2.5 Laser ablation

Laser ablation is a physical method used for the preparation of nanomaterials, in which laser energy is used for ablating a solid target. In this method the nanoparticles are formed by condensation of laser-vaporized species in different environmental conditions, such as gas, vacuum, and liquid media. Without limit, this method allows the synthesis of various kinds of nanostructures, including CNTs, semiconductor nanowires/particles, quantum dots, and core–shell nanostructures by laser ablation from graphite and semiconductor solid crystals [61–65]. The laser wavelength, energy, repetition rate, ablation medium, and temperature are the controlled parameters for the synthesis of CNOs by laser ablation. Large-size (100–200 nm) CNOs were produced by ablating a Ni- and Co-contained graphite target with a KrF excimer UV laser ($\lambda = 248$ nm) in O_2 gas medium at room temperature [66, 67]. CNOs also produced by the high temperature (900 °C) ablation of graphite target using the same laser source in Ar medium [68]. The hollow CNOs are produced using CO_2 laser ablation-induced decomposition of acetylene (C_2H_2) gas in N_2 flow [69]. Seung *et al* obtained amorphous iron carbide (Fe–C) and cobalt carbide (Co–C) nanoparticles encapsulated by CNOs (as shown in figure 8.4(e)) by irradiation of metallocene powder with ferrocene and co-baltocene impurities with Nd-YAG laser ($\lambda = 532$ nm) in Ar medium. In the model, a ferromagnetic (FM = Ir and Co) core surrounded by a superparamagnetic (SPM = Fe–C and Co–C) amorphous shell encapsulated by CNOs is proposed in the magnetic core–shell structures [70]. The CNO and the related composites are also prepared by ablation in liquid media [71]. The Co_3C/CNO core/shell structure is obtained using Nd-YAG laser ($\lambda = 532$ nm) irradiation of a Co metal target placed in acetone medium. Figure 8.4(f) shows HRTEM images of different sizes of (denoted by the letters A, B, and C) amorphous Co_3C and CNO core/shell nanoparticles along with 28 layers of large CNO [71].

8.2.3 Chemical methods

8.2.3.1 Chemical vapor deposition

Chemical vapor deposition (CVD) is a material processing technology used for the production of high-quality semiconductor nanomaterials. CVD has emerged as an important method for the preparation of nanocarbon including graphene, CNTs, carbon fibers, and CNOs [72–78]. The growth process involves catalytic

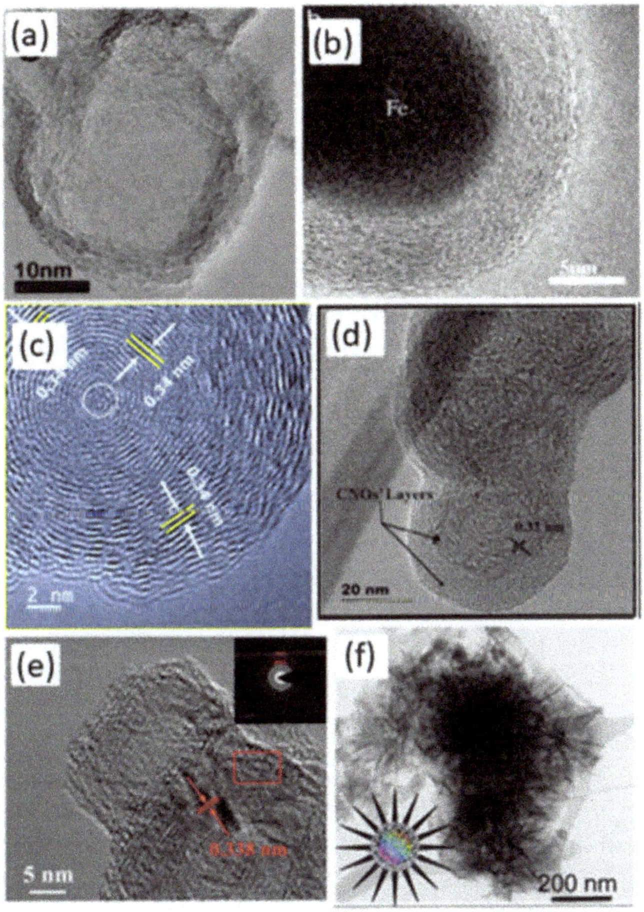

Figure 8.5. TEM images of CVD grown (a) hollow CNO with MgO supported Co catalyst at 700 °C and (b) Fe-CNO core–shell structure at 1100 °C in Ar atmosphere. TEM images of flame synthesis processed (c) N-doped CNO using acetonitrile precursor at 1000 °C and (d) CNO sample collected on a brass plate obtained by catalyst-free flame pyrolysis using ghee oil. TEM images of (e) hydrothermal synthesis processed N containing CNOs using *Lentinus edodes* and (f) core–leaf structure of CNO–MnO_2 nanocomposite. ((a) Reproduced with permission from [77]. Copyright 2020 Elsevier. (b) Reproduced with permission from [83]. Copyright 2006 Elsevier. (c) Reproduced with permission from [88]. Copyright 2020 Elsevier. (d) Reproduced with permission from [99]. Copyright 2018 Elsevier. (e) Reproduced with permission from [109]. Copyright 2021 Elsevier. (f) Reproduced with permission from [113]. Copyright 2013 Elsevier.)

decomposition of carbon precursors and forms various nanostructures in the vapor liquid solid (VLS) growth model. A metal catalyst with support or floating is generally used for the synthesis of CNOs. For CNO growth, ethylene (C_2H_4), ethane (C_2H_2), and methane (CH_4) are the most used precursors with Co, Ni, and Fe catalysts. After synthesis, the CNOs are subjected to post-synthesis purification to remove impurities and by-products. Chao *et al* have reported the gram-scale synthesis of hollow CNOs with MgO supported Co catalyst at 700 °C in Ar atmosphere using ethylene as a carbon source [77]. Figure 8.5(a) shows an HRTEM image of large hollow (50 nm) CNOs obtained at a Co and MgO precursor ratio of 5:95. In another study, Yang *et al* reported the synthesis of CNO using NaCl supported Fe catalyst by catalytic decomposition of acetylene at 400 °C in Ar atmosphere [79]. The large-sized CNOs were obtained by plasma enhanced CVD (PE-CVD) using CH_4 as a carbon source and Co as a catalyst at 300 °C in a flow of hydrogen at an applied voltage of 50 V [80]. The CNO shell is encapsulated by amorphous carbon films prepared by PE-CVD and shows superior mechanical properties in terms of a low friction coefficient of 0.01 and 92% elastic recovery in ambient conditions [81]. High-yield CNOs with an unsupported alloy (Ni–Fe) catalyst were synthesized using a CH_4 precursor at 850 °C in the flow of hydrogen. The synthesized CNOs with polyhedral shapes had a core–shell structure with Fe–Ni as the core and CNO as the shell [78]. Phosphorus doped CNOs (P-CNOs) were prepared using TiC and triphenylphosphine (TPP) as the carbon and phosphorus precursors at 870 °C in a gas mixture of H_2 and Ar. The prepared P-CNO samples exhibited surface activity, high sensitivity, and a reversible character as a NH_3 gas sensor [82]. Using ferrocene as the Fe source Fe-CNOs core–shell structures were synthesized with an acetylene (C_2H_2) precursor heated at 1100 °C [83, 84]. A TEM image of the Fe/CNO core–shell is shown in figure 8.5(b). From figures 8.5(a) and (b) it is observed that the CVD process produces various types of CNOs by changing the process parameters, such as precursor, temperature, and gas flow, etc.

8.2.3.2 Flame synthesis

The flame pyrolysis technique is a thermochemical method that is used widely for the preparation of CNTs and CNOs [85–88]. This method offers a high yield of production, feasible doping, low cost, simple operation, and ambient condition synthesis of CNOs. Mahapatra *et al* prepared CNOs and CNO/ZnO composites by one-step flame synthesis using clarified butter and zinc chloride as the precursors [89, 90]. The same group also prepared nitrogen (N-CNO) and sulfur (S-CNO) doped CNOs using acetonitrile and thiophene (C_4H_4S) as a source of C, N, and S, respectively [88, 91–93]. The TEM image (figure 8.5(c)) of N-CNO prepared by the flame synthesis in the temperature range 800 °C–1000 °C reveals that the interplanar spacing between the graphitic shells is 0.34 nm and the CNOs are free of impurities [88]. Furthermore, N-CNOs are also prepared using various combinations of N–C sources such as ethanol-benzylamine [94], pyrrole [95], and nanodiamond-urea [96]. Liquified petroleum gas (LPG) and household candles are also used as C sources to obtain high surface area CNOs by flame synthesis [97, 98]. Hou *et al* have synthesized high-yield CNOs using a diffusion flame with ethylene as the C source on a catalytic Ni substrate [85, 87]. The morphological and structural properties of CNOs prepared by catalyst-free flame pyrolysis using Ghee oil depend on the sample collecting plate (glass or brass) [99]. Figure 8.5(d) shows typical HRTEM images of CNO samples collected on a brass plate. It is observed that the samples show spherical shapes, having different sizes and lattice spacings on glass (55 nm and 0.34 nm) and brass (37 nm and 0.32 nm). Further, the sample collected on glass has a disordered structure, whereas the sample on brass was covered by an amorphous carbon layer. The thermal annealing of the sample on both collecting plates results in low disorder and lower defect density [99]. Recently, Yeon *et al* synthesized amorphous carbon encapsulated CNOs by implementing the laser pyrolysis method with an infrared CO_2 laser used to pyrolyze C_2H_2 and C_2H_4 gases [100].

8.2.3.3 Hydrothermal synthesis

Hydrothermal synthesis is the most popular method of preparation of various types of nanomaterials including nanocarbons [101, 102], semiconductor nanostructures [103–105], and nanocomposites [106, 107]. Unlike other methods, this method has the advantages of low temperature and low-cost synthesis of nanostructures with well-controlled morphology, composition, and structure. There are some reports that exist on the hydrothermal synthesis of CNOs with various sources of carbon precursor. For example, Sang *et al* have synthesized CNO using citric acid at 180 °C [108]. N containing CNOs were prepared using *Lentinus edodes* as a source of N and C with Fe $(NO_3)_3$ as the N fixing agent at 200 °C. The corresponding HRTEM image of the CNO is shown in figure 8.5(e) and has a polyhedron shape with 10–30 concentric rings with an interplanar spacing of 0.338 nm [109]. The electron diffraction image of the sample is shown in the inset of the figure. Spherically shaped nano-onions with an average size of 25 nm were prepared on a graphene sheet using glucose and glycerol as the C sources at 200 °C [110]. The hydrogenated CNOs obtained using $CHCl_3$ precursor required a longer time of reaction at 100 °C [111]. The hydrothermal method is further used for post-synthesis surface modification/functionalization of CNOs obtained by different methods (thermal annealing and carbonization of phenolic-formaldehyde). The CNO-made composite with MnO_2 (OLC/MnO_2) by hydrothermal synthesis using $KMnO_4$ is applied for energy storage applications [112–114]. Figure 8.5(f) is a low-resolution TEM image of the core–leaf structure of CNO–MnO_2 nanocomposite with clear evidence of MnO_2 flakes spreading out from the OLC core [113]. Some work has also been reported on Bi_2WO_6/CNO nanocomposite using Bi $(NO_3)_3{\cdot}5H_2O$ and $Na_2WO_4{\cdot}2H_2O$ as precursors for Bi and W by hydrothermal synthesis [115, 116]. In addition to the above-discussed methods, CNO and their composites are occasionally prepared using other methods such as electrochemical synthesis [117–119] and aerosol methods [120].

8.3 Applications of CNOs

8.3.1 Electrochemical energy conversion and storage applications

Over the past two decades, non-noble metal and nanocarbon-based materials have been used widely as electrodes for electrochemical energy conversion and storage applications [123–126]. Figure 8.6 illustrates the use of a CNO catalyst for the electrochemical energy conversion technologies of water splitting (converting electrical energy into chemical energy) and fuel cells (converting chemical energy into electrical energy). CNOs are also used as electrode materials for energy storage devices such as supercapacitors and batteries.

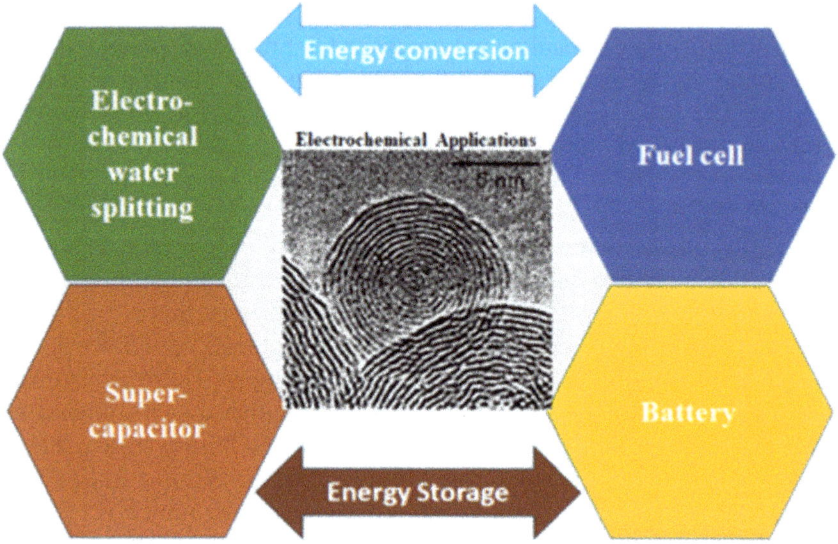

Figure 8.6. Schematic of CNO applications for the electrochemical energy conversion technologies of water splitting and fuel cells, and the energy storage devices of the supercapacitor and battery. (Reproduced with permission from [33]. Copyright 1995 Elsevier.)

8.3.1.1 Electrochemical water splitting

CNOs derived from ZIF-8 and embedded in TiO_2/Fe_2O_3 and core–shell structures of Mo–Ni/CNOs shows bifunctional electrocatalytic activity and produce hydrogen and oxygen simultaneously through a hydrogen evolution reaction (HER) and oxygen evolution reaction (OER) in 1 M KOH [46, 125]. A nitrogen-doped CNO hetero-structure with Ni/Fe_2O_3 shows good performance in the electrochemical water oxidation reaction [126]. The core–shell structure of Ni nanoparticles encapsulated with CNOs supported on CNTs shows superior electrocatalytic activity for the HER in 0.1 M H_2SO_4 [127]. Onion-like graphene sheets (OGNs) surrounded by MoS_2 nanosheets (OGN@MoS_2) prepared by a one-pot plasma-induced electrochemical method show better catalytic activity of the HER. Figure 8.7(a) shows a HRTEM image of OGN@MoS_2 with clear evidence of onion-like graphene and MoS_2 nanosheets. The interlayer spacing identified in figure 8.7(b) corresponds to the (002) plane of graphene (0.34 nm) and (100) plane of MoS_2 (0.27 nm). The inset in figure 8.7(a) shows the SAED pattern of OGN@MoS_2 showing planes corresponding to onion-like graphene and MoS_2. Figures 8.7(c) and (d) show the electrochemical HER polarization curves and Tafel slopes of individual onion-like graphene (OGN) and MoS_2 nanosheets, and composite graphene sheets with MoS_2 (GN@MoS_2) and onion-like graphene with MoS_2 (OGN@MoS_2) in 0.5 M H_2SO_4 electrolyte. Among these samples, OGN@MoS_2 shows the better catalytic activity and shows an overpotential (η) of 212 mV at a current density (j) of 10 mA cm^{-2}. The sample also shows a lower Tafel slope of 73 mV dec^{-1}. This result indicates the strong HER ability of OGN@ MoS_2 catalyst [119]. A nanodiamond derived CNO composite with metal (Mn, Co, Ni) doped Fe_3C shows efficient oxygen evaluation.

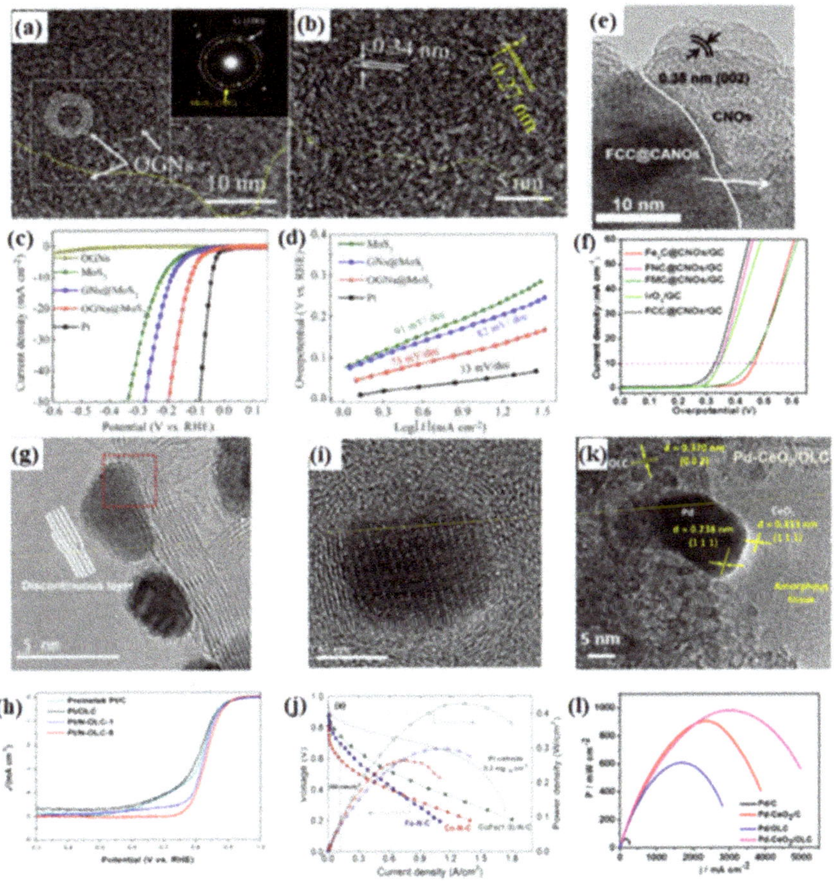

Figure 8.7. (a) HRTEM image of OGN@MoS_2 nanocomposite with (b) identified lattice planes of MoS_2 and OGN. (c) Electrochemical polarization curves and (d) corresponding Tafel plots of individual OGNs, MoS_2, and composite GNs@MoS_2, OGNs@MoS_2, and Pt. (Reproduced with permission from [119]. Copyright 2020 American Chemical Society.) (e) HRTEM image and (f) OER polarization curves of FCC@CNO along with Fe_3C@CNOs, FNC@CNOs, FMC@CNOs, and IrO_2 on a glassy carbon (GC) electrode. (Reproduced with permission from [128]. Copyright 2020 Elsevier.) (g) HRTEM image and (h) ORR polarization curve of Pt/N-OLC-5 along with Pt/OLC and Pt/N-OLC-1 catalyst. (i) and (k) HRTEM images and (j) and (l) fuel cell polarization curves of CoFe/N-CNO and Pd–CeO_2/CNO nanocomposites. ((i) and (j) Reproduced with permission from [26]. Copyright 2011 Elsevier. (k) and (l) Reproduced with permission from [134]. Copyright 2022 American Chemical Society.)

Figure 8.7(e) shows a typical TEM image of Co-doped Fe_3C and CNO composite (FCC@CNO) with clear evidence of Fe_3C nanoparticles and CNO present in the image. The comparative OER polarization curves (figure 8.7(f)) of undoped (Fe_3C@CNO) and Ni (FNC@CNO), Mn (FMC@CNO) and Co (FCC@CNO) doped Fe_3C nanoparticle composites with CNO, with commercial IrO_2 on glassy carbon electrode (GC), all show strong OER activity and the electrode with the FCC@CNO nanocomposite catalyst exhibits an overpotential of 320 mV at a current density of 10 mA cm^{-2} in 1 M KOH [128].

8.3.1.2 Fuel cells

Recently some progress has been made towards the use of nanocarbon materials (CNOs/fullerene/carbon fibers) as catalysts for fuel cell applications [129, 130]. Choi *et al* observed that nitrogen-doped nano-onions (NNOs) prepared by the pyrolysis of nanodiamond derived oxidized nano-onions (NO) showed good catalytic ORR activity with a higher current density than NDs, NOs, and oxidized NOs (ONOs) in 0.1 M KOH [96]. In another study, N-CNO was synthesized by flame synthesis and applied as a catalyst for the ORR in a microbial fuel cell in a 50 mM phosphate buffer saline (PBS) electrolyte [90]. Nitrogen and boron co-doped CNOs (NB-CNOs) prepared using a conventional thermal annealing method show good ORR activity ($J = 2.4$ mA cm^{-2} onset potential = 0.8 V) in 0.1 M KOH [38]. Gang *et al* have examined the potentials of N-doped CNOs (N-CNOs) and Pt-supported N-CNOs (Pt/N-CNOs) for fuel cell applications. The catalyst with 2% Pt in N-CNOs (Pt/N-CNOs) showed a more significant enhancement of oxidation–reduction (ORR) activity than N-CNOs. The catalyst delivered a current density of 0.14 and 0.59 mA cm^{-2} at a cell voltage of 0.8 and 0.65 V, respectively, in 0.5 M H_2SO_4 solution [131]. In another study, Kim *et al* reported the efficient catalytic activity of ORR from a catalyst made up of Pt and CNO composite (Pt/CNO) [44]. The composite samples were prepared by having 20 wt% Pt with different moles of liquid NH_3 1 M (Pt/N-OLC-1) and 5 M (Pt/N-OLC-5) using an arc discharge method. A HRTEM image of Pt/N-OLC-5 is shown in figure 8.7(g) with evidence of Pt nanoparticles presented in the OLC matrix. The ORR polarization curves shown in figure 8.7(h) indicate that the Pt/N-OLC-5 catalyst with a 2.31 mA cm^{-2} current density shows higher ORR activity compared to Pt/OLC and Pt/N-OLC-1 [44]. The hybrid nanocatalyst made up of Pd-supported N-CNOs exhibits a high current density of 17.4 mA cm^{-2} for ethanol oxidation in an alkaline ethanol fuel cell [132]. The catalyst also shows improved ORR activity in 1 M KOH and 1 M C_2H_5OH [132]. A nanodiamond derived CNO composite with Pd doped MnO_2 (CNO/Pd–MnO_2) acts as an active catalyst for glycerol oxidation with a cathode current density of 5.58 mA cm^{-2} at 5.38 V [133]. The heat treatment of hexamethylene diamine (HDA) in the presence of Co and Fe species at 900 °C under N_2 flow results in the formation of binary CoFe nanoparticle encapsulated N-doped CNO (CoFe/N-CNO) in a core–shell structure and its HRTEM image is shown in figure 8.7(i). The fuel cell polarization curves of the CoFe/N-CNO catalyst along with the Co/N-CNO catalyst presented in figure 8.7(j) reveal that the CoFe/N-CNO catalyst shows higher cell performance, having a current density of 0.05 A cm^{-2} at 0.8 V and a high power density of 0.42 W cm^{-2} [26]. Recent studies have reported that CNOs and their derivatives were utilized for proton exchange membrane fuel cell (PEMFC) applications [100, 134]. Amorphous carbon encapsulated CNO (AC/CNO) prepared by laser pyrolysis exhibited a PEMFC power density of 0.72 W cm^{-2} under low humidity conditions [100]. The composite Pd nanoparticles containing CeO_2 on CNOs (Pd–CeO_2/CNO) prepared by a simple chemical method shows excellent catalyst activity for hydrogen oxidation reaction (HOR) in 1 M KOH electrolyte. The HRTEM image of Pd–CeO_2/CNO composite shown in figure 8.7(k) has Pd nanoparticles with CeO_2 in the CNO network as identified by the respective crystal planes. From the fuel cell performance data (figure 8.7(l)), it is observed that the Pd–CeO_2/CNO catalyst delivered a higher peak power density (1.0 mW cm^{-2} at a current density of 3 A cm^{-2}) than other catalysts made without CeO_2 (Pd/CNO) and Pd–CeO_2 with C composite (Pd–CeO_2/C) [134].

8.3.1.3 Supercapacitors and batteries

Excellent research progress has been made on CNOs and their composites for electrochemical supercapacitor and battery applications [8, 23, 24, 135]. Ultrasonication (in N, N-Dimethylformamide at 60 °C for 15 min) treated OLCs prepared by flame synthesis shows a three times better specific capacity (C_p) of 647 F g^{-1} at 10 A g^{-1} current density with 97% retention over 5000 cycles in 1 M H_2SO_4 than untreated and as-prepared OLCs [136]. John *et al* have reported the high capacitance performance of OLCs prepared by thermal annealing with a charge/discharge rate of 50 V s^{-1} at 10 ms time constant compared with nanodiamonds, carbon black, and activated carbon in 1 M KOH [137]. Mesoporous carbon fullerene micro-particles supported on carbon cloth (mCF@CC) flexible electrodes show high electrochemical performance with a specific capacitance of 440 F g^{-1} at a current density of 2 A g^{-1}. The mCF@CC electrode has a good rate capability of 84.85% at 12 A g^{-1} with 98.67% capacitance retention and high stable life up to 10 000 charge/discharge cycles [138]. Activated CNOs (A_t-CNO) prepared by treating pyrolysis derived CNO using KOH solution were examined for energy storage applications [139, 140]. The specific capacitance of A_t-CNO was obtained as 71.4 F g^{-1} at 2 A g^{-1} in organic 1 M $TEABF_4$ electrolyte with a capacitance retention of 80% even after 10 000 cycles [139]. In another study, CNOs obtained by laser excitation of ethylene molecules have porosity on the outer shell after activation chemically in 6 M KOH solution. Compared to pristine sample, A_t-CNO exhibits a nearly five times specific capacitance of 122 F g^{-1}, a power density of 153 kW kg^{-1}, and an energy density of 8.5 Wh kg^{-1} in 2 M l^{-1} KNO_3 electrolytes [140]. A polyaniline (PANI) embedded CNO (CNO-PANI) composite prepared by thermal annealing and a chemical method combination shows 450 F g^{-1} specific capacitance in a symmetric cell [141]. In a separate work, it is reported that the quinone-coated OLCs, and decorated OLCs/carbon fiber composites, have better specific capacitances of 264 and 288 F g^{-1} in 1 M H_2SO_4 electrolyte [25, 142]. Monoatomic doping induces superior specific capacitance and energy density than undoped CNO in different electrolyte environments [88, 92, 143]. S-doped CNO (S-CNO) exhibits a high specific capacitance of 305 F g^{-1} at a current density of 2 A g^{-1} with high power and energy densities of 5500 W kg^{-1} and 6.1 Wh kg^{-1} in 1 M H_2SO_4 solution [92]. Similarly, the specific capacitance of N-doped CNO (N-CNO) is 88 F g^{-1} at $J = 4$ A g^{-1}, and the power and energy density values are 1850 W kg^{-1} and 12.3 Wh kg^{-1}, respectively, in 1 M Na_2SO_4 electrolyte [88]. Further, the composite N-CNO and NiO (N-CNO/NiO) shows improved supercapacitor performance in the same electrolyte. The electrochemical CV curves of a N-CNOs/NiO asymmetric supercapacitor (ASC) at different scan rates measured in 1 M Na_2SO_4 are shown in figure 8.8(a). The figure shows the high current response and capacitive nature of the N-CNOs/NiO ASC with the specific capacitance of 113 F g^{-1} at a current density of 4 A g^{-1} (at 100 mV s^{-1}). From the Ragano plot shown in figure 8.8(b), it is observed that the N-CNO/NiO ASC cell delivered an energy density of

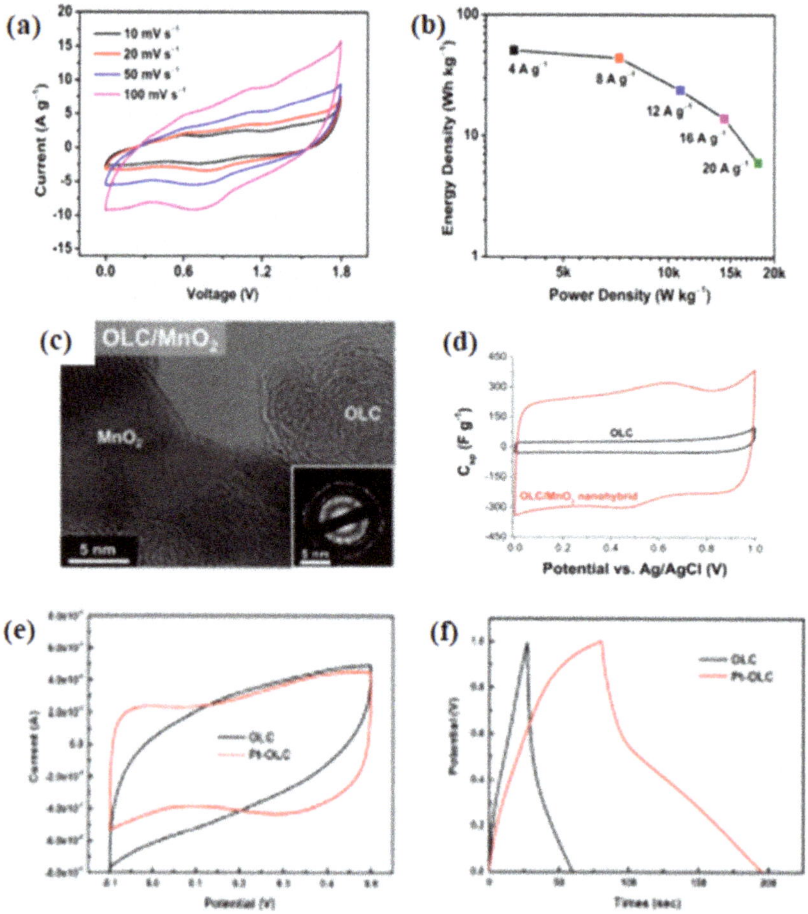

Figure 8.8. (a) CV curves at different scan rates and (b) a Ragano plot of N-CNO/NiO. (Reproduced with permission from [88]. Copyright 2020 Elsevier.) (c) HRTEM image and (d) CV curve of OLC/MnO_2 along with OLC. (Reproduced from [112]. CC BY 3.0.) (e) CV curves and (f) charge and discharge curves of Pt-OLC composite with OLC. (Reproduced from [145]. CC BY 3.0.)

51 Wh kg^{-1} and a power density of 3.6 kW kg^{-1} at a current density of 4 A g^{-1}. At a current density of 20 A g^{-1} the ASC cell shows an energy density of 6 Wh kg^{-1} and power density of 18 kW kg^{-1}. The reported values of specific capacitance, power density, and energy density of CNO/NiO are higher than for N-CNO and NiO cells. The cyclic stability and coulombic efficiency of N-CNOs/NiO ASC are 98% and 99% at the current density of 20 A g^{-1} even after 10 000 cycles, respectively [88]. A recent study observed that N-CNOs demonstrate both symmetric and asymmetric supercapacitors natures and show a high specific capacitance value of 205 F g^{-1} for asymmetric devices, and excellent capacity retention of 96% to 98% over 5000 cycles for both symmetric and asymmetric devices [143]. Two-dimensional layered δ-MnO_2 decorated on CNO cells shows temperature-dependent pseudocapacitance charge storage behavior with a specific capacitance varying from 147 to 321 F g^{-1} as the temperature rises from 4 °C

to 70 °C [144]. A symmetric pseudocapacitance nature is observed from the birnessite type MnO_2 on an OLC nanohybrid (OLC/MnO_2). Figure 8.8(c) shows a HRTEM image of OLC/MnO_2 nanohybrid with an SAED pattern (inset) having lattice planes belonging to MnO_2 and OLC. Figure 8.8(d) shows the CV curves of OLC and OLC/MnO_2 nanohybrids measured in 1 M Na_2SO_4 using Ni foam as a working electrode. Clearly, the OLC/MnO_2 nanohybrid shows a symmetric pseudocapacitance with a high specific capacity of 408 F g^{-1} (at 1 A g^{-1}), the specific energy density of 5.6 Wh kg^{-1}, power density of 74.8 kW kg^{-1}, and excellent long hours capacity retention [112]. The composite Pt nanoparticle decorated OLC (Pt-OLC) has proven to be an excellent electric-double layer supercapacitor compared to pristine OLC, as is evident from figure 8.8(e). The galvanostatic charge/discharge curve for OLC and Pt-OLC at 1 A g^{-1} current density shown in figure 8.8(f) reveals a change in linearity in Pt-OLC that signifies an electric-double layer supercapacitor. The obtained specific capacity of 115.2 F g^{-1} of Pt-OLC is higher than the reported value for noble metal-OLC (Pd-OLC, Au-OLC and Ag-OLC) composites [145]. A chemically oxidized CNO (OCNO) composite with $RuO_2{\cdot}nH_2O$ ($RuO_2{\cdot}nH_2O$/OCNO) shows a high electrochemical supercapacitor performance. The best electrode having RuO_2 wt% (67.5 wt%) exhibits an excellent specific capacitance of 334 F g^{-1}, a power density of 242.2 kW kg^{-1}, and a high energy density of 11.6 Wh kg^{-1} in 1 M H_2SO_4 electrolyte [146].

It is demonstrated that CNOs obtained from flame synthesis can become universal anode materials for metal ion (Li^+, Na^+ and K^+) batteries [147–150]. The high temperature heat treatment of candle soot carbon transforms it into CNOs that show electrochemical high performance for Li^+, Na^+, and K^+ ion batteries. A sample obtained by heat treatment at 750 °C in Ar atmosphere exhibited a high specific capacitance (at current density) of 1247 (150 mA g^{-1}), 461 (20 mA g^{-1}), and 406 (20 mA g^{-1}) for Li^+, Na^+, and K^+ ion batteries, respectively, at a low charging rate. The samples had capacity retention of 78% (Li-ion), 87% (Na-ion), and 94% (K-ion) for 1000 cycles [147]. In the other work, it is reported that CNOs formed after heat treatment of candle soot carbon at 2400 °C in N_2 showed a high discharge capacity of 534 mAh g^{-1} with an improved coulombic efficiency of 76.4% for a Li-ion battery [148]. Mesoporous and high surface area N-doped CNO catalyst improved the Li–O_2 battery performance with a specific capacitance of 12 181 mAh g^{-1} and 76% cyclic stability for nearly 200 cycles [151]. N-OLC also shows an excellent discharge capacity of 805 mAh g^{-1} at a current density of 0.1 C after 50 cycles for Li-ion batteries [152]. The metal (Sn) and N-CNO composite (Sn@N-CNO) prepared by arc discharge exhibit a high discharge capacity of 902.8 mAh g^{-1} at 0.1 A g^{-1} and capacitance is stable for 1000 cycles [153]. The Co-CNO structure was obtained by thermal treatment of Co and C precursors at 850 °C (the sample denoted as Co-N-CNO-850) and shows excellent Li-ion battery performance [154]. The CV curve of CO-N-CNO-850 for three cycles reveals the formation of a solid electrolyte interface (SEI) in the first cycle, confirmed by observing an anodic peak at 0.6 V. The absence of SEI film in the second and third CV cycle suggests Li^+ migration during the charge/discharge process and improved cell capacity. The sharp peaks at 0.01 and 0.1 V represent the insertion and extraction of Li^+ into the carbon layer [154]. The Co-N-CNO-850 anode has a reversible capacity of 774 mAh g^{-1} at a current density of 0.5 C after 500 cycles [154].

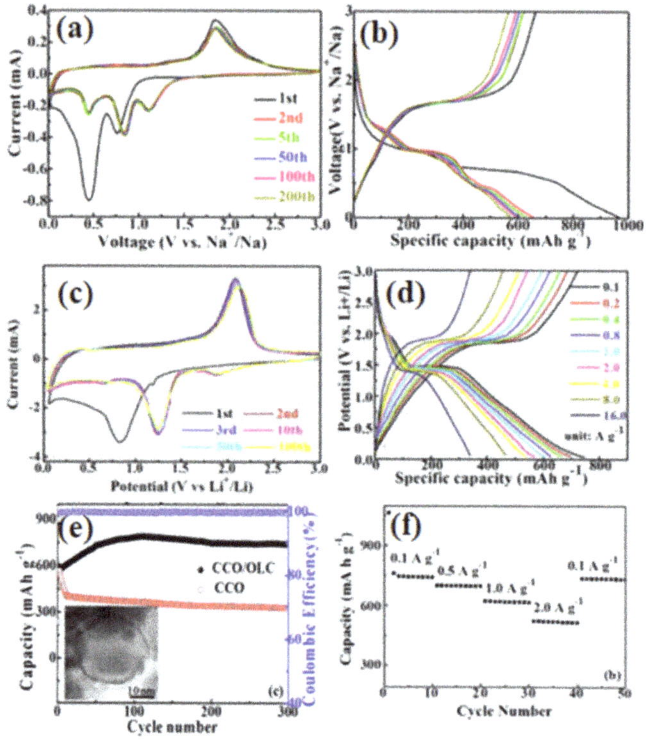

Figure 8.9. CV curves and discharge/charge curve of (a) and (b) CoS@S-CNO and (c) and (d) FeS@CNO/a-C anode materials. (e) The cycling performance of CCO/OLC composite along with bare CCO measured at 0.2 A g^{-1} with a TEM image of CCO/OLC as the inset. (f) The rate capability of Mo_2C@CNO/a-C composite under different current densities. ((a) and (b) Reproduced with permission from [155]. Copyright 2018 Elsevier. (c) and (d) Reproduced with permission from [156]. Copyright 2018 Elsevier. (e) Reproduced with permission from [157]. Copyright 2016 Elsevier. (f) Reproduced with permission from [159]. Copyright 2017 Elsevier.)

Liu *et al* have recently prepared composite CNO coated metal sulfide (CoS and FeS) nanocapsules using an arc discharge method and examined their use in Na- and Li-ion battery applications [155, 156]. The composite sulfur-doped CNO coated CoS (CoS@S-CNO) nanocapsules have become a superior anode material for the researchable Na-ion battery with a long cycle life and high rate capacities in 1 M $NaCF_3SO_3$. The first CV curve (figure 8.9(a)) shows a reduction peak at 0.44 V corresponding to SEI film formation and an oxidation peak at 1.84 V corresponding to CoS phase formation. The charge/discharge data of the CoS@S-CNO anode in figure 8.9(b) shows that the initial discharge and charge capacity are around 968 and 663 mAh g^{-1}, respectively. The discharge capacity reaches 582 mAh g^{-1} after 500 cycles at a 0.1 A g^{-1} current density [155]. Another composite CNO coated FeS nanocapsules embedded in amorphous carbon (FeS@CNO/a-C) anode material has the advantages of good cycling stability and high rate capability. The first CV curve of the FeS@CNO/a-C anode (figure 8.9(c)) and three reduction peaks at 0.7, 1.11, and 1.37 V represent the formation of SEI film between the sample and electrolyte interface, conversion reaction, and the interaction. The only anodic peak in the CV curve at 2.1 V represents the Fe oxidation reaction. In all CV curves (except the first), the reduction and oxidation peak position intensity is almost unchanged. This indicates the long-term cycle stability of the FeS@CNO/a-C anode. The discharge/charge profile measured at different rates (figure 8.9(d)) reveals that the discharge capacity is varied from 746 to 340 mAh g^{-1} as the rate changes from 0.1 to 1.6 A g^{-1}. This implies that the FeS@CNO/a-C anode has high rate stability for the Li-ion battery [156]. The anode materials made with core–shell $CuCo_2O_4$ nanoparticles encapsulated by CNO (CCO/CNO) show high cycling stability and rate capability [157]. From the cycling performance data shown in figure 8.9(e) in comparison with a bare CCO anode, the CCO/CNO anode has high stability with discharge capacities of 743.2 mAh g^{-1} at 0.2 A g^{-1} after 300 cycles with a high coulombic efficiency of 99.5% [157]. A TEM image of CCO/CNO core–shell is shown as an inset in figure 8.9(e).

Similarly, many composites such as CuO@CNO, Fe_3O_4@CNO, Mo_2C@CNO, and NiS@CNO having core–shell structures show improved Li-ion battery performance with high rate (cycling) stability and discharge capacity [158–161]. Figure 8.9(f) show the rate capability of Mo_2C@CNO embedded with amorphous carbon (Mo_2C@CNO/a-C) under different current densities. It seen that the composite exhibits good rate stability with 98.5% retention when the current densities return from 2 to 0.1 A g^{-1} [159]. The composite α-MnO_2 nanorod/CNO (α-MnO_2/CNO) prepared by the molten salt method is used as high performance cathode material for Zn-ion battery in 1 M $ZnSO_4$ electrolyte. The α-MnO_2/CNO exhibits a stable and high reversible capacity of 1.68 mAh g^{-1} with 93% capacity retention [162].

8.3.2 Magnetic memory

CNOs and their composite materials exhibit magnetic properties and have potential for magnetic memory applications. The CNO shell prevents the oxidation of metal core nanoparticles and enhances the magnetic interactions in metal/CNO core/shell structures. Here, we review the work carried out on the magnetic properties and applications of CNOs. The metal carbide (Co–C and Fe–C) nanoparticle encapsulated amorphous CNO prepared by pulsed laser ablation method form core–shell structures and the corresponding TEM images of Fe–C@a-CNO and Co–C@a-CNO are shown in figure 8.10(a) and (b). These composite samples show room temperature (RT) ferromagnetic behavior, as observed from the M–H curve shown in figures 8.10(c) and (d). Close observation of the M–H curve reveals that the samples have saturation magnetization (M_s) around 0.002 emu. Both samples show Curie temperature at 300 K as the field cooling (FC) and zero-field cooling (ZFC) M–T curve measure at an applied field of $H = 100$ Oe converges around 300 K [70].

The metal nanoparticle–CNO composites (Pd@CNO and Pt@CNO) prepared using a chemical reduction method show ferromagnetic nature at room temperature. The magnetic interactions of Pd and Pt nanoparticles with CNO gives rise to the ferromagnetic behavior in the samples. The Pd@CNO and Pt@CNO composites show saturation magnetization of 149 and 171 Oe coercive fields at room temperature [163]. Fe cores encapsulated with CNO shells were prepared by different

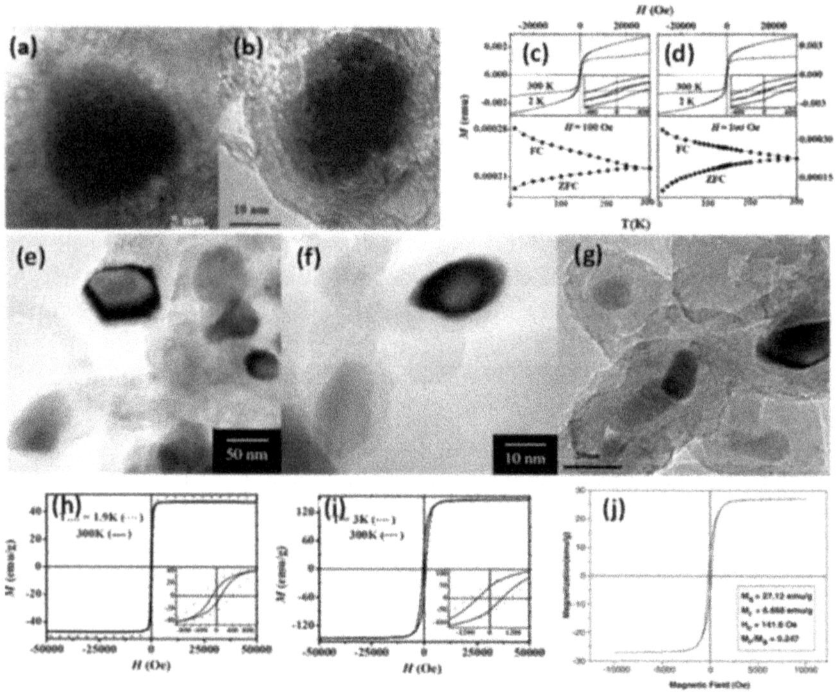

Figure 8.10. HRTEM image (a) and (c) and magnetic *M–H* and *M–T* curves (b) and (d) of Fe–C@CNO and Co–C@CNO core@shell structures. TEM image (e) and (f) and magnetic *M–H* curve (h) and (i) of Ni and Co core–fullerene shell composites. (g) TEM image and magnetic and (j) *M–H* curve of Ni/Al@CNO core–shell composite. ((a), (b), (c), and (d) Reproduced with permission from [70]. Copyright 2006 AIP Publishing. (e), (f), (h), and (i) Reproduced with permission from [166]. Copyright 2002 Elsevier. (g) and (j) Reproduced with permission from [167]. Copyright 2008 Elsevier.)

methods and showed room temperature ferromagnetism [164, 165]. Ning *et al* studied the magnetic properties of CNO encapsulated Fe nanoparticles synthesized by the detonation method. The Fe@CNO core/shell structure showed ferromagnetic behavior at room temperature with a saturation magnetization of 54.56 emu g^{-1} and remanent magnetization (M_r) of 5.35 emu g^{-1} [164]. In another work, carbon encapsulated Fe nanoparticles were processed by an arc discharge method exhibiting ferromagnetic behavior with low coercivity between 129 and 362 Oe [165]. Fullerene encapsulated ferromagnetic nanoparticles (Ni and Co) synthesized by a thermal decomposition method having core–shell structure (figures 8.10(e) and (f)) exhibit RT ferromagnetism as observed from the *M–H* curve shown in figures 8.10(h) and (i). It is seen that the Co core–fullerene shell structure has higher M_s than the Ni core–fullerene shell structure [166]. The bimetallic Ni/Al core and CNO shell (Ni/Al@CNO, figure 8.10(g)) composite also shows a strong RT ferromagnetic nature (figure 8.10(j)) [167]. CNO buckypaper filled with ferromagnetic nanoparticles Fe_5C_2/Fe_7C_3 (Fe_5C_2/Fe_7C_3@CNO) and obtained using a pyrolysis method show M_s of 47 emu g^{-1} at room temperature [168]. Table 8.1 presents a summary of CNOs and their composite materials used for different applications.

Table 8.1. Summary of CNOs and their composite materials used for different applications.

Sl. No.	Type	Size	Shape	Synthesis method	Application	Specification	Reference
1	Core–shell CNO/ TiO_2/Fe_2O_3	—	Spherical	Hydrothermal	HER and OER	$\eta = 291$ mV for HER and 290 mV for OER	[125]
2	Core–shell N-CNO/$NiFe_2O_3$	150 nm	Spherical	Pyrolysis	OER	$\eta = 296$ mV	[126]
3	OGN/MoS_2	100 nm	Spherical	Electrochemical method	HER	$\eta = 118$ mV	[119]
4	CNO/M–Fe_3C (M = Mn, Co, Ni)	100 nm	Spherical	Thermal annealing	OER	FCC@CNO $\eta = 271$ mV	[128]
5	Core–shell Pt@N-CNO	2–3 nm	Spherical	Pyrolysis	Fuel cell	$j = 0.14$ mA cm^{-2} at 0.8 V and 0.59 at 0.65 V	[131]
6	Pt/N-OLC	15–19 nm	—	Arc discharge	ORR	$j = 2.31$ mA cm^{-2}	[44]
7	Core–shell N-CNO/NiO	30 nm	Spherical	Pyrolysis	Supercapacitor	$C_p = 113$ F g^{-1}	[88]
8	Lattice shell OLC/MnO_2	—	Circular	Hydrothermal	Supercapacitor	$C_p = 408$ F g^{-1}	[112]
9	Core–shell Pt@OLC	5 nm	Spherical	Hydrothermal	Supercapacitor	$C_p = 115.2$ F g^{-1}	[145]
10.	CNO	35–40 nm	—	Flame synthesis	Li-, Na-, and K-ion batteries	Lower charging rate capacity 1247, 461, and 406 at current density 150 mA g^{-1}, 20 mA g^{-1}, and 20 mA g^{-1}	[147]
11.	CNO	10–50 nm	Spherical	Flame synthesis	Li- and K-ion batteries	Discharge capacity of 534 mAh g^{-1}	[148]
12.	N-CNO	20 nm	Quasi- spherical	Thermal annealing	Li–O_2 battery	Specific capacitance of 12 181 mAh g^{-1}	[151]
13.	Core–shell Sn@N-CNO	5–60 nm	spherical	Arc discharge	Li-ion battery	Discharge capacity 902.8 mAh g^{-1} at 0.1 A g^{-1}	[153]
14.	Co-N-CNO	12–30 nm	Spherical	Thermal treatment	Li-ion battery	Reversible capacity of 774 mAh g^{-1}	[154]
15.		—	Spherical	Arc discharge	Na-ion battery	Discharge capacity of 582 mAh g^{-1}	[155]

(*Continued*)

Table 8.1. (*Continued*)

Sl. No.	Type	Size	Shape	Synthesis method	Application	Specification	Reference
	Core–shell CoS@S-doped OLC						
16.	$CuCo_2O_4$/OLC	5–40 nm	Spherical	Arc discharge	Li-ion battery	Discharge capacity 743.2 mAh g^{-1} at a current of 0.2 A g^{-1}	[156]
17.	FeS@OLC/a-C	20–80 nm	Spherical	Arc discharge	Li-ion battery	Capacity of 719.2 mAh g^{-1} after 200 charge/discharge cycles	[157]
18.	Core–shell NiS@OLC	20–40 nm	Spherical	Arc discharge	Li-ion battery	Discharge capacity of 546 mAh g^{-1} after 100 cycles at 0.1 A g^{-1}	[160]
19.	α-MnO_2/OLC	40 nm	—	Chemical method	Zn-ion battery	High reversible capacity of 168 mAh g^{-1}	[162]
20.	Core–shell Fe@CNO	10–40 nm	Spherical	Pulsed laser ablation	Magnetic memory	Room temperature ferromagnetism	[70]
21.	Core–shell Fe@CNO	15–50 nm	Spherical	Thermal annealing	Magnetic memory	54.56 emu g^{-1} at room temperature	[164]
22.	Pd@CNO and Pt@CNO	30–40 nm	—	Pyrolysis	Magnetic memory	149 and 171 Oe room temperature	[163]
23.	Core–shell Fe@CNO	—	Spherical	Arc discharge	Magnetic memory	220 emu g^{-1} and 165 emu g^{-1} at 300 K	[165]
24.	Ni@ CNO and Co@CNO	30–60 nm and 10–30 nm	—	Chemical method	Magnetic memory	Room temperature ferromagnetism	[166]
25.	Ni/Al@CNO	10–30 nm and 5–50 nm	Hollow sphere	CVD	Magnetic memory	27.12 emu g^{-1} at room temperature	[167]
26.	Fe_5C_2/Fe_7C_3@CNO	10–30 nm	Spherical	Pyrolysis	Magnetic memory	47 emu g^{-1} at 300 K	[168]
27.	Core–shell Co@CNO	22 nm	Spherical	Laser ablation	Magnetic memory	7 emu g^{-1} at room temperature	[71]

8.4 Conclusions

In this chapter we first summarize the work reported on the synthesis of CNOs and their derivatives by physical methods (thermal annealing, arc discharge, electron beam irradiation, ion implantation, and laser ablation) and chemical methods (chemical vapor deposition, flame synthesis, and hydrothermal methods). The composites of metal oxide/CNO and metal nanoparticle@CNO core–shell structures show improved electrochemical HER, OER, and ORR performance compared to pristine CNOs. It is noted that CNO composites have gained research attention for fuel cell applications. The potential for the use of CNO composites as electrode materials for electrochemical energy storage devices, supercapacitors, and Li- and Na-ion battery applications are discussed. Finally, the magnetic properties of metal nanoparticle@CNO core–shell structures are discussed.

Acknowledgments

The authors acknowledge the STARS (Scheme for Transformational and Advanced Research in Sciences, MoE, Government of India, MoE-STARS/STARS-1/751) program and EEQ (Empowerment and Equity Opportunities for Excellence in Science, DST-SERB, Government of India, EEQ/2020/000319) for financial support.

References

[1] Kono J and O'Connell M J 2018 Carbon Nanotubes Properties and applications (Boca Raton, FL: CRC Press)
[2] Geim A K and Novoselov K S 2007 *Nat. Mater.* **6** 183
[3] Schedin F, Geim A K, Morozov S V, Hill E W, Blake P, Katsnelson M I and Novoselov K S 2007 *Nat. Mater.* **6** 652
[4] Chang H and Wu H 2013 *Energy Environ. Sci.* **6** 3483
[5] Lim S Y, Shen W and Gao Z 2015 *Chem. Soc. Rev.* **44** 362
[6] Rao C N R, Sood A K, Subrahmanyam K S and Govindaraj A 2009 *Angew. Chem. Int. Ed.* **48** 7752
[7] Dhand V, Yadav M, Kim S H and Rhee K Y 2021 *Carbon* **175** 534
[8] Pech D, Brunet M, Durou H, Huang P, Mochalin V, Gogotsi Y and Simon P 2010 *Nat. Nanotechnol.* **5** 651
[9] Ugarte D 1992 *Nature* **359** 707
[10] Krasheninnikov A V and Banhart F J N M 2007 *Nat. Mater.* **6** 723
[11] Banhart F and Ajayan P M 1997 *Adv. Mater.* **9** 261
[12] Pichler T, Knupfer M, Golden M S, Fink J and Cabioc'h T 2001 *Phys. Rev.* B **63** 155415
[13] Pérez-Ojeda M E, Castro E, Kröckel C, Lucherelli M A, Ludacka U, Kotakoski J and Abellán G 2021 *J. Am. Chem. Soc.* **143** 18997
[14] Gu W, Peters N and Yushin G 2013 *Carbon* **53** 292
[15] Plonska-Brzezinska M E 2019 *Chem. Nano Mater.* **5** 568
[16] Bartkowski M and Giordani S 2020 *Nanoscale* **12** 9352
[17] Luszczyn J, Plonska-Brzezinska M E, Palkar A, Dubis A T, Simionescu A, Simionescu D T and Echegoyen L 2010 *Chem. Eur. J.* **16** 4870

[18] Joly-Pottuz L, Matsumoto N, Kinoshita H, Vacher B, Belin M, Montagnac G and Ohmae N 2008 *Tribol. Int.* **41** 69
[19] Hirata A, Igarashi M and Kaito T 2004 *Tribol. Int.* **37** 899
[20] Mamezaki O M O, Adachi H A H, Tomita S T S, Fujii M F M and Hayashi S H S 2000 *Jpn. J. Appl. Phys.* **39** 6680
[21] Wang M S, Golberg D and Bando Y 2010 *ACS Nano* **4** 4396
[22] Okotrub A V, Bulusheva L G, Gusel'nikov A V, Kuznetsov V L and Butenko Y V 2004 *Carbon* **42** 1099
[23] Zeiger M, Jäckel N, Mochalin V N and Presser V 2016 *J. Mater. Chem.* A **4** 3172
[24] Plonska-Brzezinska M E and Echegoyen L 2013 *J. Mater. Chem.* A **1** 13703
[25] Anjos D M, McDonough J K, Perre E, Brown G M, Overbury S H, Gogotsi Y and Presser V 2013 *Nano Energy* **2** 702
[26] Wu G, Nelson M, Ma S, Meng H, Cui G and Shen P K 2011 *Carbon* **49** 3972
[27] Iijima S 1980 *J. Cryst. Growth* **50** 675
[28] Kuznetsov V L, Chuvilin A L, Butenko Y V, Mal'kov I Y and Titov V M 1994 *Chem. Phys. Lett.* **222** 343
[29] Zeiger M, Jäckel N, Aslan M, Weingarth D and Presser V 2015 *Carbon* **84** 584
[30] Mykhailiv O, Zubyk H, Brzezinski K, Gras M, Lota G, Gniadek M, Romero E, Echegoyen L and Brzezinsk M E P 2017 *Chem. Nano Mater* **3** 583
[31] Mochalin V N, Shenderova O, Ho D and Gogotsi Y 2012 *Nat. Nanotechnol.* **7** 11
[32] Kumar S, Nehra M, Kedia D, Dilbaghi N, Tankeshwar K and Ki K H 2019 *Carbon* **143** 678
[33] Ugarte D 1995 *Carbon* **33** 989
[34] Bogdanov K, Fedorov A, Osipov V, Enoki T, Takai K, Hayashi T and Baranov A 2014 *Carbon* **73** 78
[35] Zeiger M, Jäckel N, Weingarth D and Presser V 2015 *Carbon* **94** 507
[36] Liu Y and Kim D Y 2014 *Electrochim. Acta* **139** 82
[37] Thomas M P, Wanninayake N, Goonatilleke M D A, Kim D Y and Guiton B S 2020 *Nanoscale* **12** 6144
[38] Camisasca A, Sacco A, Brescia R and Giordani S 2018 *ACS Appl. Nano Mater.* **1** 5763
[39] Sano N, Wang H, Alexandrou I, Chhowalla M, Teo K B K, Amaratunga G A J and Iimura K 2002 *J. Appl. Phys.* **92** 2783
[40] Sano N, Wang H, Chhowalla M, Alexandrou I and Amaratunga G A 2001 *Nature* **414** 506
[41] Alexandrou I, Wang H, Sano N and Amaratunga G A 2004 *J. Chem. Phys.* **120** 1055
[42] Vasu K, Pramoda K, Moses K, Govindaraj A and Rao C N R 2013 *Mater. Res. Express* **1** 015001
[43] Ishigami M, Cumings J, Zettl A and Chen S 2000 *Chem. Phys. Lett.* **319** 457
[44] Kim S M, Heo Y K, Bae K T, Oh Y T, Lee M H and Lee S Y 2016 *Carbon* **101** 420
[45] Glad X, Gorry J, Cha M S and Hamdan A 2021 *Sci. Rep.* **11** 7516
[46] Chattopadhyay J, Singh S and Srivastava R 2020 *J. Appl. Electrochem.* **50** 207
[47] Banhart F 1999 *Rep. Prog. Phys.* **62** 1181
[48] Chuvilin A, Kaiser U, Bichoutskaia E, Besley N A and Khlobystov A N 2010 *Nat. Chem.* **2** 450
[49] Terrones M 2010 *Nat. Chem.* **2** 82
[50] Gan Y and Banhart F 2008 *Adv. Mater.* **20** 4751

[51] Gonzalez-Martinez I G, Bachmatiuk A, Bezugly V, Kunstmann J, Gemming T, Liu Z and Rümmeli M H 2016 *Nanoscale* **8** 11340
[52] Golberg D, Bando Y, Kurashima K and Sasaki T 1999 *Carbon* **37** 293
[53] Roddatis V V, Kuznetsov V L, Butenko Y V, Su D S and Schlögl R 2002 *Phys. Chem. Chem. Phys.* **4** 1964
[54] Hiraki J, Mori H, Taguchi E, Yasuda H, Kinoshita H and Ohmae N 2005 *Appl. Phys. Lett.* **86** 223101
[55] Lin Z, Shao G, Liu W, Wang Y, Wang H, Wang H and Zhang R 2022 *Carbon* **192** 356
[56] SCabioc'h T, Jaouen M, Thune E, Guerin P, Fayoux C and Denanot M F 2000 *Surf. Coat. Technol.* **128** 43
[57] Cabioc'h T, Jaouen M, Denanot M F and Bechet P 1998 *Appl. Phys. Lett.* **73** 3096
[58] Thune E, Cabioc'h T, Jaouen M and Bodart F 2003 *Phys. Rev.* B **68** 115434
[59] Cabioc'h T, Thune E and Jaouen M 2002 *Phys. Rev.* B **65** 132103
[60] Matsumoto N, Kinoshita H, Choi J and Kato T 2020 *Sci. Rep.* **10** 10037
[61] Zeng H, Du X W, Singh S C, Kulinich S A, Yang S, He J and Cai W 2012 *Adv. Funct. Mater.* **22** 1333
[62] Morales A M and Lieber C M 1998 *Science* **279** 208
[63] Yudasaka M, Komatsu T, Ichihashi T and Iijima S 1997 *Chem. Phys. Lett.* **278** 102
[64] Gautam U K, Vivekchand S R C, Govindaraj A, Kulkarni G U, Selvi N R and Rao C N R 2005 *J. Am. Chem. Soc.* **127** 3658
[65] Vasu K 2014 *Bull. Mater. Sci.* **37** 1383
[66] Radhakrishnan G, Adams P M and Bernstein L S 2007 *Appl. Surf. Sci.* **253** 7651
[67] Radhakrishnan G, Adams P M and Bernstein L S 2006 *Thin Solid Films* **515** 1142
[68] Dorobantu D, Bota P M, Boerasu I, Bojin D and Enachescu M 2014 *Surf. Eng. Appl. Electrochem.* **50** 390
[69] Choi M, Altman I S, Kim Y J, Pikhitsa P V, Lee S, Park G S and Yoo J B 2004 *Adv. Mater.* **16** 1721
[70] Seung H H and Nakajima A 2006 *J. Appl. Phys.* **99** 064302
[71] Zhang H, Liang C, Liu J, Tian Z and Shao G 2013 *Carbon* **55** 108
[72] Bae S *et al* 2010 *Nat. Nanotechnol.* **5** 574
[73] Zhang Y, Zhang L and Zhou C 2013 *Acc. Chem. Res.* **46** 2329
[74] Reina A, Jia X, Ho J, Nezich D, Son H, Bulovic V, Dresselhaus M S and Kong J 2009 *Nano Lett.* **9** 30
[75] Kumar M and Ando Y 2010 *J. Nanosci. Nanotechnol.* **10** 3739
[76] Che G, Lakshmi B B, Martin C R, Fisher E R and Ruoff R S 1998 *Chem. Mater.* **10** 260
[77] Ruan C, Li Z, Zhang D, Yuan X, Liang C, Chang Y, Huang H, Xu L and Chen M 2020 *Carbon* **161** 622
[78] Zhang C, Li J, Shi C, Liu E, Du X, Feng W and Zhoa N 2011 *Carbon* **49** 1151
[79] Yang Y, Liu X, Guo X, Wen H and Xu B 2011 *J. Nanopart. Res.* **13** 1979
[80] Chen X H, Deng F M, Wang J X, Yang H S, Wu G T, Zhang X B, Peng J C and Li W Z 2001 *Chem. Phys. Lett.* **336** 201
[81] Gong Z, Bai C, Qiang L, Gao K, Zhang J and Zhang B 2018 *Diam. Relat. Mater.* **87** 172
[82] Raghu A V, Karuppanan K K and Pullithadathil B 2019 *ACS Appl. Electron. Mater* **1** 2208
[83] Wang X, Xu B, Liu X, Guo J and Ichinose H 2006 *Diam. Relat. Mater.* **15** 147
[84] Wang X, Xu B, Jia H, Liu X and Ichinose H 2006 *J. Phys. Chem. Solids* **67** 871
[85] Houa S S, Chung D H and Lin T H 2009 *Carbon* **47** 938

[86] Chungb D H, Lin T H and Hou S S 2010 *Nanotechnology* **21** 435604
[87] Zahid M U, Pervaiz E, Hussain A, Shahzad M I and Niazi M B K 2018 *Mater. Res. Express* **5** 052002
[88] Mohapatra D, Muhammad O, Sayed M S, Parida S and Shim J J 2020 *Electrochim. Acta* **331** 135363
[89] Mohapatra D, Badrayyana S, Pant A and Parida S 2016 *Mater. Lett.* **183** 56
[90] Mahopatra D, Badrayyana S and Parida S 2016 *Mater. Chem. Phys.* **174** 112
[91] Han T H, Mahopatra D, Mahato N, Parida S, Shim J H, Nguyen A T N, Nguyen V Q, Cho M H and Shim J J 2020 *J. Ind. Eng. Chem.* **81** 269
[92] Mahopatra D, Dhakal G, Sayed M S, Subramanya B, Shim J J and Parida S 2019 *ACS Appl. Mater. Interfaces* **11** 8040
[93] Mohapatra D, Gowthaman N S K, Sayed M S and Shim J J 2020 *Sensors Actuators* B **304** 127325
[94] Martinez E T, Torres J A M, Salazar J V C, Reyes M R, Ruiz L F C and Sandoval R L 2018 *Carbon* **140** 171
[95] Pallavolu M R, Gaddam N, Banerjee A N, Nallapureddy R R and Joo S W 2021 *Int. J. Eng. Res.* **45** 1–16
[96] Choi E Y and Kim C K 2017 *Sci. Rep.* **7** 4178
[97] Dhand V, Prasad J S, Rao M V, Bharadwaj S, Anjaneyulu Y and Jain P K 2013 *Mater. Sci. Eng.* C **33** 758
[98] Bu I Y Y 2014 *Sol. Energy* **105** 236
[99] Mongwe T H, Matsoso B J, Mutuma B K, Coville N J and Maubane M S 2018 *Diam. Relat. Mater.* **90** 135
[100] Yeon J H, Choi J, Jang S and Choi M 2021 *Carbon* **184** 910
[101] Gogotsi Y, Joseph A L and Yoshimura M 2000 *J. Mater. Res.* **15** 2591
[102] Shen J, Zhu Y, Yang X, Zong J, Zhang J and Li C 2012 *New J. Chem.* **36** 97
[103] Liu B and Zeng C 2003 *J. Am. Chem. Soc.* **125** 4430
[104] Keerthana B G T, Muniyappan S S and Murugakoothan P 2018 *Mater. Lett.* **220** 20
[105] Thennarasi A and Vasu K 2022 *Mater. Today Commun.* **32** 103934
[106] Perera S D, Mariano R G, Vu K, Nour N, Seitz O, Chabal Y and Balkus K J 2012 *ACS Catal.* **2** 949
[107] Wang R, Han M, Zhao Q, Ren Z, Guo X, Xu C and Lu L 2017 *Sci. Rep.* **7** 44562
[108] Sang S, Yang S, Guo A and Gao X 2020 *Chem. Asian J.* **15** 3428
[109] Han L, Zhang P, Li L, Su B, An X and Lei Z 2021 *Appl. Surf. Sci.* **543** 148752
[110] Najafi A S G and Alizageh T 2022 *J. Mater. Sci. Mater. Electron.* **33** 7444
[111] Zhai C, Kong X, Hu K, Wang L and Fan X 2020 *Chem. Phys. Lett.* **738** 136866
[112] Makgopa K, Ejikeme P M, Jafta C J, Raju K, Zeiger M, Presser V and Ozoemena K I 2015 *J. Mater. Chem.* A **3** 3480
[113] Wang Y, Han Z J, Yu S F, Song R R, Song H H, Ostrikov K and Yang H Y 2013 *Carbon* **64** 230
[114] Wang Y, Yu S F, Sun C Y, Zhuc T J and Yang H Y 2012 *J. Mater. Chem.* **22** 17584
[115] Zhang W, Wang J, Yang Y, Liang Y and Gao Z 2018 *J. Colloid Interface Sci.* **531** 502
[116] Wang J, Zhang W, Zhang X, Wang F, Yang Y and Lv G 2020 *J. Alloys Compd.* **826** 154217
[117] Mykhailiv O, Imierska M, Petelczyc M, Echegoyen L and Brzezinska M E P 2015 *Chem. Eur. J.* **21** 5783

[118] Santiago D, Rodríguez-Calero G G R, Palkar A, Barraza-Jimenez D, Galvan D H, Casillas G, Mayoral A, Jose-Yacamán M J, Echegoyen L and Cabrera C R 2012 *Langmuir* **28** 17202
[119] Nguyen V T, Le P A, Hsu Y C and Wei K H 2020 *ACS Appl. Mater. Interfaces* **12** 11533
[120] Fua Y, Liua Y and Li H 2020 *Particuology* **53** 58
[121] Najafpour M M, Hosseini S M, Hołyńska M, Tomo T and Allakhverdiev S I 2015 *Int. J. Hydrog. Energy* **40** 10825
[122] Avramov-Ivić M, Štrbac S and Mitrović V 2001 *Electrochim. Acta* **46** 3175
[123] Tran N Q, Bui V Q, Le H M, Kawazoe Y and Lee H 2018 *Adv. Energy Mater.* **8** 1702139
[124] Babar P, Lokhande A, Karade V, Pawar B, Gang M G, Pawar S and Kim J H 2019 *ACS Sustain. Chem. Eng.* **7** 10035
[125] Vattikuti S P, Devarayapalli K C, Reddy Nallabala N K, Nguyen T N, Nguyen Dang N and Shim J 2021 *J. Phys. Chem. Lett.* **12** 5909
[126] Liu G, Yao R, Zhao Y, Wang M, Li N, Li Y, Bo X, Li J and Zhao C 2018 *Nanoscale* **10** 3997
[127] Xiao Y, Wang W, Li T, Mao Y and Liu C 2021 *Electrochim. Acta* **386** 138406
[128] Xu S, Wang M, Saranya G, Chen N, Zhang L, He Y and Gou H 2020 *Appl. Catal.* B **268** 118385
[129] Coro J, Suarez M, Silva L S, Eguiluz K I and Salazar-Banda G R 2016 *Int. J. Hydrog. Energy* **41** 17944
[130] Debe M K 2012 *Nature* **486** 43
[131] Wu G, Dai C, Wang D, Li D and Li N 2010 *J. Mater. Chem.* **20** 3059
[132] Sikeyi L L, Ntuli T D, Mongwe T H, Maxakato N W, Carleschi E, Doyle B P and Maubane-Nkadimeng M S 2021 *Int. J. Hydrog. Energy* **46** 10862
[133] Ejikeme P M, Makgopa K, Raju K and Ozoemena K I 2016 *Chem. Electro. Chem.* **3** 2243
[134] Ogada J J, Ipadeola A K, Mwonga P V, Haruna A B, Nichols F, Chen S and Ozoemena K I 2022 *ACS Catal.* **12** 7014
[135] McDonough J K and Gogotsi Y 2013 *Electrochem. Soc. Interface* **22** 61
[136] Ramya A V, Thomas R and Balachandran M 2021 *Ultrason. Sonochem.* **79** 105767
[137] McDonough J K, Frolov A I, Presser V, Niu J, Miller C H, Ubieto T and Gogotsi Y 2012 *Carbon* **50** 3298
[138] Khan A J, Hanif M, Javed M S, Hussain S and Liu Z 2019 *J. Mater. Sci., Mater. Electron.* **30** 8568
[139] Jung S, Myung Y, Das G S, Bhatnagar A, Park J W, Tripathi K M and Kim T 2020 *New J. Chem.* **44** 7369
[140] Gao Y, Zhou Y S, Qian M, He X N, Redepenning J, Goodman P and Lu Y F 2013 *Carbon* **51** 52
[141] Kovalenko I, Bucknall D G and Yushin G 2010 *Adv. Funct. Mater.* **20** 3979
[142] Zeiger M, Weingarth D and Presser V 2015 *Chem. Electro. Chem.* **2** 1117
[143] Pallavolu M R, Gaddam N, Banerjee A N, Nallapureddy R R and Joo S W 2022 *Int. J. Energy Res.* **46** 1234
[144] Gupta S P, Kakade B A, Sathe B R, Qiao Q, Late D J and Walke P S 2020 *ACS Appl. Energy Mater.* **3** 11398
[145] Wang C, Wu C, Chen S, He Q, Liu D, Zheng X and Song L 2017 *RSC Adv.* **7** 4667
[146] Borgohain R, Li J, Selegue J P and Cheng Y T 2012 *J. Phys. Chem.* C **116** 15068
[147] Kanakaraj R and Sudakar C 2020 *J. Power Sources* **458** 228064

[148] Pathak A D, Potphode D and Sharma C S 2022 *Mater. Adv* **3** 3610

[149] Zheng Z, Su Q, Zhang Q, Ye H and Wang Z 2018 *J. Mater. Sci.* **53** 12421

[150] Han F D, Yao B and Bai Y J 2011 *J. Phys. Chem.* C **115** 8923

[151] Shu C, Lin Y and Su D 2016 *J. Mater. Chem.* A **4** 2128

[152] Xiao M, Meng Y, Duan C, Zhu F and Zhang Y 2018 *J. Electroanal. Chem.* **827** 167

[153] Han D, Xu S, Li J, Liu B, Song G and Guo Z 2017 *Mater. Lett.* **199** 93

[154] Xiao M J, Ma B, Zhang Z Q, Xiao Q, Li X Y, Zhang Z T and Zhang H L 2021 *J. Mater. Chem.* A **9** 7227

[155] Yu J, Li X, Sun Y and Liu X 2018 *Ceram. Int.* **44** 17113

[156] Liu X, Wu Y, Li X, Yu J and Sun Y 2018 *Ceram. Int.* **44** 13654

[157] Han D, Hu H, Liu B, Song G, Yan H and Di J 2016 *Ceram. Int.* **42** 12460

[158] Bi N, Liu X, Wu N, Cui C and Sun Y 2015 *RSC Adv.* **5** 32452

[159] Liu X, Li Z, Zhang S, Long H, Wei H, Zhang H and Zhao C 2017 *Ceram. Int.* **43** 14446

[160] Han D, Xiao N, Liu B, Song G and Ding J 2017 *Mater. Lett.* **196** 119

[161] Liu X, Bi N, Feng C, Or S W, Sun Y, Jin C and Xiao F 2014 *J. Alloys Compd.* **587** 1

[162] Palaniyandy N, Kebede M A, Raju K, Ozoemena K I, Le Roux L, Mathe M K and Jayaprakasam R 2019 *Mater. Chem. Phys.* **230** 258

[163] Nouh E A S, Roy M, Nandi G and Sarkar S 2013 *Adv. Sci. Eng. Med* **5** 1181

[164] Ning L, Xiaojie L, Xiaohong W, Honghao Y, Fei M and Wei S 2009 *Compos. Sci. Technol.* **69** 2554

[165] Bystrzejewski M, Łabędź O, Kaszuwara W, Huczko A and Lange H 2013 *Powder Technol.* **246** 7

[166] Lee G H, Huh S H, Jeong J W and Ri H C 2002 *J. Magn. Magn. Mater.* **246** 404

[167] He C N, Tian F, Liu S J, Du Z J, Liu C J, Li F and Chen S Q 2008 *Mater. Lett.* **62** 3697

[168] Medranda D, Borowiec J, Zhang X, Wang S, Yan K, Zhang J and Boi F S 2018 *R. Soc. Open Sci.* **5** 170981

IOP Publishing

Nanocarbon Allotropes Beyond Graphene
Synthesis, properties and applications
Arpan Kumar Nayak and Santosh K Tiwari

Chapter 9

Synthesis and application of carbon nanotori

Maya Devi, Swetapadma Praharaj and Dibyaranjan Rout

Owing to their unique physical and chemical properties, carbon materials have gained popularity in recent years. A wide variety of structures and textures of these materials make them suitable for applications in diverse fields. Carbon materials are classified into two basic categories, classic and new carbon forms. Of the several new carbon forms, nanotubes are very significant because of their high tensile strength and improved electrical and thermal properties. The discovery of fullerene crop circles has led to a new form of carbon material called carbon nanotori. These are basically graphene sheets rolled up into carbon nanotubes whose ends are joined together to form a torus. These are thought to be very useful in next-generation electronic devices owing to their novel electromagnetic transport properties. In this chapter various synthesis methods, relevant material properties, and applications of carbon nanotori are discussed in detail.

9.1 Introduction

Of all the forms of carbon materials, carbon nanotubes (CNTs) have drawn the attention of researchers because of their high tensile strength and flexibility, and metal or semiconductor nature depending on their atomic structure. CNTs consist of one or several graphene sheets rolled up into tubes forming single-walled carbon nanotubes (SWCNTs) or multi-walled carbon nanotubes (MWCNTs). Based on the type of crystal lattice of the graphene and how it folds up in one dimension to form nanotubes, there are three varieties of carbon nanotubes: (i) armchair, (ii) zigzag, and (iii) chiral. In armchair CNTs the hexagons are arranged straight along the axis whereas in chiral CNTs the hexagons are ordered spirally along the axis, as shown in figure 9.1 [1]. It is very challenging to synthesize ideal tubes with perfect hexagon networks. Also, the existence of non-hexagonal rings such as pentagons and heptagons act as crystal defects causing bending of nanotubes [1]. These nanotubes have strong attraction (van der Waals force) between them and also with the

doi:10.1088/978-0-7503-5177-5ch9

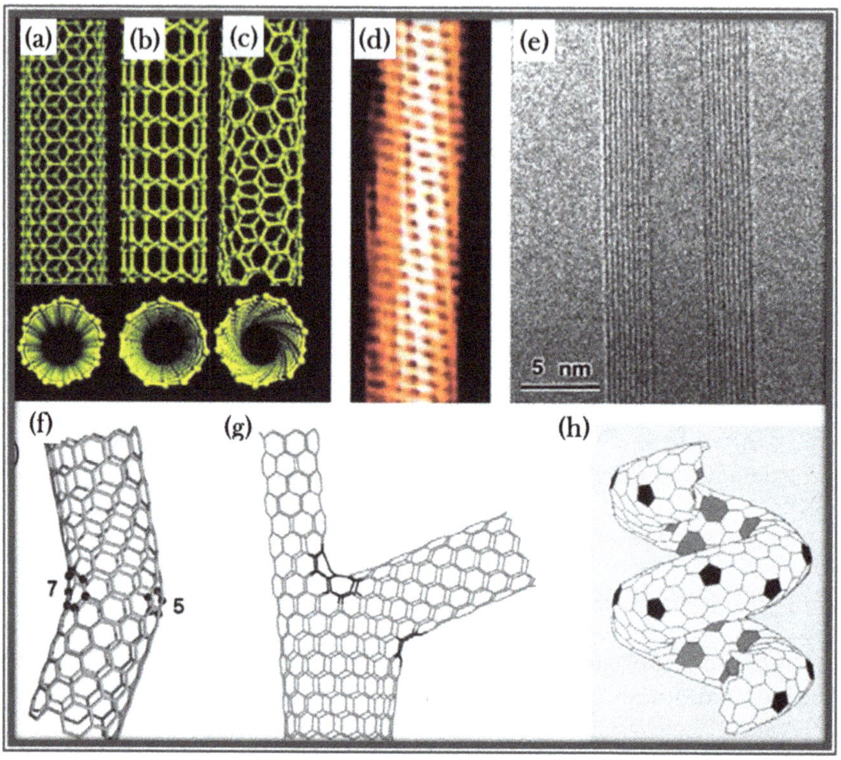

Figure 9.1. Schematic representation of (a) armchair, (b) zigzag, and (c) chiral SWNT structures; (d) SPM image of chiral SWNTs; (e) TEM image of an MWNT; (f) carbon-bone network of a kink junction between an 'armchair' and a 'zigzag' tube (5: pentagon; 7: heptagon; dark spheres: atoms); (g) a structural model of an SWNT asymmetric, zigzag Y junction. n-H rings (dark areas), and (h) helically coiled SWNT (Helix C360). (Reproduced with permission from [1]. Copyright 2009 Elsevier.)

substrate. The strong adhesive forces between the SWCNTs give rise to a self-organized structure in the form of a ring/torus called a carbon nanotorus or nanoring.

Thus nanotori are basically circular CNTs formed by connecting the ends. The structural model of a carbon nanotorus is shown in figure 9.2 [2]. The geometry of nanotori can be determined by the parameters known as the major and minor radii. Depending on the values of these parameters, the nanotori are classified into different categories, such as large and small. In large nanotori the major radius is greater than 100 nm. Before their experimental synthesis, Dunlap proposed a model for the construction of carbon nanotori by connecting two CNTs of different diameters [3]. At the same time a Japanese research group constructed a nanotorus C_{360} from C_{60} fullerene [4]. They generated a series of toroidal CNTs with the number of carbon atoms varying from 120 to 1920 using the method of Goldberg [5]. So far six different major methods have been proposed for the construction of the structural models of carbon nanotori: (i) bending a CNT and connecting its open

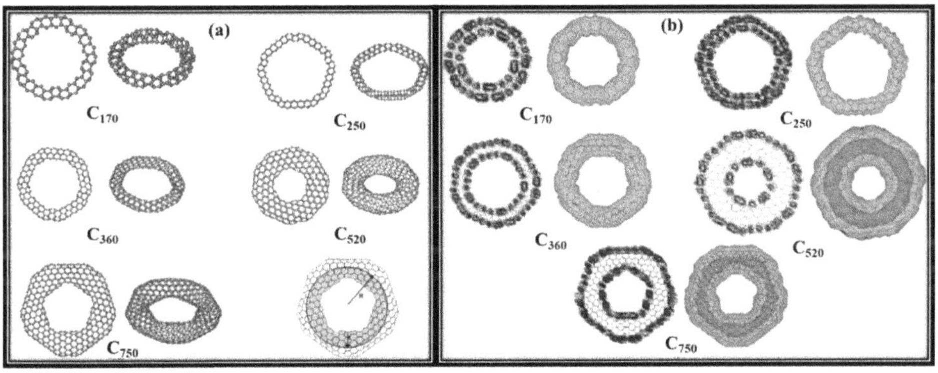

Figure 9.2. Carbon nanotori C_{170}, C_{250}, C_{360}, C_{520}, and C_{750} with five-fold symmetry and their charge distributions. (Reproduced with permission from [2]. Copyright 2004 Elsevier.)

ends together and stabilizing the structure through van der Walls interaction [6]; (ii) introducing a pentagon–heptagon pair into the honeycomb network of two carbon nanotubes [3, 7, 8]; (iii) constructing from C_{60} fullerenes using Goldberg's method [9]; (iv) connecting zigzag-edged chains of hexagons to the armchair-edged chain of hexagons found in different types of nanotubes [10]; (v) sewing the walls of single-walled, double-walled, or triple-walled carbon nanotubes at both ends [11–13]; and (vi) constructing only from pentagons and heptagons [14]. Thus far it has been difficult to obtain carbon nanotori under experimental conditions in the laboratory, in particular scaling up the production. In this chapter, the various synthesis methods of carbon nanotori are presented along with their properties and various applications.

9.2 Synthesis methods of carbon nanotori

The last few decades have witnessed unprecedented progress in the synthesis methods for producing different carbon nanostructures ranging from fullerenes to carbon nanotubes and grapheme. The predominant methods used for the synthesis of CNTs are arc discharge, laser ablation, and chemical vapor deposition (CVD). Nevertheless, in the recent years several modified techniques have been adopted for the synthesis of carbon materials; some of them are listed in figure 9.3 and discussed in this section.

9.2.1 Laser ablation method

In 1997, nanotori were first observed during the synthesis of SWCNTs by the laser ablation method. In this method, a laser source is used to vaporize a graphite target (shown schematically in figure 9.4) in a controlled environment. The temperature is maintained at 1200 °C and on a relatively cool target; the condensed material is collected in the form of nanoparticles and nanotubes. In the improved ablation method, cobalt and nickel are used as catalysts, either to dope the graphite target or to help in the alignment of growth of the carbon nanotubes. The first observed

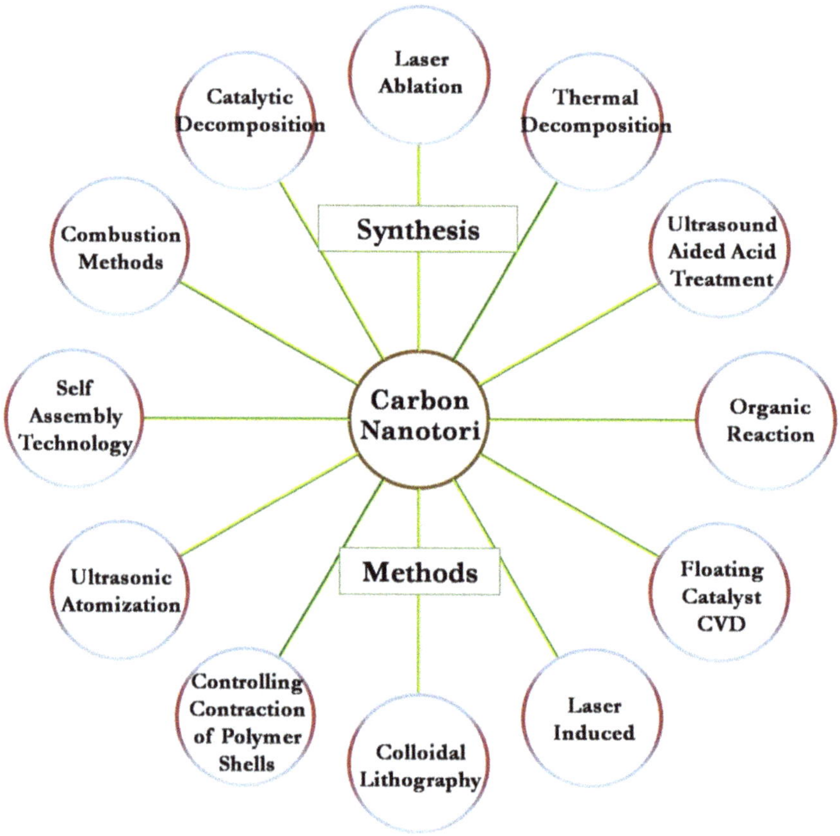

Figure 9.3. Synthesis methods for carbon nanotori.

carbon nanotori appeared as SWNT ropes consisting of 10–100 individual nanotubes aligned over the entire length and packed in 2D crystalline array. They were named crop circles and nearly 1% of the total ropes were circular with diameters from 300 to 500 nm and a width of 5–15 nm [15].

9.2.2 Thermal decomposition of hydrocarbon gas

The catalytic decomposition of acetylene was done in a flow reactor at 700 °C and atmospheric pressure. The catalyst was removed by strong acid (HF) treatment. By sonicating the pure form of nanotube materials in isopropanol they were deposited on a silicon wafer. Rings of carbon nanotubes of several hundred nanometers are formed which are interpreted as compressed sections of helical carbon nanotubes [16].

9.2.3 Ultrasound aided acid treatment

In this synthesis method, long twisted SWNTs 1.4 nm in diameter were used as the raw materials, which were obtained from laser ablation. The nanotubes were dispersed and their length was reduced by an oxidation process. Oxidation was

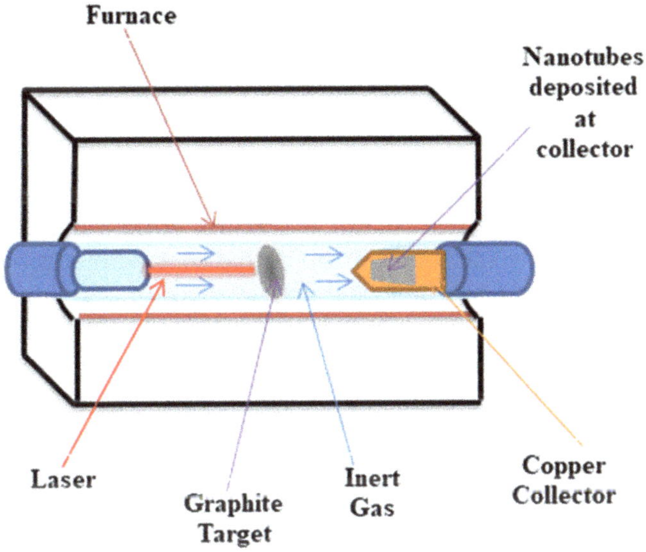

Figure 9.4. Schematic of the laser ablation method.

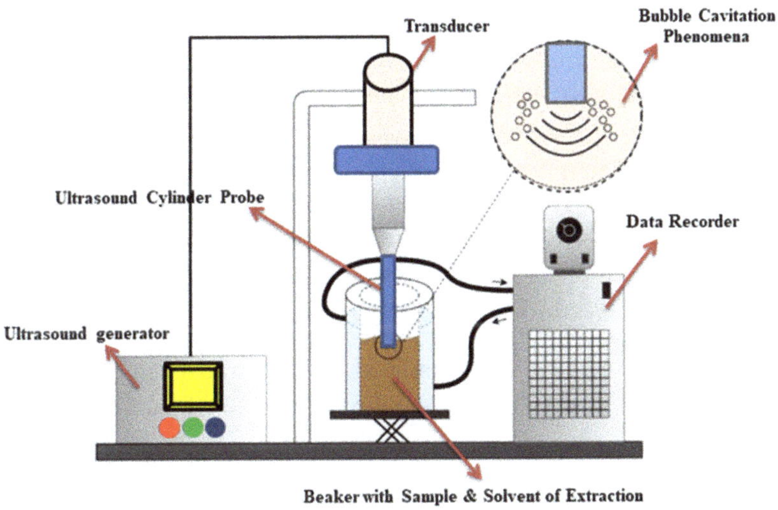

Figure 9.5. Schematic diagram of an ultrasound assisted extraction system.

carried out in a strong acidic medium and irradiated with ultrasound for 1–3 h at 40 °C–50 °C, as shown in figure 9.5. The acidic solution was prepared by adding H_2SO_4 (concentrated) with H_2O_2 in the ratio 9:1 using a sonicator. The solution was then filtered through a 0.2 μm membrane and the residue was ringed by DI water and air dried. Using sonication with low power ultrasound the nanotubes were suspended in 1,2-dichloroethane. The nanorings formed in this method have diameters in the range 300–400 nm. The yield of the nanorings is dependent on the exposure time to ultrasound and the concentration of the peroxide solution [17].

9.2.4 Organic reaction method

Pristine SWNTs were treated with concentrated H_2SO_4 or HNO_3 and large aggregates were removed by ultracentrifuge. The remaining floating solution was dispersed in dry dimethylformamide (DMF). Excess dicyclohexylcarbodiimide (DCC) was added to the solution and stirred overnight at room temperature. Finally, the solid material was collected and washed on a Teflon filter. Here the ring closure was achieved with a number of organic reactions with the activated oxygen containing group. This process yields rings with a diameter of 540 nm. These formed rings are very stable and do not reopen easily [18]. The synthesis of self-assembled SWNT rings occurs due to the formation of noncovalent interactions of SWNTs and porphyrins. The porphyrins are synthesized by phenyl ring para-substituted tetraphenylporphine (TPP). The noncovalent interaction between the substituted TPP and SWNTs is facilitated by the electron donating properties of the substituent group. The self-assembly of the SWNT rings is very much dependent on (i) the mass ratio of SWNTs to porphyrin, (ii) the concentration of the porphyrin hybrid, and (iii) the nature of the substrate surface. SWNT rings of diameter 0.5–1.2 μm and width 15 nm are reproducible by suitably adjusting the above parameters [19].

9.2.5 Floating catalyst chemical vapor deposition (FCCVD)

A high yield of bundles of SWNT rings was achieved by the thermal decomposition of acetylene at 1100 °C in a floating iron catalyst system. The synthesis is performed in a two-stage furnace system fitted with special quartz tubes. A quartz tube of diameter 10 mm is embedded in another quartz tube of diameter 30 mm. The catalyst (ferrocene + sulfur) was sublimed at 50 °C–65 °C in the first furnace and then carried to a second furnace with the help of flowing argon and acetylene mixture through a small inserted quartz tube, as shown in figure 9.6. The reaction temperature and pressure were maintained at 1100 °C and 1 atm, respectively. After 30 min of reaction, the product was collected on substrates placed at a lower

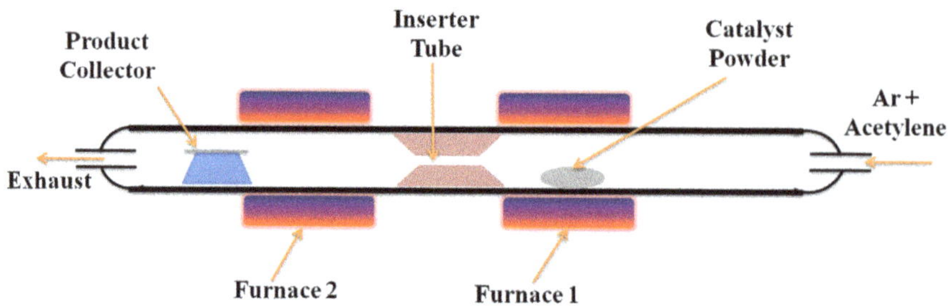

Figure 9.6. A schematic diagram of an FCCVD reactor.

temperature between 200 °C and 300 °C. The rings formed in this process primarily consist of SWNT toroids having diameters in the range of 100–300 nm, which is relatively smaller than that obtained in other methods. The thickness of the film ranges from 1 to 8 nm. The proportion of SWNT rings formed depends on (i) the amount of sublimed catalyst and (ii) the reactant gas flow, and (iii) the diameter and configuration of the quartz reactor. The amount of sublimed catalyst can be manipulated by adjusting the sublimation temperature. Below a 50 °C sublimation temperature, no rings or linear SWNT products are formed. The reactant flow rate also affects the SWNT ring formation. The optimum flow rate ranges between 200 and 600 sccm with an acetylene to argon ratio ranging from 0.02 to 0.1 volume percent [20, 21].

9.2.6 Carbon toroids from fullerene using a laser induced method

In this method, 248 nm laser output from a KrF laser operating at 10 Hz is incident on a mixed fullerene target consisting of 76% C_{60} and 22% C_{70} kept inside a vacuum chamber. The ablated target is irradiated with a 532 nm laser produced by the second harmonic of a Nd:YAG laser operating at 10 Hz. The ejected substance is deposited on a carbon/copper grid kept at 1.5 cm from the fullerene target. Toroids of diameter ranging from 170 nm to 0.5 μm and thickness ranging from 35 nm to 140 nm are formed [22].

9.2.7 Colloidal lithography

Polystyrene spheres in aqueous suspension were deposited on n-type Si wafer to form a hexagonal monolayer of mask. Using a sonicator, the SWNTs were dispersed thoroughly in methanol and allowed to settle. Then they were deposited on the colloidal mask, allowed to dry, and heated in an oven at 90 °C for 5 min. The liquid retracted toward the bottom of sphere by capillary forces. Thus CNT circular ring structures were formed around each sphere. Removing these spheres using adhesive tape, an ordered array of rings was formed on the substrate. The diameter of the rings ranged from 203 ± 21 nm to 97 ± 14 nm formed by 780 nm and 450 nm colloidal spheres [23].

9.2.8 Controlling the contraction of polymer shells

Single-walled CNT bundles were encapsulated in polystyrene block polyacrylic acid (PSPAA). The mixture was sonicated until a transparent black solution was obtained. Then water was added drop wise, and the mixture was sonicated and cooled down to room temperature. Again, 1,2-dichlorobenzene (DCB) was added to this solution and it was centrifuged to collect a concentrated sample. These concentrated samples contained polymer encapsulated CNT rings. The polymer cells were removed from the CNT rings and purified. The expansion and compression cycles of the CNT/PSPAA rings were repeated and by controlling the concentration of polymer shell the carbon nanotubes could be coiled into rings [24].

9.2.9 Ultrasonic atomization for isolation of toroidal aggregates of SWCNTs

A suspension of SWCNTs prepared by the arc discharge method was taken and purified through acid treatment and annealing in air. The purified sample was then sonicated at 20 °C under carrier gas flow. A black solid was obtained after blowing on the droplets generated on the surface of SWCNT suspension. The analysis of this showed a high concentration of toroidal SWCNTs (t-SWNTs). From trials with different solvents it was observed that with a suspension of n-heptane the purity of the t-SWCNTs improves. Different shapes of tori ranging in the diameter 0.5–3.0 μm were observed. Finally, through ultrasonic atomization, the t-SWCNTs were separated from the fabricated structure [11].

9.2.10 Self-assembly technology of CNT rings based on wet chemistry

SWNTs of diameter 1.4 nm were prepared by the arc discharge method and purified. The purified fibers were acid treated with a mixture of H_2SO_4 and H_2O_2 in a ration of 9:1. Then they were ultrasonicated to reduce the length to 1–3 μm. The solution was filtered through 0.2 μm membrane and the residue was suspended ultrasonically in 1,2-dichloroethane. Droplets of nanotube solution were dropped onto SiO_2/Si substrate. Circular and non-circular rings of various sizes were formed depending on the competition of surface tension force, strain force, and van der Waals force of attraction [25].

9.2.11 Combustion method

Carbon samples were prepared by reducing $Mg_{0.9}Co_{0.1}O_3$ solid solution in a highly reducing type atmosphere (H_2–CH_4 with 18 mol% CH_4, 250 sccm) at 1000 °C. Taking high pure metal nitrates and urea in a stoichiometric ratio, the solid solution was synthesized by the combustion method. The prepared carbon material was extracted by dissolution of MgO and part of the Co treatment in HCl aqueous solution. In this method, 80% of the nanotubes have either one or two walls with a torus diamete of the order of 250 nm and the concentration of the rings is 0.01%–0.1% [12].

9.2.12 Catalytic decomposition

The catalytic decomposition of acetylene (C_2H_2) on an MgO/Fe catalyst in a quartz fluidized bed reactor at 900 °C–1000 °C gave rise to ring-like structures. The use of such a reactor facilitates the rapid and complete contact of catalyst with the carbon source which enhances the decomposition rate. Even without any additional energy from outside, ring-like structures of CNTs were formed, which was confirmed by SEM images. These structures may be formed due to confinement in the catalyst pores which would have allowed bending and coiling of CNTs. A HRTEM image confirmed that the bundles consisted of TWNTs which have uniform size (2.04 nm outer diameter and 0.75 nm inner diameters) and ordered arrangement [13].

9.2.13 Pulsed (high voltage) discharge in ethanol vapor

The experimental set-up was an assembly of a reactor chamber, two electronic control units, a forepump, and a vapor–gas mixture feed system. A high voltage electric discharge was applied between pointed metal electrodes in a carbon containing vapor medium. The optimal condition for carbon torus synthesis was a discharge frequency of 120 Hz and an inter-electrode distance of 8 mm. The reaction product deposited on a silicon substrate by pulsed arc discharge plasma in ethanol vapor was studied by AFM. It confirmed the formation of tori of radius 100–500 nm and thickness 1–8 nm [26].

9.3 Properties of carbon nanotori

As carbon nanotori are very useful for various applications, their properties are studied by molecular dynamics using mainly tight binding models. A few of the properties obtained from these theoretical calculations are given below.

9.3.1 Structural and electronic properties of carbon nanotori

The electronic properties of nanotori are very much dependent on the curvature, hybridization, and disorder. The average value of σ- and π-bond charge per atom provides adequate information to understand the relation between the mechanical deformation and corresponding electronic properties. When the nanotorus is under increasing strain, a slight reduction in σ- and π-bond charges and the circumference associated with the bond occurs. Effectively, no pivotal change is observed in the nature of the electronic structure resulting in no substantial change in the conductance of an SWNT under delocalized deformations up to a radius of curvature $\approx$ 13.3 Å. However, at a critical strain value (the wall collapsing value), a notable increase in σ-bond charge and decrease in the π-bond charge occurs. At the locations where the walls collapse, a transition from sp_2 to sp_3 bonding configuration takes place. It is unlikely that the same mechanism is seen in the case of an SWNT at relatively small bending angles under a localized deformation. Mostly σ-electrons are localized, whereas the π-electrons are delocalized and largely responsible for conduction. This results in a dramatic reduction in π-bond charge and substantial increase in σ-bond charge which will lead to a remarkable decrease in conductance value for a nonlocalized deformation $R \leqslant 13.3$ Å. The factors responsible for the dramatic reduction of the conductance of metallic SWNTs are the combined effect of (i) sp_2- to sp_3-bonding configuration transition and (ii) the extent of the bending region where the transition occurs [27]. The metallic or semiconducting nature of toroidal CNTs depends on chirality. The electronic properties are varied by bending and intrusion of pentagons and heptagons. The electronic property is very much dependent on the chiral and translational vectors, given as $C(n, m)$ and $T(p, q)$ [28]. The various types of nanotori obtained with different relations are presented in table 9.1.

The electronic properties are very much dependent on the size of the nanotorus. With an increase in the size of the nanotorus, the energy gap exhibits a well-defined

Table 9.1. Types of nanotori obtained with different relations.

Nature of carbon nanotorus	Relation between m and n	Relation between p and q
Metallic	$m - n = 3i$	$p - q = 3i$
Semiconducting	$m - n = 3i$	$p - q \neq 3i$
Semiconducting	$m - n \neq 3i$	$p - q = 3i$
Non-existing	$m - n \neq 3i$	$p - q \neq 3i$

oscillating feature. With increasing size of the nanotorus, the electronic structure shows a transition from semiconductor to metal due to a change in the characteristic density of state function from zero dimensions to quasi one dimension. The variation tendency in the energy gap is dependent on the disorder state. The semiconductor–metal state transition is predicted to occur at strong disorder strength [29]. From a study of the conductance of nanotori contacted with carbon nanotubes by the Landauer–Büttiker formula, it is observed that conductance is very much dependent on the chemistry and geometry of the contacts. The difference in the contacts reduces the constructive quantum interference of transmission in comparison to similar types of contacts which affects the transport properties of nanotori [30].

9.3.2 Magnetic properties of nanotori

The nanotori respond to an applied magnetic field due to the interplay between delocalized π electrons and the geometrical structure of the torus. According to the geometry the carbon nanotorus can be either armchair–zigzag (armchair along the transverse direction and zigzag along the longitudinal direction) or zigzag–armchair (vice versa). In the absence of magnetic flux, there are three possible types of electronic structures of nanotori: (i) zigzag–armchair with a large energy gap; (ii) armchair–zigzag and zigzag–armchair with a zero energy gap; and (iii) armchair–zigzag with a small energy gap. From the tight binding model it is found that in type (i) the electronic structure is independent of magnetic flux but in types (ii) and (iii) the electronic structure is very much affected by magnetic flux. The energy gap is a linear periodic function of the magnetic flux (φ). With variation of φ, the electronic structure changes from metallic to semiconductor or vice versa. It is also found that the energy gap is inversely proportional to the toroid radius and is independent of the height and width of the toroid. The curvature effect, due to misorientation of π electron orbitals, plays a pivotal role in the low energy electronic structure of thin armchair–zigzag carbon nanotori. They exhibit the periodic Aharonov–Bohm oscillation (AB oscillation) except at high magnetic flux. At large magnetic flux, Zeeman splitting has a strong effect on the magnetic properties due to spin B interaction. Thus metal–semiconductor transition occurs very frequently [31]. The magnetoelectronic properties are dependent on changes in the radius, chirality, temperature, direction, and magnitude of the magnetic field. The dependence of magnetization on the toroidal radius is strong for a parallel magnetic field in comparison to a perpendicular magnetic field.

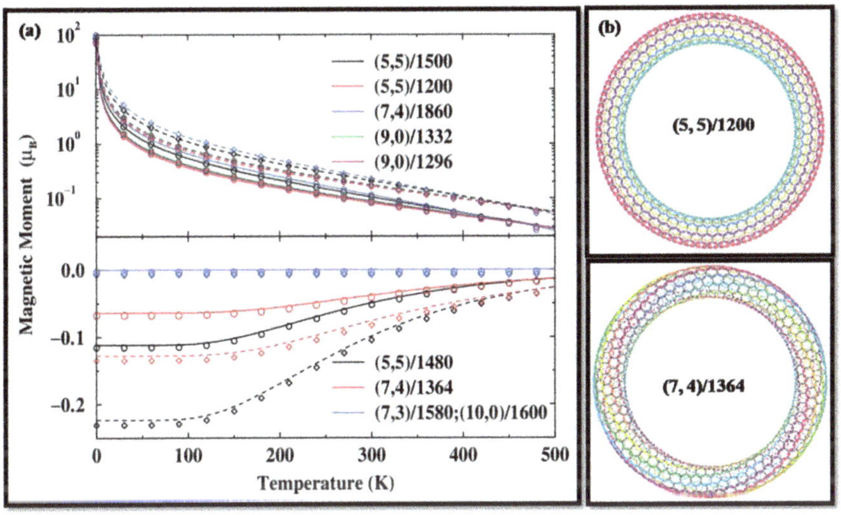

Figure 9.7. (a) Temperature dependent magnetic moments of various carbon nanotori. (b) The geometric structure and corresponding ring current of a (5,5)/1200 metallic nanotorus and (7,4)/1364 semiconducting nanotorus, respectively. (Reproduced with permission from [33]. Copyright 2002 the American Physical Society.)

The armchair nanotori show a transition from diamagnetism to paramagnetism when the direction of the applied magnetic field is close to the symmetry axis. The paramagnetic transition occurs above the critical angle 30° in armchair nanotori, but the other class of nanotori is paramagnetic for any field direction [32]. The unusual paramagnetic response is observed in a metallic nanotorus in comparison to a small gap semiconducting nanotorus. This colossal paramagnetic moment in metallic nanotori is explained by Liu *et al* [33]. Figure 9.7(a) shows the temperature dependent induced magnetic moment for different types of nanotori that are formed from armchair, chiral, and zigzag nanotubes exhibiting both metallic and semiconducting behaviors, while figure 9.7(b) shows the ring current structures. In a metallic nanotorus, as the nanotube is a ballistic conductor, so the electrons move freely along the circumference of the torus in opposite directions. This makes the net dipole moment zero in the absence of the external magnetic field. But when the magnetic field is applied perpendicular to its plane, the dipole moments align themselves along the direction of the magnetic field which gives rise to a paramagnetic moment. The magnetic response of a nanotorus is determined by the combined effect of magnetic moments associated with the motion of electrons in the torus and the diamagnetic response of electrons.

Temperature also affects the magnetic moment induced in the system. It affects the magnetic moment in two different ways. In the first process it mixes the contributions of states both below and above the Fermi level which reduces the magnetic moment and in the second it reduces the phase coherence length. So for a coherence length less than the circumference of the torus, the magnetic moment reduces exponentially. The disorder and curvature affect the magnetic moment in nanotori. In armchair nanotori the diamagnetic response is observed at low temperatures near the zero magnetic field due to the curvature effect, but the zigzag nanotori show a paramagnetic response. The curvature effect and high temperature reduce the amplitude of AB oscillation. Disorder also plays a significant role in the reduction of AB oscillation and with increased disorder a phase transition occurs from diamagnetism to paramagnetism. A strong disorder state along with high temperature destroys the AB oscillation and there is an oscillation from a diamagnetic moment to a paramagnetic moment and vice versa [34]. From the study of magnetic properties of various carbon nanotori formed by coalescing of C_{60} molecules along the three symmetry axes, it is found that a combination of +ve and −ve Gaussian curves due to presence of pentagonal, heptagonal, and octagonal carbon rings play a significant role in magnetic behavior. These carbon structures show ferromagnetism in the absence of any defects, however, this is very sensitive to the introduction of defects or rupture of symmetry [35].

9.3.3 Optical properties of nanotori

The study of the low frequency absorption spectra of carbon tori show that the spectrum intensity decreases rapidly with an increase of frequency. The spectral function shows many peaks. These peaks are due to the excitation of electrons from occupied to unoccupied states. The energy difference between two neighboring channels increases with frequency and the threshold energy is nearly equal to the energy gap. The absorption spectra are observed to reflect the characteristics of low energy states. The threshold energy is inversely proportional to the square of width or the width of the toroid depending on its type. The optical measurement of threshold energy gives an idea of the curvature effect on low energy states [31]. The study of magneto-optical excitation by cross polarized light shows their dependence on the magnitude and direction of the magnetic field. The toroid geometry also affects these properties. It is observed that the absorption spectra of armchair toroids are independent of the magnitude and direction of magnetic field, whereas those of zigzag toroids are very much dependent on the magnitude and direction of the magnetic field. For different field directions, the threshold excitation energy shows periodic AB oscillation or with an increase in the magnitude of the field the excitation energy shows a monotonous decrease. The curvature effect, which is very much dependent on the height of the tori, leads to changes in the excitation energy [36].

9.3.4 Thermal properties of nanotori

As the electronic structure of carbon tori depends on the toroidal geometry, temperature, and magnetic flux, so the electronic specific heat also shows magnetic flux and temperature dependence. The semiconductor–metal transition occurs with variation of the magnetic flux. The temperature dependence shows nonlinear variation of the specific heat (C_V) at low temperatures, whereas it is linear at high temperatures. The electronic specific heat is defined as variation of the mean energy with temperature, reflecting the characteristic of electronic structure near the Fermi level. At low temperatures, the flux dependent specific heat $C_V(\varphi)$ shows two peak structures and the C_V value reduces to zero at a flux value corresponding to the semiconductor–metal transition. The toroid shows oscillatory behavior and the amplitude of oscillation remains the same at all temperatures. The peak positions show linear temperature dependence. With increasing magnetic flux the two peak structures of $C_V(\varphi)$ are doubled because the Zeeman splitting make the spin down and spin up states of angular momentum cross the Fermi level. This causes the semiconductor–metal transition to occur frequently. This leads to double two peak structures in C_V. With an increase or decrease in temperature, the two peak structures at low temperature $C_V(\varphi)$ are broadened, but on increasing the temperature beyond 4 K this two peak structure completely vanishes due to the contribution from more degenerate states. The flux value at which the semiconductor–metal transition occurs is very much dependent on the toroid height and radii, and the chiral angles. The temperature dependant specific heat $C_V(T)$ for armchair carbon tori shows a nonlinear behavior at $T<10$ K. It shows a peak structure. The temperature (T_m) corresponding to the local maximum of $C_V(T)$can be calculated, from which the energy gap can be obtained easily [37]. The magnetic flux changes the state energy and state degeneracy which gives rise to a change in peak structure. The temperature dependence of $C_V(T)$ changes from nonlinear to linear as the temperature increases. Above 20 K the temperature, as thermal energy K_BT, is comparable to the typical energy level spacing, and the discrete energy levels behave as a continuous one-dimensional band. Thus the value of $C_V(T)$ can be calculated from the temperature and density of states [37].

9.3.5 Transport properties

The transport properties of a thin carbon toroid threaded by magnetic flux φ and weakly coupled to two leads is studied, when a bias voltage 'v' is applied and a temperature difference ΔT is given between two reservoirs. The electrons flow from one lead to the other by tunneling through the thin carbon toroid. At low temperatures the special transport properties, such as conductance and thermopower, are dependent on states close to the chemical potential and the symmetric and antisymmetric functions of the π-electron state energy. At zero magnetic flux the transport property is dependent on the chemical potential and resonance is absent. At low temperatures, when the chemical potential has the required value, the conductance shows a peak. The height of the resonant peaks suggests π-electron state degeneracy. At resonance the thermopower shows an antisymmetric structure which has a peak followed by a dip. It does not give any suggestion of state degeneracy. The toroid structure, radius, and height affect the transport properties [38]. At zero temperature, electric conductivity relates to the number of π-electron states touching the chemical potential, but with an increase in temperature the conductance value reduces. The thermopower increases with increasing temperature and the special resonance curves are visible at very low temperatures. The magneto-transport properties show periodic AB oscillations. Both conductivity and thermopower are periodic in φ. In the presence of magnetic flux, when the state degeneracy reduces to ½, the amplitude of the conductivity also decreases, but the amplitude of the themopower does not decrease, it becomes +ve and −ve as the flux varies. The period of AB oscillation is destroyed by Zeeman splitting at a large magnetic flux.

9.4 Applications of carbon nanotori

The unique mechanical and electronic properties of carbon nanotori make them suitable for various applications.

9.4.1 Gigahertz oscillators

One such application is the nanoscale oscillator. Zheng and Jiang proposed the concept of a nano-oscillator in which an inner shell oscillates inside the outer shell of an MWCNT [39]. The nano-oscillator has advantages over micromechanical oscillators as it can have an oscillation frequency in the range of GHz. Due this they find applications in ultrafast optical fibers and nano antennae sensitive to high frequencies. They can also be used in nanoscale sensors, actuators, resonators, and memories. From calculations it is found that the frequency of oscillation can be increased by decreasing the length of the inner tube. This gives the maximum possible frequency of oscillation by using C_{60} fullerene or bucky balls as the core of the oscillator. Liu *et al* studied the oscillation of a C_{60} fullerene in a carbon nanotube and by molecular dynamics it is predicted that the oscillation frequency can be of the order of 74 GHz [40]. Cox *et al* proposed a suitable acceptance energy at which the bucky ball can be sucked to the nanotube by van der Waals forces [41]. Hilder and Hill proposed an oscillator in which a nanotorus is sucked into an SWCNT and oscillates along its length [42]. It is calculated that a nanotorus with a ring radius greater than the nanotube radius + nanotorus tube radius + interlayer distance of graphite layer will be accepted onto a carbon nanotube. The radius of the nanotube and nanotorus must be very close to each other for maximum acceptance energy and maximum frequency of oscillation. Hilder, using a mathematical model, determined the equilibrium configuration for a number of carbon nanotori and nanosector oscillators and calculated their operating frequency [43]. The values of oscillating frequencies for various nanotorus–nanosector systems are presented in table 9.2 [43], where *a* and *b* are the tube radii of the inner torus and outer torus, respectively, and *c*

Table 9.2. Oscillating frequencies for various nanotorus–nanosector systems.

a (Å)	*b* (Å)	*c* (Å)	*β* (°)	Frequency (GHz)
5	8.45	1500	1.8	0.962
5	8.45	1500	30	0.984
5	8.45	1500	45	0.984
6.5	9.97	110	45	12.9
6.78	10.225	200	45	7.09
7	10.431	150	45	9.41

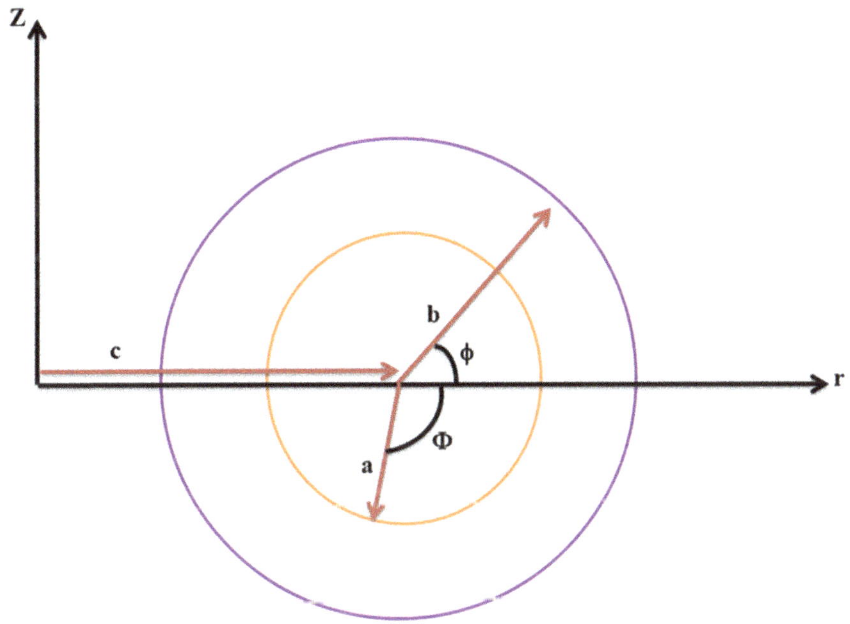

Figure 9.8. Sector of torus oscillating inside a nanotorus.

is the ring radius, which is the same for both the tori presented in figure 9.8. β is the sector angle.

These values of a, b, and c correspond to a symmetrical location of the nanosector in the nanotorus corresponding to equilibrium energy. Thus from numerical calculations it is confirmed that the operating frequency of the nanotorus and CNT oscillator is in the range of GHz.

9.4.2 Reinforcing material for lubricants

New lubricants prepared by dispersing carbon nanotori in distilled water show enhanced load carrying capacity [44]. It improved by nearly 950% with 2 wt% nanotori. The wear mass loss and coefficient of friction were reduced by 96% and 76%, respectively, with this concentration. These enhanced properties were due to the high elastic modulus and hollow toroidal geometry. Due to these properties they could withstand the test loads and hence limited metal–metal contact. Thus the tribological properties of nanolubricants were enhanced due to an increase in the concentration of nanotori. At extreme pressure carbon nanotori provide a load bearing effect due to a reduction in metal–metal contacts, showing a reduction in surface roughness and an increase in load carrying capacity. At lower contact pressure, carbon nanotori may provide a rolling/sliding mechanism as shown in figure 9.9 [44].

Thus carbon nanotori may be used as extreme pressure, antiwear additives for water-based lubricants in metal forming processes. The introduction of nanostructures increased the viscosity of conventional lubricants, which makes the carbon nanotori a suitable reinforcing material for lubricants in industrial applications [45].

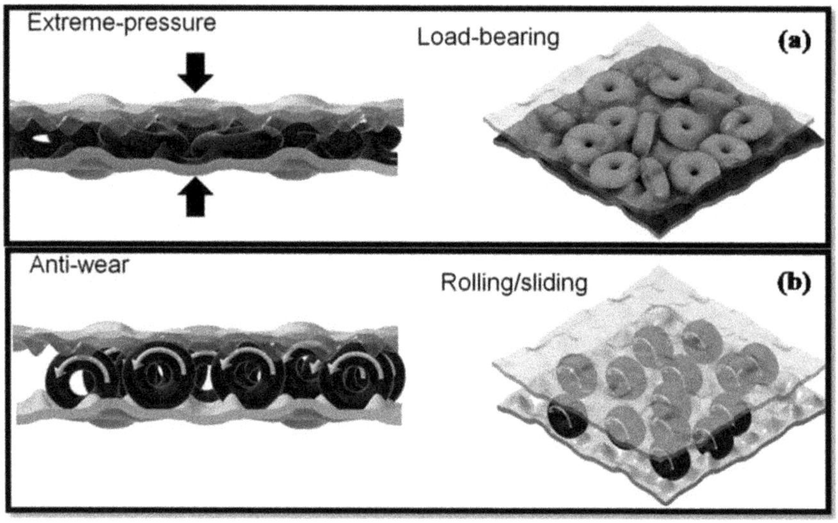

Figure 9.9. Tribological mechanism of carbon nanotori: (a) load bearing under extreme pressure and (b) rolling/sliding for antiwear conditions. (Reproduced with permission from [44]. Copyright 2019 Elsevier.)

9.4.3 Hydrogen storage

A theoretical study of the adsorption of hydrogen on carbon nanotorus C_{120} was done by Torres *et al* [42]. The geometry optimization was performed for H atoms from 1 to 15. The charge densities, HOMO–LUMO gap, Mulliken population, and binding energy were calculated. From a study of binding energy as a function of the number of H atoms, it is concluded that a few hydrogen atoms for $n = 14$ and 15 can be chemisorped on the inner surface of the carbon nanotorus. From geometry optimization it is observed that all H_2 molecules are uniformly distributed and aligned in a trajectory equidistant to all points around the inner surface of the nanotorus. This is independent of the orientation of the H_2 molecule to the axis of the nanotorus. It is found that as the number of H_2 molecules increases, there is a decrease in H_2 binding energy. The binding energy of H_2 inside the nanotorus is less than that of gaseous H_2, which is due to the repulsive energy between the walls of the nanotorus, which puts a limit on the maximum H_2 storage capacity. From Mulliken charge analysis it is observed that electron transfer occurs from H_2 to the nanotorus which is observed from table 9.3. Variation of H_2–H_2 and H_2–C_{120} repulsive energies with the number of H_2 molecules confined in a C_{120} nanotorus was calculated. From all these results, it is found that H_2 storage is possible [46].

9.4.4 Carbon nanotori act as traps for atoms and ions

Chan *et al* studied the possibility of carbon nanotori acting as traps for atoms and ions [47], as it is observed that the magnetic moment of some metallic nanotori of minor and major radii is a thousand times more than for other rings. This may be due to the unusual behavior of electrons when they move in a ring shaped structure.

Table 9.3. Mulliken charge analysis for n = 1–15.

No. of H_2 molecules	H–H average distance (Å)	Average distance between two adjacent H_2 molecules	HOMO (eV)	LUMO (eV)	ΔE (eV)	Average Mulliken charges	
						C atoms	H atoms
1	0.753	–	−5.19	−4.24	0.95	−0.0004	0.0235
2	0.756	1.784	−5.13	−4.27	0.86	−0.0007	0.0235
3	0.758	1.681	−5.10	−4.30	0.80	−0.0013	0.0268
4	0.758	1.662	−5.12	−4.33	0.79	−0.0020	0.0299
5	0.760	1.621	−5.12	−4.36	0.76	−0.0026	0.0323
6	0.761	1.648	−5.13	−4.41	0.72	−0.0033	0.0335
7	0.760	1.668	−5.15	−4.44	0.70	−0.0037	0.0324
8	0.764	1.679	−5.13	−4.48	0.65	−0.0049	0.0370
9	0.766	1.659	−5.11	−4.49	0.62	−0.0058	0.0393
10	0.765	1.710	−5.19	−4.52	0.67	−0.0060	0.0360
11	0.766	1.477	−5.23	−4.52	0.71	−0.0064	0.0353
12	0.770	1.340	−5.08	−4.57	0.52	−0.0076	0.0381
13	0.770	1.286	−5.07	−4.61	0.46	−0.0094	0.0435
14	0.772	1.516	−4.89	−4.77	0.12	−0.0157	0.0673
15	0.760	1.849	−4.95	−4.75	0.20	−0.0191	0.0763

When different nanotori are subjected to an external magnetic field (0.1 T), the spins of electrons line up to produce the magnetic moment. The radius of the torus and the area enclosed by it affect the magnetic moment. This colossal magnetic moment exhibited by metallic nanotori along with the predicted geometrical reversibility of toroidal CNTs with strain in a colloid [24] can be used for fabricating ultra-sensitive sensors.

9.5 Conclusion

Mostly theoretical model calculations proposed by many researchers suggested the idea of a new structure of carbon which was initially called crop circles. Gradually, from a few experiments, structures were obtained, although in very small amounts. Large scale production of this material has not yet been achieved, although it is showing many novel properties in theoretical studies. Due to these properties it is also predicted that carbon nanotori might be a potential candidate for several applications, so efforts are still ongoing for the large scale synthesis of this new and novel material.

References

[1] Zhang M and Li J 2009 Carbon nanotube in different shapes *Mater. Today* **12** 12–8

[2] Yazgan E, Taşci E, Malcioğlu O B and Erkoç Ş 2004 Electronic properties of carbon nanotoroidal structures *J. Mol. Struct. THEOCHEM* **681** 231–4

[3] Dunlap B I 1992 Connecting carbon tubules *Phys. Rev.* B **46** 1933

[4] Itoh S, Ihara S and Kitakami J I 1993 Toroidal form of carbon C_{360} *Phys. Rev.* B **47** 1703

[5] Ihara S, Itoh S and Kitakami J I 1993 Toroidal forms of graphitic carbon *Phys. Rev.* B **47** 12908

[6] Kirby E C, Mallion R B and Pollak P 1993 Toroidal polyhexes *J. Chem. Soc., Faraday Trans.* **89** 1945–53

[7] Baowan D, Cox B J and Hill J M 2008 Toroidal molecules formed from three distinct carbon nanotubes *J. Math. Chem.* **44** 515–27

[8] Liu L, Zhang L, Gao H and Zhao J 2011 Structure, energetics, and heteroatom doping of armchair carbon nanotori *Carbon* **49** 4518–23

[9] Klein D J, Seitz W A and Schmalz T G 1986 Icosahedral symmetry carbon cage molecules *Nature* **323** 703–6

[10] Itoh S and Ihara S 1994 Isomers of the toroidal forms of graphitic carbon *Phys. Rev.* B **49** 13970

[11] Komatsu N, Shimawaki T, Aonuma S and Kimura T 2006 Ultrasonic isolation of toroidal aggregates of single-walled carbon nanotubes *Carbon* **44** 2091–2

[12] Colomer J F, Henrard L, Flahaut E, Van Tendeloo G, Lucas A A and Lambin P 2003 Rings of double-walled carbon nanotube bundles *Nano Lett.* **3** 685–9

[13] Yu H, Zhang Q, Luo G and Wei F 2006 Rings of triple-walled carbon nanotube bundles *Appl. Phys. Lett.* **89** 223106

[14] László I and Rassat A 2001 Toroidal and spherical fullerene-like molecules with only pentagonal and heptagonal faces *Int. J. Quantum Chem.* **84** 136–9

[15] Liu J, Dai H, Hafner J H, Colbert D T, Smalley R E, Tans S J and Dekker C 1997 Fullerene crop circles *Nature* **385** 780–1

[16] Ahlskog M, Seynaeve E, Vullers R J M, Van Haesendonck C, Fonseca A, Hernadi K and Nagy J B 1999 Ring formations from catalytically synthesized carbon nanotubes *Chem. Phys. Lett.* **300** 202–6

[17] Martel R, Shea H R and Avouris P 1999 Rings of single-walled carbon nanotubes *Nature* **398** 299–9

[18] Sano M, Kamino A, Okamura J and Shinkai S 2001 Ring closure of carbon nanotubes *Science* **293** 1299–301

[19] Geng J, Ko Y K, Youn S C, Kim Y H, Kim S A, Jung D H and Jung H T 2008 Synthesis of SWNT rings by noncovalent hybridization of porphyrins and single-walled carbon nanotubes *J. Phys. Chem.* C **112** 12264–71

[20] Song L *et al* 2006 Large-scale synthesis of rings of bundled single-walled carbon nanotubes by floating chemical vapor deposition *Adv. Mater.* **18** 1817–21

[21] Zhou Z, Wan D, Bai Y, Dou X, Song L, Zhou W, Mo Y and Xie S 2006 Ring formation from the direct floating catalytic chemical vapor deposition *Physica* E **33** 24–7

[22] Lyn M E, He J and Koplitz B 2005 Laser-induced production of large carbon-based toroids *Appl. Surf. Sci.* **246** 44–7

[23] Motavas S, Omrane B and Papadopoulos C 2009 Large-area patterning of carbon nanotube ring arrays *Langmuir* **25** 4655–8

[24] Chen L, Wang H, Xu J, Shen X, Yao L, Zhu L, Zeng Z, Zhang H and Chen H 2011 Controlling reversible elastic deformation of carbon nanotube rings *J. Am. Chem. Soc.* **133** 9654–7
[25] Guo A, Fu Y, Guan L, Zhang Z, Wu W, Chen J, Shi Z, Gu Z, Huang R and Zhang X 2007 Spontaneously formed closed rings of single-wall carbon nanotube bundles and their physical mechanism *J. Phys. Chem.* C **111** 3555–9
[26] Glukhova O E, Kondrashov V A, Nevolin V K, Bobrinetsky I I, Savostyanov G V and Slepchenkov M M 2016 Prediction of the stability and electronic properties of carbon nanotori synthesized by a high-voltage pulsed discharge in ethanol vapor *Semiconductors* **50** 502–7
[27] Liu L, Jayanthi C S and Wu S Y 2001 Structural and electronic properties of a carbon nanotorus: effects of delocalized and localized deformations *Phys. Rev.* B **64** 033412
[28] Dresselhaus G, Dresselhaus M S and Saito R 1998 *Physical Properties of Carbon Nanotubes* (Singapore: World Scientific)
[29] Liu C P and Ding J W 2006 Electronic structure of carbon nanotori: the roles of curvature, hybridization, and disorder *J. Phys.: Condens. Matter* **18** 4077
[30] Chou Y Y, Guo G Y, Liu L, Jayanthi C S and Wu S Y 2004 Electrical conductance of carbon nanotori in contact with single-wall carbon nanotubes *J. Appl. Phys.* **96** 2249–53
[31] Lin M F and Chuu D S 1998 Electronic states of toroidal carbon nanotubes *J. Phys. Soc. Jpn.* **67** 259–63
[32] Tsai C C, Shyu F L, Chiu C W, Chang C P, Chen R B and Lin M F 2004 Magnetization of armchair carbon tori *Phys. Rev.* B **70** 075411
[33] Liu L, Guo G Y, Jayanthi C S and Wu S Y 2002 Colossal paramagnetic moments in metallic carbon nanotori *Phys. Rev. Lett.* **88** 217206
[34] Liu C P, Chen H B and Ding J W 2007 Magnetic response of carbon nanotori: the importance of curvature and disorder *J. Phys. Condens. Matter* **20** 015206
[35] Rodríguez-Manzo J A, Lopez-Urias F, Terrones M and Terrones H 2004 Magnetism in corrugated carbon nanotori: the importance of symmetry, defects, and negative curvature *Nano Lett.* **4** 2179–83
[36] Shyu F L 2005 Magneto-optical properties of carbon toroids: influence of geometry and magnetic field *Phys. Rev.* B **72** 045424
[37] Lin M F 1999 Thermal properties of carbon toroids *J. Phys. Soc. Jpn.* **68** 3585–91
[38] Lin M F 1999 Transport properties of carbon toroids *J. Phys. Soc. Jpn.* **68** 3744–5
[39] Zheng Q and Jiang Q 2002 Multiwalled carbon nanotubes as gigahertz oscillators *Phys. Rev. Lett.* **88** 045503
[40] Liu P, Zhang Y and Lu C 2005 Oscillatory behavior of C60-nanotube oscillators: a molecular-dynamics study *J. Appl. Phys.* **97** 094313
[41] Cox B J, Thamwattana N and Hill J M 2007 Mechanics of atoms and fullerenes in single-walled carbon nanotubes. I. Acceptance and suction energies *Proc. R. Soc.* A **463** 461–77
[42] Hilder T A and Hill J M 2007 Oscillating carbon nanotori along carbon nanotubes *Phys. Rev.* B **75** 125415
[43] Hilder T A and Hill J M 2007 Orbiting nanosectors inside carbon nanotori *Micro Nano Lett.* **2** 50–3
[44] Peña-Parás L, Maldonado-Cortés D, Kharissova O V, Saldívar K I, Contreras L, Arquieta P and Castaños B 2019 Novel carbon nanotori additives for lubricants with superior anti-wear and extreme pressure properties *Tribol. Int.* **131** 488–95

[45] Taha-Tijerina J J, Martínez J M, Euresti D and Arquieta-Guillén P Y 2022 Carbon nanotori reinforced lubricants in plastic deformation processes *Lubricants* **10** 74

[46] Cruz-Torres A, Castillo-Alvarado F D L, Ortíz-López J and Arellano J S 2010 Hydrogen storage inside a toroidal carbon nanostructure C_{120}: density functional theory computer simulation *Int. J. Quantum Chem.* **110** 2495–508

[47] Chan Y, Cox B J and Hill J M 2012 Carbon nanotori as traps for atoms and ions *Physica* B **407** 3479–83

IOP Publishing

Nanocarbon Allotropes Beyond Graphene
Synthesis, properties and applications
Arpan Kumar Nayak and Santosh K Tiwari

Chapter 10

Nanocarbons: commercialization, shortcomings, and future prospects

Srikanta Moharana, Nannan Wang, Bibhuti B Sahu, Bhim Prasad Kafle and Santosh K Tiwari

Carbon nanotubes, graphene, carbon quantum dots, and other nanocarbon-based materials are becoming more popular for making new types of composites that could be used to further technological progress. The superior intrinsic qualities and combined effects of nanocarbon-based materials have, in particular, been found to augment or modify the characteristics of the individual composites. The use of carbon materials in environmental remediation is an established and tested approach. In current times, there has been a major shift in interest toward the environmentally responsible synthesis of nanocarbons for use in environmental remediation and powering a range of devices. In order to make nanocarbons that are not harmful for the environment, researchers have looked into new and environmentally friendly ways to make them. A new category of biocompatible, non-hazardous materials has emerged as a result of the development of nanocomposite materials, which are formed by combining nanocarbons and polymers. This chapter focuses on the current advancements in the investigation of nanocarbon-based materials and their composites for use in technological applications, including their commercialization, challenges, and prospects for their use.

10.1 Introduction

In recent decades, both developed and developing nations have become more dependent on chemical goods, including artificial-detergents, drugs, dyes, fertilizers, polymers and materials, food extracts, and chemicals related to the agricultural sector [1]. Their use has increased exponentially, which has increased pollution from operational processes, burning fuels, and waste disposal. The extensive use and overexploitation of conventional energy sources such as coal and other fossil fuels remain the main contributors to the alarming increase in pollution levels throughout

doi:10.1088/978-0-7503-5177-5ch10

the biosphere [2]. Also, their waste includes hazardous carbon, such as carbon soot, which interacts with ecosystems and damages the environment [3, 4]. The creation of an eco-friendly environment inside the biosphere requires the study and manufacture of ecologically friendly materials that can keep up with rapidly evolving technological advancements [5]. Industries are particularly concerned with avoiding waste and using effective resources to achieve the goal of zero waste. Thus, green chemistry is crucial to achieving these objectives, which are necessary for a flourishing society [6, 7]. Green chemistry has gained popularity as a method of chemical synthesis since it uses fewer potentially harmful chemicals to create useful chemical products [8]. Owing to the inimitable physical, chemical, thermodynamic, mechanical, dielectric, and optical characteristics of carbon-based materials, including fullerenes, carbon nanotubes (CNTs), carbon quantum dots (CQDs), graphene, graphyne, and MXene, have attracted a lot of interest in recent years [9–11]. Carbon nanomaterials have been touted as potential 'wonder materials' that could be used either as-is or with some adjustments for the aforesaid uses [12]. Graphene and CNTs are both two-dimensional carbon nanomaterials, but graphene has advantages over CNTs owing to the low cost of synthesis and the widespread obtainability of graphite as a natural precursor. There are various applications, such as sensing, batteries, photovoltaic cells, actuators, supercapacitors, catalysis, organic light-emitting diodes (LEDs), field-effect transistors (FETs), optical/electrochemical devices, etc, that all benefit from the use of graphitic materials. It has recently been revealed that these graphitic composites may be employed as excellent nanoscale-fillers in polymer matrices, greatly improving the characteristics of polymers even at very modest loads. However, agglomeration is still a major issue for both CNTs and graphene, since it is caused by contact between tubes and the restacking of graphene in the form of layers. Another problem is that they do not mix well with polymer matrices and many popular organic solvents. Strong filler–polymer interactions may help improve filler dispersion by counteracting the restacking propensity of these graphitic fillers on their surfaces [13–17]. Other composites of carbon have been studied in the growth of polymer nanomaterials include graphite oxide [5, 15] and nanofibers of carbon [16, 17]. More recently, three-dimensional materials obtained by hybridizing 1D and 2D graphene have been utilized as amalgam fillers because of their improved dispersal in contrast to the limited in dispersal presented by the individual fillers in polymer matrices [18, 19].

The fabrication of nanocarbon composites has received a lot of attention recently, and much research has been dedicated to developing new nanocarbon-based composite materials [20]. The aforementioned studies intend to deliver a wider impression of emerging nanocarbons that are being employed for various purposes. There are various types of nanocarbon-based materials that are of immense interest among scientists, academic researchers, and industry, such as CNTs, graphene, graphene oxide, CQDs, graphene quantum dots, onion-like carbon, 3D graphene, graphene nanoribbons, fullerene, nanotubes, graphyne, graphdiyne, carbon nanobuds, graphene nanowires, nanotube moirés, diamanoid, and diamanes (figure 10.1). In this chapter we have focused on some of these nanocarbon-based materials, which

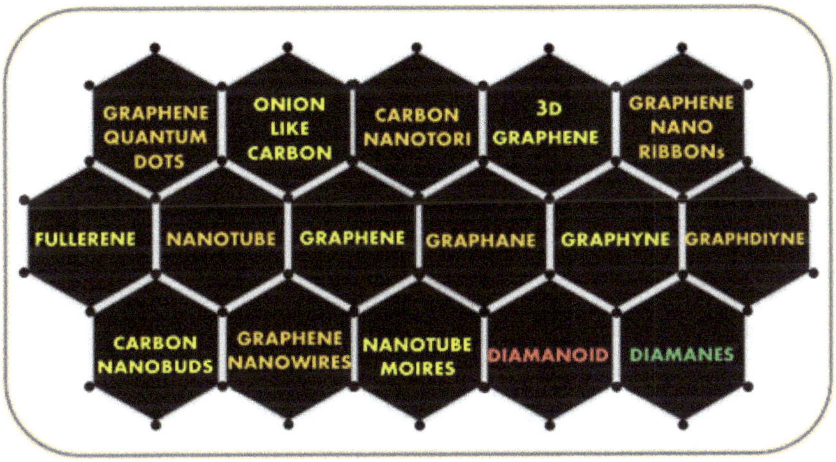

Figure 10.1. Diagrammatic representation demonstrating various kinds of nanocarbons. (Reproduced with permission from [20]. Copyright 2016 Taylor and Francis.)

have future applications in many fields of science and technology for the creation of improved efficiency nanocarbon composites.

10.2 Overview of nanocarbon-based materials

There are numerous possible carbon nanostructures that have been used effectively in synthetic or composite materials. The different materials are classified into categories such as fullerenes (CFu), CNTs, carbon nanofibers (CNFs), graphene, and graphene-like substances (e.g. extended graphite (EG), graphene oxide (GO), and graphene nanoplatelets (GNPs)). Each class represents a whole family of structures with various tunable attributes. Carbon, a single element, may be utilized to make a huge range of different 3D, 2D, and even 1D materials. The conventional 3D composites, such as carbon black, artificial and natural graphite, and carbon fibers are generally reinforced in a thermosetting polymeric matrix [19]. Metal-based filler particles show high conductivity, yet a metal filler combines strength. Moreover, the porous nature of sponges makes them susceptible for disintegration. which prevents them from being used in the majority of supercapacitor electrodes. Materials, despite their excellent performance and scalability, have minimal charge–dischargecycle instability, and high performance pores. [21]. When compared to metal fillers, carbon fillers have additional positive effects on ion transport properties, including the ability to grow, get along well with others, couple well electrically, and exhibit robust stability over charge–discharge cycles. [21]. Thus, materials based on nanoscale carbon structures provide encouraging potential for supercapacitor electrodes. Meng and co-workers [22] have fabricated nanocarbon-based flexible multifunctional composite electrodes. Figure 10.2 shows CNT-sponge electrode materials with a 0.08 wt%. It is observed that carbon nanotubes are attached to the sponge skeleton, and the linked porous network structure of the sponge is shown in figures 10.2(a1)–(a3). The features in figure 10.2(a2) are expanded upon in

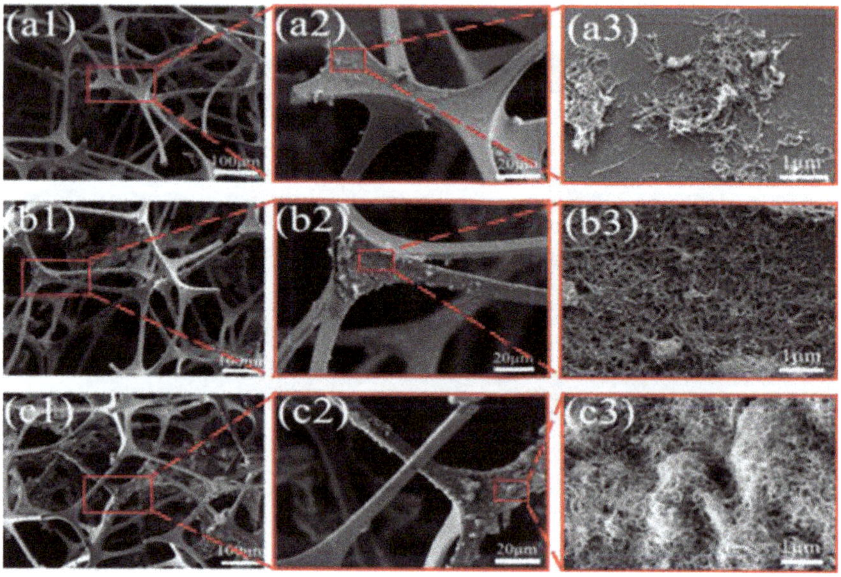

Figure 10.2. SEM pictures of (a1)–(a3) 0.08 wt%, (b1)–(b3) 0.11 wt%, and (c1)–(c3) 0.14 wt% of CNT-sponge material sample. (Reproduced with permission from [22]. Copyright 2021 American Chemical Society.)

figure 10.2(a3). In figure 10.2(a2), CNTs are distributed sparsely across the sponge pore surface, creating a conductive electrical channel. However, the overall conductivity of the conductive network may not be maximized with only a few carbon nanotubes. Carbon nanotube sponge material electrodes with a filling proportion of 0.11 wt% are shown in figures 10.2(b1)–(b3). The area shown by a red circle in figure 10.2(b1) has been enlarged for clarity in figure 10.2(b2). Expansion of the data seen in figure 10.2(b2) is shown in figure 10.2(b3). It is expected that the CNT would be evenly dispersed across the exterior of the sponge apertures, creating electrical channels of high conductivity. There is a significant improvement in the conductive network's capacity to transport electrons and ions due to the more uniform distribution and higher density of carbon nanotubes on the sponge skeleton exhibited in figure 10.2(b2) compared to that displayed in figure 10.2(a2). Further, carbon nanotube sponge composite electrodes with a filling percentage of 0.14 wt% are shown in figures 10.1(c1)–(c3). Figure 10.2(c2) is an enlargement of a key area in figure 10.2(c1). The features in figure 10.2(c2) are expanded upon in figure 10.2(c3). The CNTs are observed to be highly concentrated on the skeletal surface of the sponges, with numerous carbon nanotubes packed in a highly agglomerated fashion. When the carbon nanotubes join together, the interior confrontation of the conductive system rises, and the electron conduit becomes less efficient. Carbon nanotubes are overrepresented and distributed excessively densely on the sponge skeleton, as seen in figure 10.2(c2), in comparison to figure 10.2(b2). Carbon nanotube aggregation decreases the overall conductive network's capacity to transport electrons and ions. Figure 10.2 shows that the most consistent attachment to the sponge pore structure occurs at a filling percentage of 0.11 wt%, suggesting that the conducting system created may have the greatest capacity for transporting electrons as well as ions.

Han and co-workers [23] fabricated materials based on cement with CNT-nanocarbon black materials. The influence of water on the piezoelectric characteristics of materials based on cement has been reported. Jiang and co-workers [24] recently reported diamond-carbon nanowall (D-CNW) films associating large sp^2-C amounts for applicability in the electrochemical sensing domain. It was observed that, at 1065 °C on the protruding cylinder, microwave plasma assisted chemical vapour deposition (MPCVD) yielded films with a wide range of D-CNW morphologies and CH4 contents. Figures 10.3(a)–(h) show that the D-CNW films with different CH_4/H_2 ratios exhibit a network-like three-dimensional structure. In figure 10.3, the red arrows indicate sp^3 bonded diamond, which is intense and straight like nanoplatelets. The yellow arrows indicate that the dark, curved sheets are sp^2 bound CNWs. Small diamond nanoplatelets and CNWs, linked among diamond ounces are visible in the D-CNWs-3% film (figure 10.3(a)). The area (specific surface) of the film may increase greatly if CNWs are present. The vertical diamond flakes expand to 1 nm in size when the CH_4/H_2ratio is increased to 5% (figure 10.3(b)), and the number of CNWs affixed to the upright diamond nanoplatelets similarly increases. While CNWs with a large proportion of edges are significantly more compact in appearance, upright diamond nanoplatelets over the silicon substrate become extra dispersed when the concentration of CH_4 rises from 0% to 7% (figure 10.3(c)). At a CH_4 concentration of 9%, the width of diamond nanoplatelets grows to 2 nm, and the bulk of CNWs increases, as illustrated in figure 10.3(d). Moreover, the D-CNW morphological evolution shows that increasing methane concentration promotes diamond nanoplatelet expansion and CNW abundance. Therefore, the nucleation and development of diamond and CNWs could be controlled and the surface morphology of the D-CNW films could be modulated by altering the CH_4/H_2 ratio in an assortment of gases. Furthermore, as shown in figures 10.3(e)–(h), the growth of D/CNW films is sped up by rising CH_4 concentration, as shown by the progressive increase in film thickness from 1.03 to 2.40 nm with the growing ratio of CH_4/H_2.

Feng and co-workers [25] have fabricated nanocarbon–polyaniline (PANI)–graphene nanosheet (GNS) composites through an *in situ* polymerization technique. Figure 10.4 depicts micro-structural views of GNSs, CNTs, carbon black (CB), and their respective compounds after *in situ* polymerization of aniline. The average GNS is a few micrometers in size, and has a corrugated, layered structure (figure 10.4(a)). In a core–shell configuration, layered PANI was deposited uniformly and firmly over the exterior of CNTs or CB (figures 10.4(e) and (f) and figures 10.4(c,g) and (d,h)). In figures 10.4(c) and (d), high resolution TEM (HRTEM) images give the impression of a PANI layer deposited uniformly on the exterior of CNTs and CBs, with widths of 8–10 nm and 3–5 nm for CNT–PANI and CB–PANI materials, respectively. As illustrated in figures 10.4(a) and (b), *in situ* polymerization resulted in a homogenous PANI coating on the GNS surface or filling the spaces between the GNSs. Graphene and PANI coatings alternate in the GNS–PANI composite's layered structure (figures 10.4(a) and (b)).

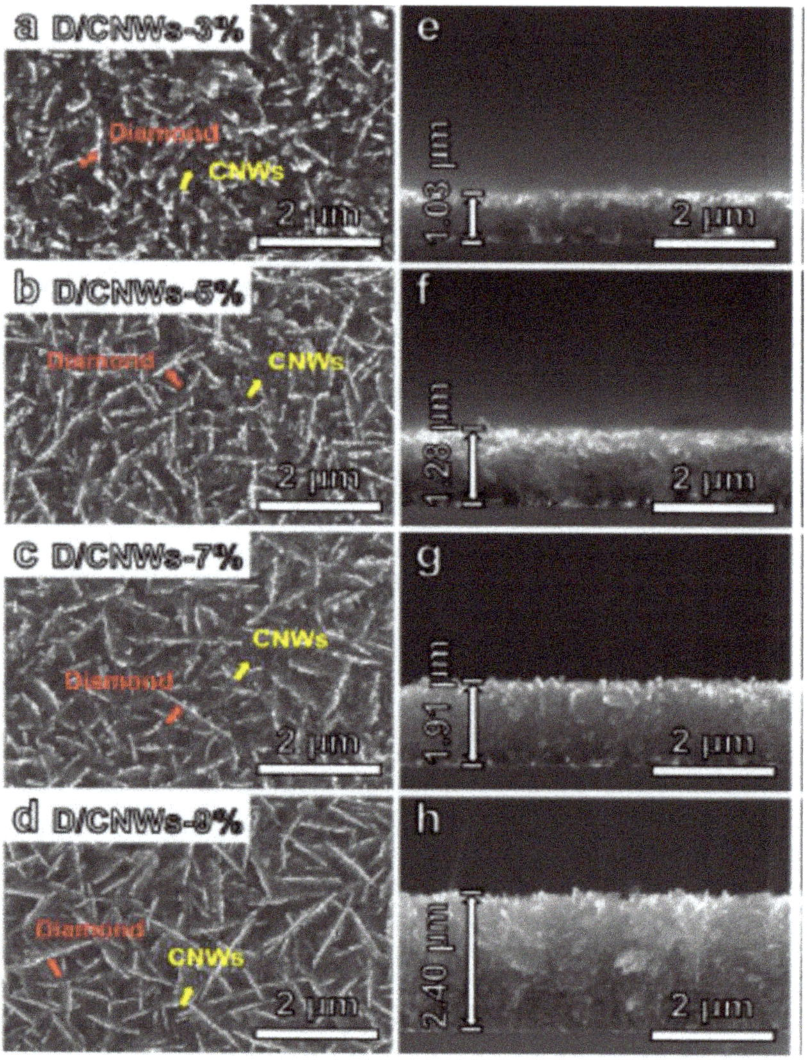

Figure 10.3. Surface and cross-section images of D/CNWs films made at 1065 °C with various CH_4/H_2 ratios of (a), (e) 3%, (b), (f) 5%, (c), (g) 7%, and (d), (h) 9%. (Reproduced with permission from [24]. Copyright 2021 Elsevier.)

10.3 Classification

The unusual electrical structure and lower size of carbon compared to the four most abundant elements makes it capable of bonding with other elements, and this gives rise to the element's propensity for polymerizing into large molecular weight composites with elongated chains [26]. Since it is found mostly in the surface biosphere, rather than in the Earth's crust, where silicon ranks second, carbon is the 15th most plentiful element on Earth. Carbon's four valence electrons make it a prime candidate owing to covalent bond formation with both metals as well as

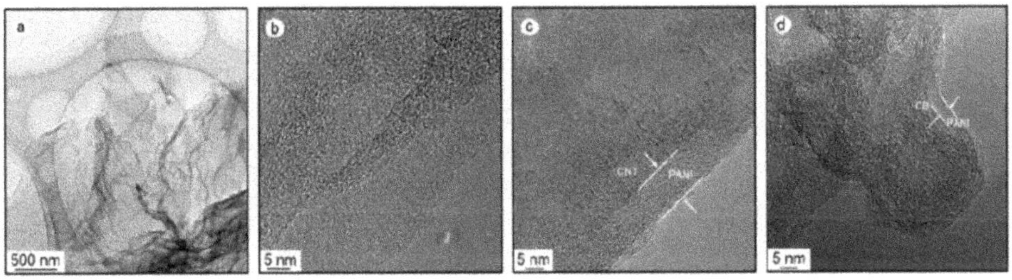

Figure 10.4. Archetypal TEM images of (a) GNS–PANI composites and HRTEM images of (b) GNS–PANI, (c) CNT–PANI, and (d) CB–PANI composites. (Reproduced with permission from [25]. Copyright 2011 Elsevier.)

non-metals. As a result of its properties, compounds based on carbon may take various molecular structures, and the same type of atom can present a wide variety of characteristics depending on its particular arrangement [27]. There are several carbon allotropes that may be produced from natural carbon sources, graphite and diamond being only two examples. The top-down and bottom-up approaches for achieving nanotechnology are both used in the development of the novel family of carbon nanomaterials that have been dubbed 'wonder materials' [28]. In the recent past, several forms of carbon-based nanomaterials have undergone extensive experimental testing and been produced successfully as cutting-edge materials in a broad spectrum of technological applications [29–31]. Generally, carbon-based nanostructured materials with dimensions in the range of 1–100 nm are excellent with suitability for many engineering applications because of their exceptional electrical, mechanical, and thermal properties at the nanoscale, and their smaller size compared to of other materials [32]. In addition, the shapes and geometrical structures of carbon-based nanostructured composites are categorized into various classes, which include carbon nanohorns, carbon nanotubes, graphene, fullerene, carbon fibers, and carbon quantum dots, etc [15, 33, 34]. Nanocarbon-based composites are commonly fabricated making use of the special qualities of individual nanocarbons, such as the anisotropy of 1D tubular CNTs or the dimensionality-breaking capabilities of 2D planar graphene. These carbon-based composites offer a broad range of practical applications, including nano-electronics, paints, conductive polymers, battery technology, energy storage, textiles and coatings [35, 36]. However, there are several other superior properties of carbon-based materials, including elevated aqueous solubility, low-cost processing, reduced toxicity, ample surface functional groups, and a greater dynamic exterior area for functionalization. Carbon quantum dots (CQDs) have also been shown to exhibit fluorescence, making them a novel biocompatible material [37].

10.3.1 Fullerene

Fullerene, often referred to as buckminsterfullerene (C_{60}) (as shown in figures 10.5(a) and (j)), was initially discovered by Kroto, Curl, and Smalley in 1985, who were recognized with the 1996 Nobel Prize in Chemistry. It was a new member of the

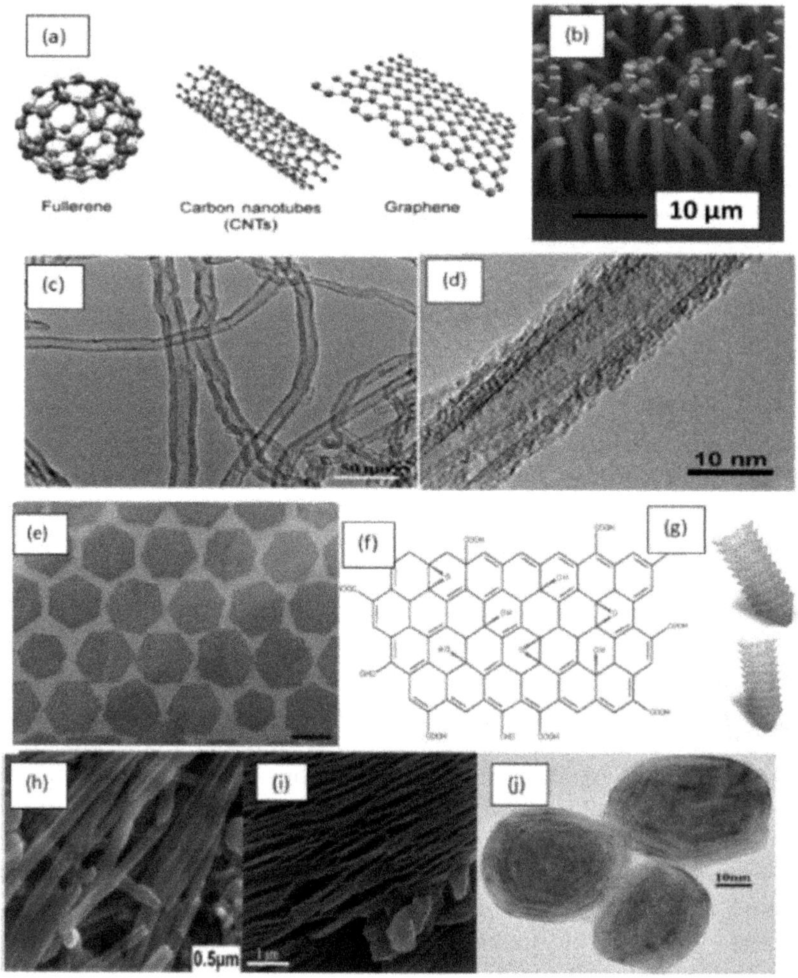

Figure 10.5. (a) Diagrams of CNTs, graphene, and fullerene. (b) SEM micrograph of aligned bundles of CNTs. (c) TEM image of CNTs. (e) A TEM image of a graphene grain array. (f) A simplified illustration of the GO structure. SEM images of (g) CNFs, (h) herringbone-like CNFs, and (i) multilayered MXene. (j) TEM image of onion-like fullerene. (Reproduced with permission from [52]. Copyright 2016 Elsevier.)

carbon family. Since then, nanoscience and nanotechnology have been on an exciting path of discovery that was recently extended with the announcement of the finding of C_{60} ions in outer space. In this carbon allotrope, the C_n repeating unit ($n > 20$) is clustered to form a spherical surface with a hollow center, giving it qualities intermediate between those of graphite and diamond [38]. At the corners of the pentagons and hexagons, carbon atoms repeatedly form sp^2-hybridizing covalent bonds. In general, for the two bond lengths of C_{60}, the double bonds for 6:6 ring bonds are shorter than those for 6:5 ring bonds. One distinguishing characteristic of C_{60} is the poor electron delocalization that the pentagonal rings of the molecule cause. As a result, C_{60} interacts readily with electron-rich species, as one would anticipate from electron-deficient alkenes. The density of the black solid C_{60} is 1.65 g cm^{-3}, the standard heat of formation is 9.08 kcal mol^{-1}, and the boiling temperature is 800 K. Researchers have examined the intrinsic qualities of fullerenes to investigate whether they may be used in applications including heat-resistant materials, superconductivity, light-activated antibacterial agents, and biocompatibility in the medical industry. The standard route of creating fullerenes involves either an arc or combustion process, followed by purification of the reaction by-products. This yields fullerenes that can be integrated into polymers. It is not common for fullerenes to be evenly distributed in polymer matrices due to their poor adherence and limited miscibility with traditional organic solvents. Fullerenes may be functionalized with hydrophilic and hydrophobic groups, which improves their physicochemical characteristics and promotes their dispersion and strong adhesion in polymer matrices, allowing them to overcome these challenges. Miscibility with host polymers is crucial for many uses, and fullerene derivatives offer better miscibility with little agglomeration [38].

10.3.2 Graphene

Graphene has received a lot of attention since its discovery in 2004, when its unusual qualities were first highlighted [39]. Graphene is an exciting substance that might be used in energy, catalysis, and sensing. It has a 2D hexagonal symmetry and is a single-atom thick (0.345 nm) sheet of carbon atoms (as shown in figure 10.5(e)). It has a massive surface area of 2630, a huge Y value (Young's modulus) of 1100 GPa, a rupture strength of 125 GPa, and a large conductivity (thermal) of 5000 W m^{-1} K^{-1}. Since graphene was first found byisolated by breaking graphite sheets mechanically, academic researchers and scientists have become more interested in this material [39]. Graphene is a key component of more complicated materials such as graphite, fullerene, and carbon nanotubes (CNTs). Graphene has an interplanar distance of 0.335 nm between its separate layers, making it an allotrope of carbon with bond length of 0.142 nm between the surrounding atoms of carbon. The distribution of graphene in graphite is limited by the force from the van der Waals interaction [40]. Graphene's few layers have substantially higher thermal stability, mechanical rigidity, and electrical conductivity than its three-dimensional equivalents. Because of its excellent thermal conductivity, graphene is also an efficient heat conductor,making it a preferable option to more traditional materials such as graphite and diamond [39–41]. There are several experimental and theoretical reports in the literature on the probable applicability of graphene in spintronics, photonics, plasmonic, and electronics that have focused on quantum confinement and graphene's electrical, magnetic, optical, and thermodynamic properties [42]. Furthermore, graphene oxide (GO) is a kind of graphene in which O- and H-atoms have been added to the C-atoms in the structure (figure 10.5(f)). These oxide-based materials are typically oxidized with strong acids and oxidizers and introduce numerous groups for functionalization (hydroxyl, epoxy, and carboxyl) into the lattice of graphene during the oxidation process. Since GO nanosheets (O- and H-based functionalization groups) are commonly dissolved easily in aqueous media and different solvents (organic), they are suitable for the fabrication of different GO-founded composite materials.

10.3.3 Carbon nanotubes

CNTs are a noteworthy allotrope of carbon-built nanostructures noted for their outstanding mechanical and electrical behavior. CNTs were discovered in 1991 by a Japanese scientist called S Ijimain, and they are lightweight, high-strength materials with nondimensional properties equivalent to steel [43, 44]. Because of their remarkable optical characteristics and high specific surface area, nanocarbons are being used as fillers for nanocarbon–polymer composites and as a support material in a variety of technological applications [45]. CNTs are made up of layers or sheets of C-atoms which have been folded into hollow, even cylinder shaped tubes (as shown in figures 10.5(a)–(d)). Rolled sheets of graphene produce cylindrical CNTs, while capped fullerene structures appear as half-spheres. CNTs are commonly classified into two types based on their precise geometrical compositions as revealed by high-voltage electron microscopy: single-walled carbon nanotubes (SWCNTs) and multi-walled carbon nanotubes (MWCNTs). These nanotubes are made by rolling and coiling sheets of graphene, which is a single layer of carbon atoms [46, 47]. CNTs range in size from 1 to 3 nm in diameter and a few millimeters in length, while MWCNTs are made up of concentric layers of graphene sheets with an inter-space distance of 0.34 nm [48]. The most common CNT synthesis techniques are arc discharge, laser ablation, and chemical vapor deposition (CVD) [49]. CNT fibers are undergoing fast development due to their potential use in body and vehicle armor, transmission line cables, interlaced draperies, and textiles. Since CNTs have begun to be mass manufactured, they have found uses in a wide variety of industries [50]. Figure 10.5 shows various carbon-based nanomaterials including CNTs, GO, CNFs, methyl xanthene (MXene), and carbon from carbide (CNC). These carbon-based nanomaterials have been studied extensively in chemistry and materials science because of their distinctive qualities, which include strong chemical, thermal, and mechanical strength, conductivity, optical properties, and low density [51]. However, due to their construction of single or double layered graphite placed in parallel with or at an inclination to the fiber axis, CNFs have a hollow core. The stacked layers are adjacent to one another and may have a variety of arrangements, such as a parallel stack, a cup stack, or a herringbone stack [52]. An example of a CNF with a herringbone structure is shown in figures 10.5(g) and (h).

10.3.4 MXene

MXene is in fact not part of the carbon family but many of its properties are similar to graphene. Therefore, we are adding an overview of Mxenes and their derivatives in this section for better readability. MXenes are a class of new 2D materials that consist of transition metal carbides and nitrides. In order to create MXenes, the 'A' layers of laminated MAX phases are removed ($M_{n+1}AX_n$, n = 1–4), where M represents transition metals such as Ti, V, Cr, and Nb, and A is an A-group element (for example, Al, Si, Sn, or In), and X may be C- or N- [53]. These chemicals have thus been dubbed 'MXenes'. Since the 2011 discovery of $Ti_3C_2T_x$ MXene, various synthesis routes have been developed to prepare MXene. Among these MAX phases, hexagonal carbides and nitrides may be found in a stacked configuration. Miniscule films of transition metal carbide or carbon nitride make up MXenes. Figure 10.5(i) shows a SEM image of 2D layered MXene. Because of the presence of transition metal carbides, compounds in the A-group-free MAX phase are hydrophilic and exhibit metallic conduciveness [52–54]. Carbon compounds generated from either carbide (CDC) or boron carbide (CBC) are referred to as 'tunable nano-permeable carbon', binary carbide precursors (e.g. TiC or SiC), or ternary carbide precursors (MAX phases; e.g. Ti_2AlC or Ti_3SiC_2) [28]. CDCs may also be made out of polymer-derived ceramics or carbonitrides such as Si–N–C. MXenes have been the subject of extensive research in many different areas, including those of energy storage, electrocatalysts, biomedicine, EMI shielding, and others [53].

10.3.5 Carbon quantum dots

CQDs are an important member of the category of composites based on carbon because they not only share the same physicochemical characteristics as representative carbon-based composites, but also display exceptional luminescence behavior, including truncated toxicity, photostability, biocompatibility, multiphoton excitation, up-converted photoluminescence, and chemiluminescence [55, 56]. They has various useful applications; CQDs have been used widely in sensing and biomedicine and have excellent performance for electrochemical processes because of their cheapness and unique electron-transfer properties. They also have a high specific surface area. In addition, the CQDs' abundant interface functionalization groups (-alcohol, amine, etc) may serve as better securing sites and dynamic sites in the production of multi-component and highly performing composites [57]. For example, Xu and co-workers [58] have reported carbon-based nanoparticles with a distinctive luminous feature when purifying SWCNTs. Sun and colleagues were the first to report that CQDs are particles smaller than 10 nm in size and hold great promise as materials for use in bio-imaging, photo-degradation, and catalysis due to their unique ability to absorb and emit light in a wide range of colors [59]. In the past few years, a wide variety of chemical ingredients has been used to manufacture CQDs. These include citric acid, ammonium citrate, ethylene glycol, benzene, phenylenediamine, phytic acid, and thiourea [60, 61]. Furthermore, the synthesis of fluorescent CQDs has been achieved via various chemical routes, which include hydrothermal, solvothermal, electrochemical, microwave aided pyrolysis, ultrasonication, and chemical oxidation, methods, etc [62].

10.3.6 Carbon nanoribbons

The creation of carbon nano-ribbon networks (figure 10.6) may result from the thermally induced breakdown of carbon-containing Si–C N–O–B preceramic polymers, which may precipitate three-dimensionally interconnected nanocarbon clusters. These collections of turbostratic carbon precipitate at 800 °C (polyphenylsiloxane) and 1400 °C (polydimethylsiloxane) due to the decay of hydrocarbon

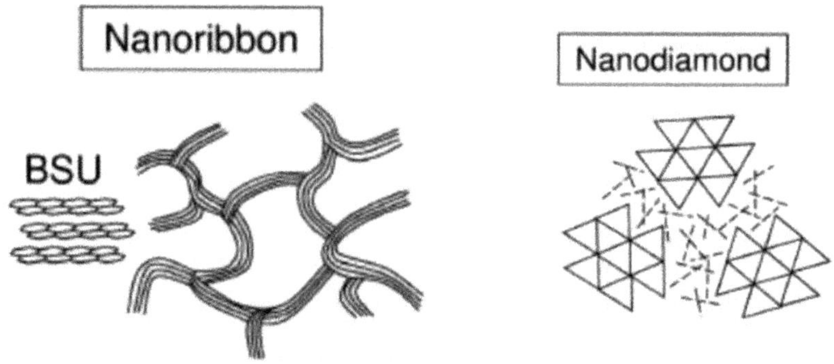

Figure 10.6. Structures of allotropes of nanocarbon.

trashes arising out of splitting of the functionalization groups linked to Si (polymethylsiloxane). A percolating network (bunch size to infinity) of turbostratic C-ribbons forms when these carbon collections expand and might combine by interface-to-interface linking of surrounding basic units of structure as the temperature increases. While semiconducting activity is ascribed to thermal-fluctuation-induced tunneling of electrons between equally dispersed charge transporter groups in annebulous Si–O–C–H, beyond the percolation threshold direct interaction of carbon particles enables the transmission of electrons [43, 63–65].

10.3.7 Nanodiamonds

Nanodiamonds (NDs) (figure 10.6) are one of the complex types of nanocarbons, which also include nano-dimensioned nebulous carbon, fullerenes, and diamondoids. Owing to their cost effectiveness, extensive synthesis via the detonation of carbon-contained explosives, tiny primary particle dimensions in nanometers with a narrow size distribution, ease of interfacial functionalization including bio-conjugation, and highly biocompatibility, NDs have gained worldwide attention since their in-depth study began in Russia in the 1960s [66]. Nanocrystalline (50–100 nm) diamonds were found in meteorites, and ultra-nanocrystalline diamonds (UNCDs) (2–5 nm) might be created using plasma-aided CVD. Contrasting carbon nanotubes and graphene, when it comes to the electrical structure of nano-diamond particles (those 4 nm or bigger), quantum confinement has no effect [43]. Nano-diamond films might be used as high-efficiency light emitters and low-voltage (cold) cathodes, whereas diamondoids could be used as electron emitters. Like carbon nanoparticles, quantum dots, and metallic nanoparticles, NDs are expected to be used as beneficial mediators in diagnostic probes, drug transport vehicles, gene rehabilitation, antiviral and antibacterial therapies, tissue frameworks, and the creation of innovative therapeutic devices such as nanorobots [66].

10.3.8 Other nanocarbon-based materials

Carbon onions (figure 10.7) are nanoparticles with a near-spherical shape and a fullerene-like carbon core surrounded by concentric graphitic shells. Because of their

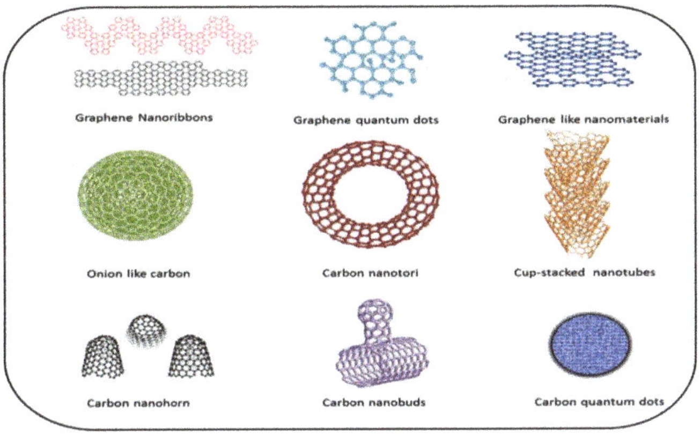

Figure 10.7. Schematic presentation of nanocarbon based materials.

extremely symmetric structure, they may be useful in ways that other carbon nanostructures, such as graphene, nanodiamonds, and nanotubes, are not. The outer graphite layers of carbon onions may shield the substances inside and provide a template for the attachment of appropriate functional groups, allowing these structures to be used as nano-capsules for medication delivery systems. Nanotori, also known as nano-rings (figure 10.7), are innovative material structures made up of CNTs connected at either end to create a circle. There is a close relationship between the well-known spherical nanostructures and the toroidal-like structures. Research has demonstrated that they have excellent characteristics and potential uses. In the metal-mechanics sector, carbon nanotorus structures have recently been tested as reinforcements for regular lubricants and fluids, resulting in less wear and friction [67]. A new kind of CNT known as a cup-stacked carbon nanotube (CSCNT) (also known as a 'herringbone' structure) [68] is well-suited for catalytic support and other energy applications. CSCNTs are made up of truncated conical graphene layers (cups), displaying a significant amount of exposed and reactive edges both on their outer surface and in their interior channel, which is very helpful for the deposition of catalysts. The unique form and physicochemical features [69] of carbon nanohorns (CNHs), discovered in 1998, make them an intriguing nanomaterial. CNHs are typically conical in shape and are made up of sp^2 hybridized carbon atoms. Single-graphene tubules, with strongly strained conical ends, are the building blocks of CNHs, and as they are aggregated into larger spherical angled superstructures CNHs are held together by van der Waals forces, and the aggregated superstructure's micro- and mesopores provide additional advantages to the CNHs. In addition, C_{60} molecules were covalently bonded to the sidewall of single-walled carbon nanotubes (SWCNTs) in 2007, creating another hybrid carbon nanostructure [70]. To describe this novel nanostructure the term 'carbon nanobud' (CNB) was coined. The electronic characteristics of CNBs are distinct from those of fullerenes and CNTs. When compared to pure SWCNTs, CNBs have several advantageous properties, including lower field thresholds, larger current densities, and better cold electron field emission efficiencies [71]. A GNR is a thin strip of graphene. Compared to graphene sheets, the more well-known two-dimensional analogue of GNRs and the quasi-one-dimensional character of GNRs lead to significant benefits. In conductive films and polymer composites, for example, a GNR's large aspect ratio considerably reduces the percolation threshold, making the materials amenable to spinning fibers [72]. As zero-dimensional derivatives of graphene, GQDs have attracted a great deal of attention from scientists in recent years [73–75]. GQDs have remarkable optical and electrical features as a result of their substantial quantum confinement and edge effects. When compared to conventional semiconductor QDs (e.g. ZnS, TiO, CdSe, CdS, and CdTe), GQDs are a viable alternative due to their low cost, high water solubility, sustained fluorescence, variable bandgap, low toxicity, and strong biocompatibility [74, 75]. Schematic views of various types of nanocarbon based materials are shown in figure 10.7.

10.4 Synthesis of natural nanocarbon

Nature contains nanocarbons, but only at a very small scale. The natural synthesis approach is novel in that it allows for the modification of the amount of layers of graphite, and hence the chemical as well as physical characteristics of the nanoscale carbon, which might have important consequences in engineering. According to earlier research, natural synthesis may produce a diverse spectrum of nanocarbons [76]. For example, Velasco-Santos and colleagues reported that carbon nanotube preparations were observed in a combination of coal and petroleum [77]. As a substrate and catalyst for CVD based SWCNT synthesis, volcanic lava containing particles of metal oxides was employed by Su and Mracek [78–80]. The scientists hypothesized that this mechanism might prove the existence of nanotubes in nature, namely in situations when temperatures reach exceptionally high levels, such as volcanic explosions. Fullerenes, like CNTs and SWCNTs, may be found in a variety of ecological materials, such as the natural mineral shungite from Karelia, where it occurs in low quantities (2% w/w) [81], and in meteorite samples of cosmological origin [82]. Chitin is a nanomaterial derived from carbohydrate polymer and is found in nature. They are generally synthesized with a variety of properties including thermal dimensional stability, dispersibility, mechanical reinforcement, antibacterial activity, etc. Depending on factors such as processing, recycling, shipping costs, etc, polymer waste may be a viable source of natural nanomaterials [80]. The flowchart shown in figure 10.8 demonstrates a variety of approaches for the synthesis of nanocarbon hybrids.

10.5 Chemical functionalization of nanocarbon

Carbon-based nanomaterials with a wide range of functionality have been implemented effectively in a wide range of biomedical, tissue engineering, and energy storage fields [82, 83]. Moreover, the surface functionalization refers to the addition of groups to carbon-based nanomaterials, which alters their chemical and physical characteristics [84]. The surface functionalization of carbon-based nanomaterials occurs in various ways: through oxidation, ionic or non-ionic aliphatic aqueous

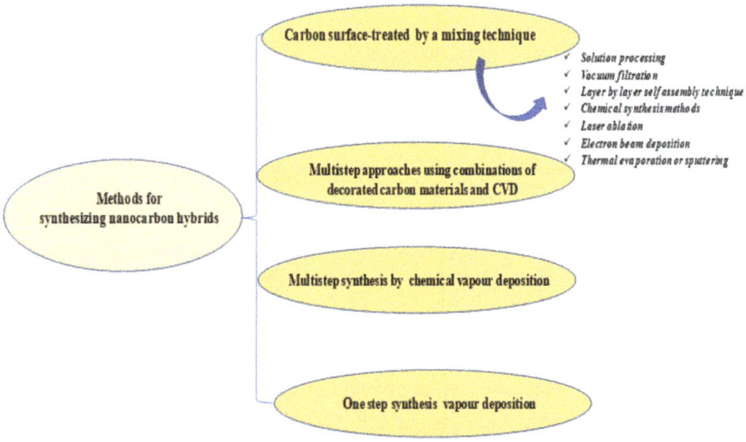

Figure 10.8. Flowchart showing different techniques for the synthesis of nanocarbon hybrids.

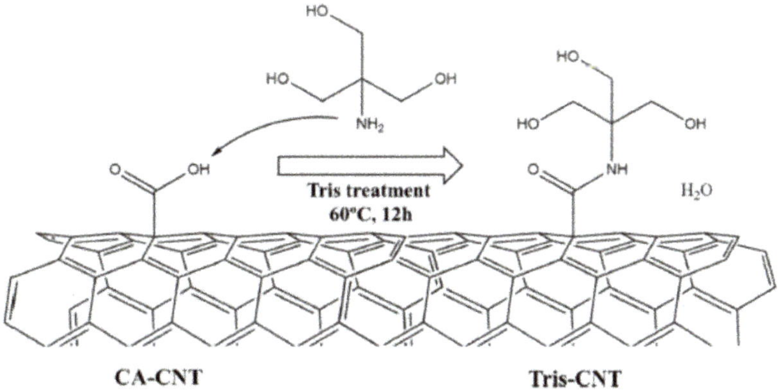

Figure 10.9. The creation of a Tris-CNT catalyst is shown schematically. (Reproduced with permission from [86]. Copyright 2021 Elsevier.)

(hydrophobic), ionic or non-ionic aromatic (piling), van der Waals' force, casing, doping, and straight deposition. The surface of the carbon materials is modified on the basis of specific applications and levels of different functional agents. Thus, various materials have been created using various methods, such as non-covalent and covalent bonds, electrostatic force, hydrogen bonds, van der Waals forces, and so on [85]. For instance, Kwon and co-workers [86] have reported a tris(hydroxy-methane) aminomethane functionalized carbon nanotube catalyst. The Tris-CNT is demonstrated in figure 10.9 to be synthesized by the Tris treatment. Condensation between the hydroxyl (–OH) group in the carboxyl (COOH) group of CA-CNT and the amine (NH_2) group (as shown in figure 10.9) is believed to be responsible for the attachment of Tris to the CA-CNT surface, resulting in Tris-CNT. Qic and co-workers [87] fabricated a MoS_2 nanosheet bonded CNTs molecule with improved photoelectrical and catalytic properties. Figure 10.10 shows the creation of core–shell structured MoS_2–CNTs–SDS hybrid materials. In this work, they suggested

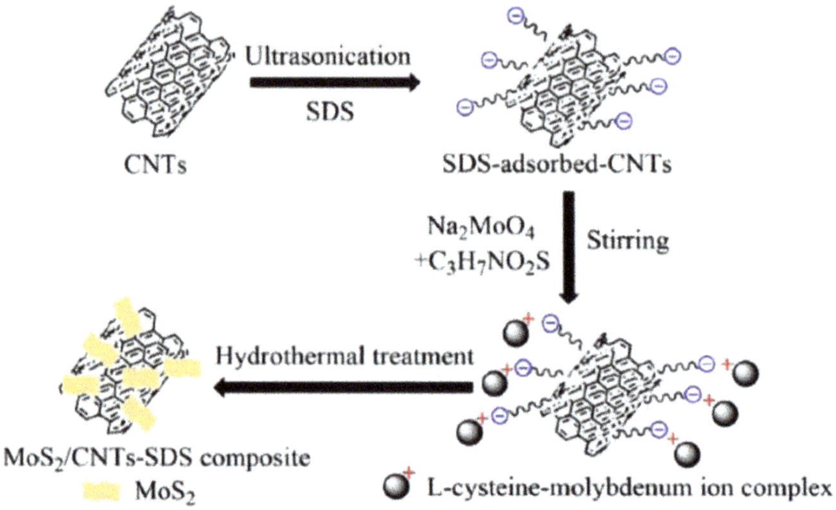

Figure 10.10. Synthesis of a MoS_2–CNTs–SDS composite, shown as a schematic. (Reproduced with permission from [87]. Copyright 2020 Elsevier.)

THAT L-multifunctional cysteine groups may react with inorganic cations and metal to create complexes, making it a useful self-assembly reagent in the production of different sulfide-based composites. Moreover, the positively charged amine groups in L-cysteine led to the formation of a complex between molybdenum ions and a molybdenum cation. Furthermore, SDS, an anionic dispersion, may produce an absorbing layer adjacent to the CNTs' outer wall [87]. Furthermore, the oxidation of CNTs is used for functionalizing carbon-based nanoparticles. Carboxylic functional groups (–COOH) are attached to the nanotubes' walls by ultrasonic treatment in a mixture of acids (figure 10.9). Because of their oxidation, CNTs become soluble in water, but their mechanical and electrical characteristics are unaffected. Additional functionality may be added by attaching carboxylic groups to the nanotubes' surface. Fullerenes and graphene that have been modified with new functions exist as well [80].

10.6 Applications

Nanocarbon's outstanding physiochemical characteristics have generated immense interest among scientists and researchers due to their potential applications in the fields of adsorption, photocatalysis, fertilization, nanobiotechnological goods, environment correlated materials, renewable energy, etc [88, 89]. De Volder and co-workers [90] reported the manufacture of nanocarbon-based materials on an industrial scale in quantities of several thousand metric tons. These materials have superior thermal and mechanical stability, which makes them a promising replacement for traditional fillers in nanocomposites with a high aspect ratio. As a result, these nanocomposites exhibit significant improvements in their mechanical and other performances over those of the basic materials used in their preparation.

Nasibulin and colleagues reported that the incorporation of nanocarbon into composite materials based on a cement matrix increased their strength [91]. These composites are lightweight, high-strength materials with improved efficiency and reduced power usage. Furthermore, nanocarbon based composite materials have excellent properties including larger tensile strength, flexibility, and superior electrical conductivity, making them preferable for energy applications [92, 93]. During the past few years, carbon-based materials such as graphene and its composites have attracted tremendous attention due to their huge range of potential applications. Similarly, when compared to typical nanocarbon materials, the polyvinyl alcohol–graphene–carbon nanotube-based composites demonstrates better stiffness, strength, and ductility. Polyvinyl alcohol-based graphene and carbon nanotube composites show high resistance, and they are still formable using epoxy polymers, so they may be used in smart, stretchy insulator devices [94].

10.6.1 Nanocarbon–polymer composite energy applications

If a nation were able to generate and store energy, particularly from renewable sources, it is feasible that the economy of that nation would see a tremendous improvement [95]. Nanomaterials based on carbon and the composites made from them show a lot of promise in a number of different fields, including energy storage and harvesting. It is well known that a huge majority of sp^2 hybridized nanocarbons has significant properties, such as a broad range of pore sizes, surface area, and enhanced mechanical and electrical performances. The improvement in the development of energy storage and energy saving are both linked with the use of nanocarbons and their composites. Photovoltaic (PV) cells (solar cells) are an approach for converting sunlight into electricity. Generally, there are two main types of PV cells (i) thin-film and (ii) crystalline silicon. Moreover, semiconductors such as silicon (Si), cadmium (Cd), and copper (Cu) are used in energy storage systems [96]. Pt based semiconductors are usually used in thin-film PV cells for their large band width in specific uses, but they have a high cost and limited availability. Carbon-based materials and their composites may be a suitable alternatives to platinum-based materials [97]. Nanomaterials based on graphene are increasingly being used for enhanced electron transport and the conversion of solar energy. Because of its favorable features, graphene may be used in fuel cells and batteries [98]. Owing to its superior qualities, such as huge-power density, extremely quick charging duration, and superior cycle stability, supercapacitors are utilized for photovoltaic storage devices in electrical cars, hybrid electric vehicles, power cells, and portable electronic device. However, the energy storage systems of supercapacitors are mostly based on pseudocapacitance and electrochemical bi-layer capacitance. A pseudocapacitor transfers charges via a faradic response method. For example, various conductors such as metals, oxides, and conductive polymers are commonly used in pseudocapacitance. Charges are collected at the boundary by the adsorption or desorption procedure of electrolyte on a wide surface area probe material during electrochemical double-layer capacitance operations. Therefore, carbon-based nanomaterials may play a considerable role in supercapacitor fabrication. Nanocarbon-based supercapacitors are better than their traditional (metal-based) counterparts in a number of significant ways, including improved cycle stability, power density, and reduced energy density for use in batteries [97–100].

Nanocarbon-based materials (carbon nanotubes) have excellent mechanical and electrical characteristics, and their exposed surface is amenable to functionalization, making them appropriate for energy storage. However, there are drawbacks, such as the low density of nanomaterials, which results in normal capacitance [101]. The lithium-ion battery (LIB) is an alternative to traditional chemical energy storage methods. In comparison to capacitors, it offers several benefits, including a higher power density and the potential for reduced emissions of greenhouse gases [102]. Since the edifice of nanocarbon-based substantial, typically expresses few usual parameters, such as the quantity of Li that is reversibly combined into the carbon lattice, the faradic losses throughout the first charge–discharge cycle, and the voltage profile through charging and discharging, these materials are used in lithium batteries. Nanomaterials made of carbon, such as carbon nanotubes, triggered carbons, and graphene-based nanolayers, are well-suited for use in long-term energy storage devices [103]. These carbon materials benefit from a number of desirable characteristics, including low cost, ease of processing, adaptable porosity, and simplicity of chemical modification. Typically, the power transport degree and energy storage capability of electrochemical capacitance are improved by nanocarbon structures due to their larger specific surface area and uniform aperture magnitude dispersal.

10.6.2 Nanocarbon–polymer composite environmental applications

The rise of pollutants due to population and industrial growth has prompted the development of effective and low-cost strategies for protecting the planet. There is a need to boost the effectiveness of standard procedures or offer novel ways of pollution remediation in order to ensure the long-term viability of our planet's ecosystems. Because of their large surface area, nanotechnology and carbon-based nanomaterials in particular could potentially make significant contributions in this field [80]. It has been suggested that nanocarbon may efficiently filter out a wide range of contaminants from air and water. In this chapter, attention has been focused on eco-friendly nanocarbon based materials. There are many hazardous substances, such as heavy metal ions and cholorbenzene compounds, that might be present in wastewater from industrial and agricultural sources. The long-term exposure to these chemicals may lead to a variety of health issues including poisoning, cancer, and nervous system issues. The sorption behavior of nanomaterials is crucial for the removal of these pollutants. The adsorption capacity of carbon nanotubes for poisons is quite high [104]. The chemical and physical interactions of hazardous chemicals with nanoparticles influence the sorption behavior of nanocomposites [105–107]. The nanocarbon surface functioning is also crucial in this respect. As a result, CNTs and graphene might be used as sensing components to track the presence of harmful chemicals in the atmosphere [108, 109]. High sensitivity, high power consumption, and optimal working temperature are the advantages that the nanocarbon-based composite gas sensors have over traditional metal oxide gas sensors [110–112]. Nanocarbon is one kind of nanomaterial that may be widely dispersed in nature. Additionally, it is found in the combustion of fuel gases as particulate matter [113, 114]. It has been shown that the nanocarbon found in the environment may be absorbed and deposited in the lungs of humans [115]. Diseases of the heart and lungs, inflammation, and cytotoxicity have all been linked to pollution, namely nanoparticles. Nanocomposites via environmentally friendly engineering is one of the most efficient ways to obtain the advantages of nanocarbons without exposing people to their toxicity. When used in nanocomposites at just the right concentration, nanocarbon toxicity may be mitigated thanks to advances in physical and chemical interaction. Since the creation of the matrix-nanofiller bonded structure, it has prevented the release of nanocarbon from the nanocomposite into the environment. Nanomaterials based on carbon have antibacterial capabilities as well, albeit the mechanism by which this property is exerted is not yet fully understood. Disinfectant applications and antimicrobial surface coatings using CNTs have been proposed in certain research areas. In particular, silver (Ag)-coated CNT hybrid nanoparticles have shown antimicrobial activity, leading the authors to speculate that these materials could one day be used in antibacterial control systems and medicinal equipment [116]. Water disinfection using microwave interactions with functionalized CNTs was reported by Al-Hakami *et al* [117]. Since long carbon chains led to a higher absorption of microwaves by CNTs, the researchers claimed that CNTs functionalized with aliphatic alcohol 1-octadecanol ($C_{18}H_{38}O$) exhibited exceptional antibacterial characteristics.

10.7 Advantages of nanocarbon composites

The benefits of nanocarbon are expected to be substantial. Industries producing tires, automobiles, printed materials, pens, pencils, laptops, desktops, copiers, and laboratory tables use nanocarbon composites (as shown in figures 10.11). One of the most useful and readily replenished elements in chemistry is hydrogen, and hydrogen can be used as a clean burning fuel that produces only water. The advancement of nanotechnology has made it urgent to design a practical hydrogen storage system that adheres to stringent volume and weight constraints [118, 119].

Gas phase storage and electrochemical adsorption are the two most fashionable methods of hydrogen storage. Capillary effect theory suggests that CNTs, due to their hollow cylindrical shape and nanometer-scale diameter, might be used to store liquids or gases in their interior cores. The hydrogen fuel cell industry must meet the minimum storage standards of 6.5% by weight established as per the proper strandard prescribed. By using gas phase adsorption, it has been shown that SWNTs may achieve and even surpass this threshold (physisorption). However, it is impossible to evaluate the claims of enormous storage capacity based on experimental data since most of these studies are much disputed and, because of this, evaluating the potential for applicability is challenging. Electrochemical storage is a different option for storing hydrogen. Chemisorption describes the adsorption of

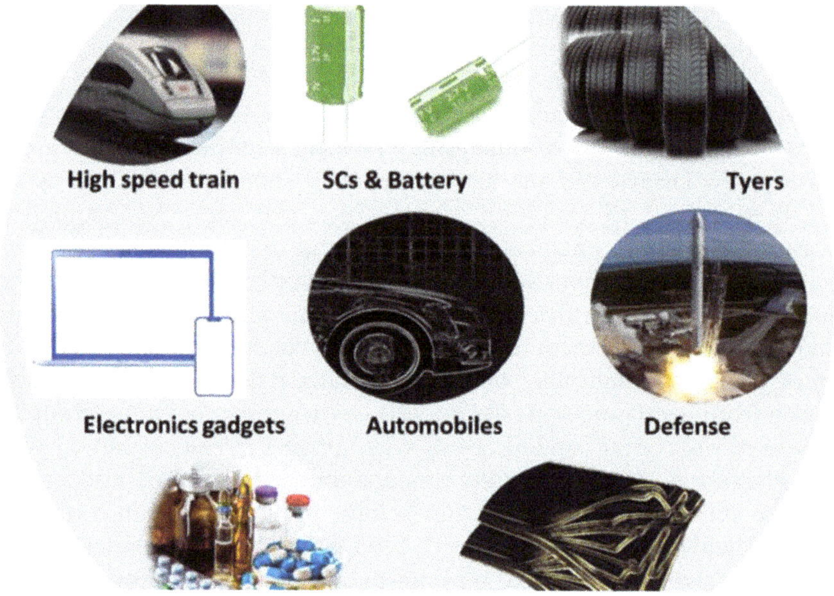

Figure 10.11. Applications of nanocarbons in different sectors of science and technology.

a single hydrogen atom rather than a hydrogen molecule. The key restriction is that new technology or challenges for hydrogen storage may be brought about by more complete knowledge of the hydrogen storage mechanism and the influence of materials processing and its mechanism, which has been a significant concern for decades and the nanocarbon process [118, 119].

10.8 Bottlenecks in the commercialization of nanocarbons

There has been a lot of enthusiasm and new dawn in the province of nanocarbons in the previous, but that is beginning to transformation. In 2022 the industrial use of carbon nanotubes (CNTs) and graphene will enter a new, sustained development phase. SWCNTs and the broader class of graphene materials are still to achieve high-volume use, but that day is fast approaching as industrial engagement continues to increase. MWCNTs are already been there because of their principal utilization in lithium-ion batteries, but they will get been there soon. Research into graphene films and wafers for electrical applications is continuing, despite the fact that most commercial interest is concentrated on graphene's use as a conductive additive. Carbon's single-layer atomic structure has the potential to support a wide variety of functional materials and cutting-edge technologies, from batteries to polymer nanocomposites, with previously unimaginable performance parameters due to its exceptional thermal, optical, electrical, and mechanical properties. Unfortunately, the high price of graphene means that global production is now below 120 metric tonnes per year [120]. In contrast, recently marketed chemical and physical methods of exfoliating graphite have paved the way for mainstream adoption of graphene as the 'enabler' of a wide range of critical technologies, such as better power storage. Since Geim, Novoselov, and collaborators revealed in 2004 that they were able to exfoliate graphite using adhesive tape, the term 'graphene' has been used to describe a single ultrathin (0.3 nm) sheet of carbon atoms structured in a 2D honeycomb lattice [121]. Despite its low density, graphene has a mechanical strength that is 50 times that of steel. Not only is it transparent to the naked eye, but it can also be stretched. More than ten thousand patents have been submitted for technologies utilizing graphene since the mechanical exfoliation process was developed in the early 2000s. Small-scale industrial manufacturing of high-surface-area graphene sheets and films has begun in a number of countries (in Europe, the US, and Asia). Sales are expected to reach $12 million in 2013, and global production has not even reached 120 metric tonnes, so the market is not quite ready for industrialization just yet. A total of 4000 metric tonnes of carbon nanotubes, which can originate from either single- or multi-walled graphene cylinders, were manufactured worldwide in 2011 [120]. Graphene's commercialization is 'within reach in a number of fields', thus it is important to learn as much as possible about graphene-based technologies so that one can evaluate the unique material's market potential. More specifically, graphene has the potential to play a game-changing role in the strategic storage and effective usage of clean power generated from renewable sources such as the Sun and wind. This is the most pressing technological requirement for dealing with the seasonal variability of local resources such as solar radiation and wind. Although typical technical yields are between 8% and 15% depending on the location of production, the most stable fullerenes have been produced at concentrations as high as 23% wt% of the total soot collected. About 2%–4% by weight of the soot produced by an arc is made up of higher-order fullerenes which are the most frequent. Without regard to run conditions or graphite rod thickness, the molar ratio is always 5.06 in a helium gas atmosphere for any carbon arc operation [122].

When used to stabilize smaller NPs, nanocarbons improve performance, and their thermal and mechanical stability also makes them more recyclable. Graphene is an improved candidate for heterogeneous catalysis support due to its larger surface area. For these products to enter the marketplace, large quantities of cheap graphene are required. Currently, GO, also known as thermally expanded graphite, is capable of the highest production rates and demonstrates promising catalytic effects. While MWCNTs are mechanically and thermally stable up to 600 °C, GO begins to decompose in air at just 400 °C. Because of this, future research should prioritize making comparisons with commercially available catalytic support materials [122, 123]. According to the most recent research, there is a huge worldwide nanocarbon materials market that is associated with several growth factors. These growth factors include the most current trends and developments in the sectors around the globe.

10.9 Shortcomings

Although there have been many exciting discoveries and uses for many of the nanocarbons, some fundamental limitations remain, such as the mixing problem and 'unsynthesised' forms. For example, 1D nanocarbons such as CNTs and GNRs have suffered from mixing issues due to the physical synthesis approaches being unable to obtain structurally uniform CNTs and GNRs. Nanocarbons are analogous to origami in the field of chemical sciences. Changes in shape and function may result from quite slight alterations. In terms of 'unsynthesized' forms, some expected nanocarbon structures have yet to be synthesized despite being predicted by theorists and mathematicians. Synthesizing these nanocarbons and manipulating their shape and structure precisely down to the atomic level present formidable difficulties. Three-dimensional carbon crystals with negative Gaussian curvatures and carbon nanotori are two examples of these fantastic but as-yet-unrealized nanocarbon structures. Many of the exact features of these frameworks are still up for debate, but past studies of nanocarbons have demonstrated that the development of new carbon geometries and morphologies leads to the discovery of functions and applications that were not previously expected. Therefore, we think that creating 'unsynthesized' nanocarbons is highly important for future scientific and technological advances. Some substantial progress has been achieved using organic synthesis methods, despite the fact that there is currently no reasonable methodology for the synthesis of these unknown carbon structures [124–126].

In this perspective, we have identified two issues with existing nanocarbon research and shown the considerable progress and breakthroughs toward addressing both of these issues. The ultimate aims of gaining access to structurally homogeneous nanocarbons and developing novel nanocarbon forms are closer and closer to being reached as we apply the idea of precision organic synthesis. However, advancing this field faces a number of obstacles. The size and form options for molecular nanocarbons are currently somewhat restricted, and their production often necessitates time-consuming synthesis techniques and complex experimental procedures. Furthermore, determining the atomic details of the architectures of molecular nanocarbons might be difficult, particularly in the case of novel scaffolds. This means that developing novel approaches to molecular synthesis and structural analysis will continue to play a crucial role in overcoming limitations in the future.

10.10 Future prospects

Considerable progress has been made in the fabrication of nanocarbon-based materials and composites with exceptional mechanical and physical properties. There is still a lot of research to be done in this area before these extraordinary properties can be fully realized in a macroscopic nanometric medium. However, any nanocarbon based composites will need to be broken down or recycled into something new; hence, investigations are needed to determine these recycling processes. In addition, there has not been much progress made in the scientific community's efforts to create naturally improved chemicals employing biologically degradable polymer matrices, since the development of such 'green composites' presents a number of challenges. They are recognized as the most environmentally friendly materials, but issues such as poor adhesion to the fiber matrix, difficulty in fiber orientation, and achieving nanoscale dimensions prevent them from finding relevant applications. Moreover, the isolation of individual nanofiber composites so that their tensile properties may be evaluated and they can be stuck to a load tester remains a challenge. Predicting the mechanical behavior of CNTs implanted in electro-spun fibers using atomistic modeling and continuum mechanism techniques remains a significant issue for theoretical approaches. It may be anticipated that in the next 10–20 years, molecular nanocarbon research will make substantial strides in a variety of applications, including optoelectronic devices and biological functions, in addition to their production. Although a number of molecular nanocarbons are available, there is currently no dependable means for fine-tuning their functions to ensure they are appropriate for each and every use. Also, there is some expectation that new sectors and applications connected to the unanticipated features and functions of nanocarbons will emerge as a result of the integration of methods for synthesizing, analysing, and modifying nanocarbons. With their increasing popularity, molecular nanocarbons have become not only a new benchmark for realizing carbon materials' full potential, but also a chance for a wide range of scientists to confirm the basic relevance and influence of doing science with molecules once again. Structures that are pure, faultless, and frequently attractive will give rise to novel functions. Comparable to the discovery of the DNA double helix, which led to revolutionary advancements in molecular biology, the advent of molecular nanocarbon science promises to have far-reaching consequences and open up exciting new avenues of research. Further, the fields of nanobioimaging, nanomedicine, water and air filtration, catalysis, energy storage, photovoltaics, and sensors stand to benefit greatly from the introduction of nanocarbons in the near future.

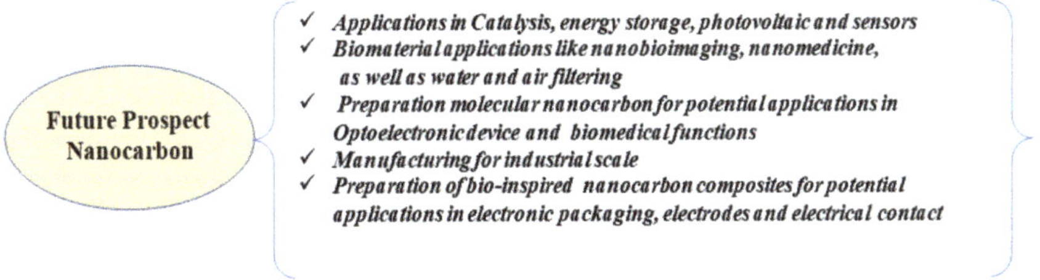

10.11 Conclusion

In materials science, the discovery of nanocarbon has been dubbed the 'breakthrough of the century'. The use of nanocarbon-based materials to improve their performance has garnered a lot of attention and research in recent years. Nanomaterials based on carbon are extraordinary technical tools due to their distinctive features (high mechanical strength, conductivity, smart optical properties, chemical versatility, etc). Graphene and carbon nanotubes are two of the most popular and widely used materials in analytical chemistry. Nanocarbon-based materials are versatile because of their many desirable characteristics. These qualities include high pore quality, a large surface area, a malleable porous structure, thermal and chemical stability, and low sensitivity to deformation. As a whole, the functional group associated with nanocarbon connected to certain materials or metals improves the composites' electrical, thermal, and other desired qualities. The improved electronic properties of modified nanocarbon-based materials may benefit energy use (for storage, conduction, radiation, and so on) and the environment (for detecting pollution parameters). These nanocarbon-based materials, which have a greater surface area than traditional materials, may be useful in the field of pollution remediation. In summary, these materials show potential, and there is a possibility that one day they may act as an interesting alternative in a variety of scenarios. The molecular nanocarbon synthesis and application of nanocarbon in nanobioimaging, nanomedicine, water and air filtration, catalysis, energy storage, photovoltaics, sensors, environment, and technology is an important subject for future research. However, analyses of environmental health and safety relevant to nanotechnology, as well as the reinforcing of nanocarbon in matrices, will need to be investigated. More research and exposure are needed to promote green nanocomposites as a reliable intermediate phase in the manufacture of vehicles, aeroplanes, solar cells, gas sensors, and water purification membranes.

References

[1] Singh H, Sharma A, Bhardwaj S K, Arya S K, Bhardwaj N and Khatri M 2021 Recent advances in the applications of nano-agrochemicals for sustainable agricultural development *Environ. Sci.: Process. Impacts* **23** 213–39

[2] Ferreira S, Monteiro E, Brito P and Vilarinho C 2017 Biomass resources in Portugal: current status and prospects *Renew. Sustain. Energy Rev.* **78** 1221–35

[3] Boncel S, Kyzioł-Komosińska J, Krzyżewska I and Czupioł J 2015 Interactions of carbon nanotubes with aqueous/aquatic media containing organic/inorganic contaminants and selected organisms of aquatic ecosystems—a review *Chemosphere* **136** 211–21

[4] Soffian M S, Halim F Z A, Aziz F, Rahman M A, Amin M A M and Chee D N A 2022 Carbon-based material derived from biomass waste for wastewater treatment *Environ. Adv.* **9** 100259

[5] Bijina V, Jandas P J, Joseph S, Gopu J, Abhitha K and John H 2022 Recent trends in industrial and academic developments of green tyre technology *Polym. Bull.* 1–30

[6] Alcántara A R, Domínguez de María P, Littlechild J A, Schürmann M, Sheldon R A and Wohlgemuth R 2022 Biocatalysis as key to sustainable industrial chemistry *Chem. Sus. Chem.* **15** e202102709

[7] Dutta P, McGranaghan A, Keller I, Patil Y, Mulholland N, Murudi V and Godineau E 2022 A case study in green chemistry: the reduction of hazardous solvents in an industrial R&D environment *Green Chem.* **24** 3943–56

[8] Sternberg J, Sequerth O and Pilla S 2021 Green chemistry design in polymers derived from lignin: review and perspective *Prog. Polym. Sci.* **113** 101344

[9] Li Y, Huang S, Wei C, Wu C and Mochalin V N 2019 Adhesion of two-dimensional titanium carbides (MXenes) and graphene to silicon *Nat. Commun.* **10** 3014

[10] Yang D C, Eglitis R I, Yi Z J, Liu C S and Jia R 2022 Overall direct photocatalytic water-splitting on *C2mm*-graphyne: a novel two-dimensional carbon allotrope *J. Mater. Chem.* C **10** 10843–52

[11] Desmond L J, Phan A N and Gentile P 2021 Critical overview on the green synthesis of carbon quantum dots and their application for cancer therapy *Environ. Sci. Nano* **8** 848–62

[12] Low F W, Lai C W, Samsudin N A, Yusoff Y, Goh S M, Chau C F and Tiong S K 2021 Graphene and its derivatives, synthesis route, and mechanism for photovoltaic solar cell applications *Sustainable Materials for Next Generation Energy Devices* (Amsterdam: Elsevier), pp 103–32

[13] Dawood A, Ahmad J, Ali S, Ullah S, Asghar Z and Shah M 2022 Metal oxide–carbon composites and their applications in optoelectronics and electrochemical energy devices *Metal Oxide—Carbon Hybrid Materials* (Amsterdam: Elsevier), pp 309–39

[14] Ahmad S O A, Ashfaq A, Akbar M U, Ikram M, Khan K, Wang F and Mahmood A 2021 Application of two-dimensional materials in perovskite solar cells: recent progress, challenges, and prospective solutions *J. Mater. Chem.* C **9** 14065–92

[15] Maduraiveeran G and Jin W 2021 Carbon nanomaterials: synthesis, properties and applications in electrochemical sensors and energy conversion systems *Mater. Sci. Eng.* B **272** 115341

[16] Li C, Liu J, Gao R, Ouyang H, Huang J, Huang Q and Liu Y 2022 Tailoring CoNi alloy embedded carbon nano-fibers by thiourea for enhanced hydrogen evolution reaction *Int. J. Hydrog. Energy* **47** 34471–82

[17] Zhang Y, Suhir E and Gu C 2021 Physical properties and mechanical behavior of carbon nano-tubes (CNTs) and carbon nano-fibers (CNFs) as thermal interface materials (TIMs) for high-power integrated circuit (IC) packages: review and extension *Nano-Bio-Electronic, Photonic and MEMS Packaging* springer, pp 137–56

[18] Zhang X, Xu J, Yuan H, Zhang S, Ouyang Q, Zhu C and Chen Y 2019 Large-scale synthesis of three-dimensional reduced graphene oxide/nitrogen-doped carbon nanotube heteronanostructures as highly efficient electromagnetic wave absorbing materials *ACS Appl. Mater. Interfaces* **11** 39100–8

[19] Laraba S R, Luo W, Rezzoug A, Zahra Q U A, Zhang S, Wu B and Li Y 2022 Graphene-based composites for biomedical applications *Green Chem. Lett. Rev.* **15** 724–48

[20] Tiwari S K, Kumar V, Huczko A, Oraon R, Adhikari A D and Nayak G C 2016 Magical allotropes of carbon: prospects and applications *Crit. Rev. Solid State Mater. Sci.* **41** 257–317

[21] Liu S, Li B, Zhou Y, Xu X, Yang R, Wang Q and Li J 2021 Hierarchical N-doped holey three-dimensional reduced graphene oxide with high performance capacitive deionization *J. Mater. Res. Technol.* **15** 1996–2006

[22] Cui X, Tian J, Zhang C, Cai R, Ma J, Yang Z and Meng Q 2021 Comparative study of nanocarbon-based flexible multifunctional composite electrodes *ACS Omega* **6** 2526–41

[23] Zhang L, Ding S, Han B, Yu X and Ni Y Q 2019 Effect of water content on the piezoresistive property of smart cement-based materials with carbon nanotube/nanocarbon black composite filler *Composites* A **119** 8–20

[24] Zhou M, Zhai Z, Liu L, Zhang C, Yuan Z, Lu Z and Jiang X 2021 Controllable synthesized diamond/CNWs film as a novel nanocarbon electrode with wide potential window and enhanced S/B ratio for electrochemical sensing *Appl. Surf. Sci.* **551** 149418

[25] Zhou G M, Wang D W, Feng L, Zhang L L, Zhe W E N G and Cheng H M 2011 The effect of carbon particle morphology on the electrochemical properties of nanocarbon/polyaniline composites in supercapacitors *New Carbon Mater.* **26** 180–6

[26] Dresselhaus M S, Dresselhaus G, Saito R and Jorio A 2005 Raman spectroscopy of carbon nanotubes *Phys. Rep.* **409** 47–99

[27] Bayatsarmadi B, Zheng Y, Vasileff A and Qiao S Z 2017 Recent advances in atomic metal doping of carbon-based nanomaterials for energy conversion *Small* **13** 1700191
[28] Garg P, Ghatmale P, Tarwadi K and Chavan S 2017 Influence of nanotechnology and the role of nanostructures in biomimetic studies and their potential applications *Biomimetics* **2** 7
[29] Arduini F, Cinti S, Scognamiglio V, Moscone D and Palleschi G 2017 How cutting-edge technologies impact the design of electrochemical (bio) sensors for environmental analysis. A review *Anal. Chim. Acta* **959** 15–42
[30] Levchenko I, Xu S, Teel G, Mariotti D, Walker M L R and Keidar M 2018 Recent progress and perspectives of space electric propulsion systems based on smart nanomaterials *Nat. Commun.* **9** 879
[31] Wu Q, Yan X, Jia Y and Yao X 2021 Defective carbon-based materials: controllable synthesis and electrochemical applications *Energy Chem.* **3** 100059
[32] Rathinavel S, Priyadharshini K and Panda D 2021 A review on carbon nanotube: an overview of synthesis, properties, functionalization, characterization, and the application *Mater. Sci. Eng.* B **268** 115095
[33] Baig N, Kammakakam I and Falath W 2021 Nanomaterials: a review of synthesis methods, properties, recent progress, and challenges *Mater. Adv.* **2** 1821–71
[34] Gaur M, Misra C, Yadav A B, Swaroop S, Maolmhuaidh F Ó, Bechelany M and Barhoum A 2021 Biomedical applications of carbon nanomaterials: fullerenes, quantum dots, nanotubes, nanofibers, and graphene *Materials* **14** 5978
[35] Islam M R, Afroj S, Novoselov K S and Karim N 2022 Smart electronic textile-based wearable supercapacitors *Adv. Sci.* **9** 2203856
[36] Chen Y, Li S, Lin S, Chen M, Tang C and Liu X 2022 Promising energy-storage applications by flotation of graphite ores: a review *Chem. Eng. J.* **454** 139994
[37] Rasal A S, Yadav S, Yadav A, Kashale A A, Manjunatha S T, Altaee A and Chang J Y 2021 Carbon quantum dots for energy applications: a review *ACS Appl. Nano Mater.* **4** 6515–41
[38] Penkova A V, Acquah S F, Piotrovskiy L B, Markelov D A, Semisalova A S and Kroto H W 2017 Fullerene derivatives as nano-additives in polymer composites *Russ. Chem. Rev.* **86** 530
[39] Greil P 2015 Perspectives of nano-carbon based engineering materials *Adv. Eng. Mater.* **17** 124–37
[40] Anwar A, Chang T P and Chen C T 2022 Graphene oxide synthesis using a top–down approach and discrete characterization techniques: a holistic review *Carbon Lett.* **32** 1–38
[41] Azadmanjiri J, Srivastava V K, Kumar P, Nikzad M, Wang J and Yu A 2018 Two-and three-dimensional graphene-based hybrid composites for advanced energy storage and conversion devices *J. Mater. Chem.* A **6** 702–34
[42] Schultz B J, Dennis R V, Lee V and Banerjee S 2014 An electronic structure perspective of graphene interfaces *Nanoscale* **6** 3444–66
[43] Khare P, Patel R K and Shankar R 2019 Recent developments in nanocarbon–polymer composites for environmental and energy applications *Green Chemistry Application* (London: Rijeka)
[44] Srivastava V, Gusain D and Sharma Y C 2015 Critical review on the toxicity of some widely used engineered nanoparticles *Ind. Eng. Chem. Res.* **54** 6209–33
[45] Ehsani M, Rahimi P and Joseph Y 2021 Structure–function relationships of nanocarbon/polymer composites for chemiresistive sensing: a review *Sensors* **21** 3291

[46] Gerasimenko A Y, Kuksin A V, Shaman Y P, Kitsyuk E P, Fedorova Y O, Murashko D T and Glukhova O E 2022 Hybrid carbon nanotubes–graphene nanostructures: modeling, formation, characterization *Nanomaterials* **12** 2812

[47] Baghel P, Sakhiya A K and Kaushal P 2022 Ultrafast growth of carbon nanotubes using microwave irradiation: characterization and its potential applications *Heliyon* **8** e10943

[48] Mirzaei S, Ahmadpour A, Shao Z and Arami-Niya A 2022 Rational design of carbon-based materials for purification and storage of energy carrier gases of methane and hydrogen arXiv: 2204.11600

[49] Shoukat R and Khan M I 2021 Carbon nanotubes: a review on properties, synthesis methods and applications in micro and nanotechnology *Microsyst. Technol.* **27** 4183–92

[50] Zhang F, Jiang W, Song X, Kang J, Cao Y and Xiang M 2021 Effects of hyperbranched polyester-modified carbon nanotubes on the crystallization kinetics of polylactic acid *ACS Omega* **6** 10362–70

[51] Grillo R, Rosa A H and Fraceto L F 2015 Engineered nanoparticles and organic matter: a review of the state-of-the-art *Chemosphere* **119** 608–19

[52] Manawi Y, Kochkodan V, Hussein M A, Khaleel M A, Khraisheh M and Hilal N 2016 Can carbon-based nanomaterials revolutionize membrane fabrication for water treatment and desalination? *Desalination* **391** 69–88

[53] Zhao Q, Zhu Q, Liu Y and Xu B 2021 Status and prospects of MXene-based lithium sulfur batteries *Adv. Funct. Mater.* **31** 2100457

[54] Xiong D, Li X, Bai Z and Lu S 2018 Recent advances in layered $Ti_3C_2T_x$ MXene for electrochemical energy storage *Small* **14** 1703419

[55] Han Y, Tang D, Yang Y, Li C, Kong W, Huang H and Kang Z 2015 Non-metal single/dual doped carbon quantum dots: a general flame synthetic method and electro-catalytic properties *Nanoscale* **7** 5955–62

[56] Hu C, Li M, Qiu J and Sun Y P 2019 Design and fabrication of carbon dots for energy conversion and storage *Chem. Soc. Rev.* **48** 2315–37

[57] Tian L, Li Z, Wang P, Zhai X, Wang X and Li T 2021 Carbon quantum dots for advanced electrocatalysis *J. Energy Chem.* **55** 279–94

[58] Xu Q, Li W, Ding L, Yang W, Xiao H and Ong W J 2019 Function-driven engineering of 1D carbon nanotubes and 0D carbon dots: mechanism, properties and applications *Nanoscale* **11** 1475–504

[59] Sun Y P, Zhou B, Lin Y, Wang W, Fernando K S, Pathak P and Xie S Y 2006 Quantum-sized carbon dots for bright and colorful photoluminescence *J. Am. Chem. Soc.* **128** 7756–7

[60] Sharma V, Tiwari P and Mobin S M 2017 Sustainable carbon-dots: recent advances in green carbon dots for sensing and bioimaging *J. Mater. Chem.* B **5** 8904–24

[61] Caglayan M O, Mindivan F and Şahin S 2022 Sensor and bioimaging studies based on carbon quantum dots: the green chemistry approach *Crit. Rev. Anal. Chem.* **52** 814–47

[62] Lu H, Li W, Dong H and Wei M 2019 Graphene quantum dots for optical bioimaging *Small* **15** 1902136

[63] Colombo P, Mera G, Riedel R and Soraru G D 2013 Polymer-derived ceramics: 40 years of research and innovation in advanced ceramics *J. Am. Ceram. Soc.* **4** 245–320

[64] Cordelair J and Greil P 2000 Electrical conductivity measurements as a microprobe for structure transitions in polysiloxane derived Si–O–C ceramics *J. Eur. Ceram. Soc.* **20** 1947–57

[65] Monthioux M and Delverdier O 1996 Thermal behavior of (organosilicon) polymer-derived ceramics. V: main facts and trends *J. Eur. Ceram. Soc.* **16** 721–37

[66] Schrand A M, Hens S A C and Shenderova O A 2009 Nanodiamond particles: properties and perspectives for bioapplications *Crit. Rev. Solid State Mater. Sci.* **34** 18–74

[67] Mykhailiv O, Zubyk H and Plonska-Brzezinska M E 2017 Carbon nano-onions: unique carbon nanostructures with fascinating properties and their potential applications *Inorg. Chim. Acta* **468** 49–66

[68] Oshida K, Murata M, Fujiwara K, Itaya T, Yanagisawa T, Kimura K and Yang K S 2013 Structural analysis of nano structured carbon by transmission electron microscopy and image processing *Appl. Surf. Sci.* **275** 409–12

[69] Huang Y, Li Z, Jin S, Zhang S, Wang H, Hiralal P and Zhou H 2020 Carbon nanohorns/nanotubes: an effective binary conductive additive in the cathode of high energy-density zinc-ion rechargeable batteries *Carbon* **167** 431–8

[70] Ahangari M G, Ganji M D and Montazar F 2015 Mechanical and electronic properties of carbon nanobuds: first-principles study *Solid State Commun.* **203** 58–62

[71] Qin Q, Chen J, Wu C, Wang Y, Li Y and Song M 2022 *In situ* growth of novel carbon nanobuds and nanoballs on graphene nanosheets by the electrochemical method *RSC Adv.* **12** 17937–43

[72] Wang H, Wang H S, Ma C, Chen L, Jiang C, Chen C and Wang X 2021 Graphene nanoribbons for quantum electronics *Nat. Rev. Phys.* **3** 791–802

[73] Abbas A, Tabish T A, Bull S J, Lim T M and Phan A N 2020 High yield synthesis of graphene quantum dots from biomass waste as a highly selective probe for Fe^{3+} sensing *Sci. Rep.* **10** 21262

[74] Tajik S, Dourandish Z, Zhang K, Beitollahi H, Van Le Q, Jang H W and Shokouhimehr M 2020 Carbon and graphene quantum dots: a review on syntheses, characterization, biological and sensing applications for neurotransmitter determination *RSC Adv.* **10** 15406–29

[75] Younis M R, He G, Lin J and Huang P 2020 Recent advances on graphene quantum dots for bioimaging applications *Front. Chem.* **8** 424

[76] Li Z, Liu Z, Sun H and Gao C 2015 Superstructured assembly of nanocarbons: fullerenes, nanotubes, and graphene *Chem. Rev.* **115** 7046–117

[77] Velasco-Santos C, Martınez-Hernández A L, Consultchi A, Rodrıguez R and Castaño V M 2003 Naturally produced carbon nanotubes *Chem. Phys. Lett.* **373** 272–6

[78] Su D S and Chen X W 2007 Natural lavas as catalysts for efficient production of carbon nanotubes and nanofibers *Angew. Chem. Int. Ed. Engl.* **119** 1855–6

[79] Mracek J, Fagan D, Stengelin M R and Hesjedal T 2011 Are carbon nanotubes a naturally occurring material? Hints from methane CVD using lava as a catalyst *Curr. Nanosci.* **7** 294–6

[80] Zaytseva O and Neumann G 2016 Carbon nanomaterials: production, impact on plant development, agricultural and environmental applications *Chem. Biol. Technol. Agric.* **3** 17

[81] Becker L, Bada J L, Winans R E and Bunch T E 1994 Fullerenes in Allende meteorite *Nature* **372** 507

[82] Innocenzi P and Stagi L 2020 Carbon-based antiviral nanomaterials: graphene, C-dots, and fullerenes. A perspective *Chem. Sci.* **11** 6606–22

[83] Zheng Y, Jiao Y and Qiao S Z 2015 Engineering of carbon-based electrocatalysts for emerging energy conversion: from fundamentality to functionality *Adv. Mater.* **27** 5372–8

[84] Zhang Y, Petibone D, Xu Y, Mahmood M, Karmakar A, Casciano D and Biris A S 2014 Toxicity and efficacy of carbon nanotubes and graphene: the utility of carbon-based nanoparticles in nanomedicine *Drug Metab. Rev.* **46** 232–46
[85] Liu J, Tang J and Gooding J J 2012 Strategies for chemical modification of graphene and applications of chemically modified graphene *J. Mater. Chem.* **22** 12435–52
[86] Shin M, Noh C, Chung Y, Kim D H and Kwon Y 2021 Vanadium redox flow battery working even at a high current density by the adoption of tris (hydroxymethyl) aminomethane functionalized acidified carbon nanotube catalyst *Appl. Surf. Sci.* **550** 148977
[87] Jiang L, Lin L, Li P, Li X, Zhong Y, Yu Y and Qiu Y 2020 Core–shell MoS_2 nanosheet-bonded carbon nanotubes through dispersant molecules for enhanced photoelectrical and photocatalytic performances *J. Solid State Chem.* **284** 121211
[88] Bhadouria R, Mishra D, Singh V K, Singh P, Srivastava P, Tripathi S and Singh R 2020 Nanocatalyst types and their potential impacts in agroecosystems: an overview *Nano-Materials as Photocatalysts for Degradation of Environmental Pollutants* Elsevier, pp 323–44
[89] Sabantina L 2021 Nanocarbons-based textiles for flexible energy storage *Nanosensors and Nanodevices for Smart Multifunctional Textiles* (Amsterdam: Elsevier), pp 163–88
[90] De Volder M F, Tawfick S H, Baughman R H and Hart A J 2013 Carbon nanotubes: present and future commercial applications *Science* **339** 535–9
[91] Nasibulin A G, Shandakov S D, Nasibulina L I, Cwirzen A, Mudimela P R, Habermehl-Cwirzen K and Kauppinen E I 2009 A novel cement-based hybrid material *New J. Phys.* **11** 023013
[92] Ni J and Li Y 2016 Carbon nanomaterials in different dimensions for electrochemical energy storage *Adv. Energy Mater.* **6** 1600278
[93] Lv T, Liu M, Zhu D, Gan L and Chen T 2018 Nanocarbon-based materials for flexible all-solid-state supercapacitors *Adv. Mater.* **30** 1705489
[94] Watanabe R, Matsuzaki R, Endo H and Koyanagi J 2018 Stretchable and insulating characteristics of chemically bonded graphene and carbon nanotube composite materials *J. Mater. Sci.* **53** 1148–56
[95] Tiwari S K, Bystrzejewski M, De Adhikari A, Huczko A and Wang N 2022 Methods for the conversion of biomass waste into value-added carbon nanomaterials: recent progress and applications *Prog. Energy Combust. Sci.* **92** 101023
[96] Deokate R J 2019 The recent research and growth in energy efficiency in Cu_2ZnSnS_4 (CZTS) solar cells *Nanostructured Materials for Next-Generation Energy Storage and Conversion* Springer, pp 137–64
[97] Ibrahim I D, Sadiku E R, Jamiru T, Hamam Y, Alayli Y and Eze A A 2020 Prospects of nanostructured composite materials for energy harvesting and storage *J. King Saud Univ. Sci.* **32** 758–64
[98] Zheng X and Zhang L 2016 Photonic nanostructures for solar energy conversion *Energy Environ. Sci.* **9** 2511–32
[99] Wu Z, Li L, Yan J M and Zhang X B 2017 Materials design and system construction for conventional and new-concept supercapacitors *Adv. Sci.* **4** 1600382
[100] Lukatskaya M R, Dunn B and Gogotsi Y 2016 Multidimensional materials and device architectures for future hybrid energy storage *Nat. Commun.* **7** 12647
[101] Saba N, Jawaid M, Fouad H and Alothman O Y 2019 Nanocarbon: preparation, properties, and applications *Nanocarbon and Its Composites* Elsevier, pp 327–54

[102] Zheng J S, Zhang L, Shellikeri A, Cao W, Wu Q and Zheng J P 2017 A hybrid electrochemical device based on a synergetic inner combination of Li ion battery and Li ion capacitor for energy storage *Sci. Rep.* **7** 41910
[103] Wang J and Kaskel S 2012 KOH activation of carbon-based materials for energy storage *J. Mater. Chem.* **22** 23710–25
[104] Kausar A 2015 Effect of sol–gel coating on microscopic, thermal and water absorption behavior of aramid/nylon 6/6/nanodiamondbased fibers *Am. J. Curr. Org. Chem.* **1** 60–8
[105] Kausar A 2015 Novel nanofiltration membranes of poly (4-chlorostyrene)-grafted-1-(4-thio carbamoylaminophenyl–sulfonylphenyl) thiourea and gold/polystyrene composite nanoparticle *Am. J. Polym. Sci. Eng.* **3** 33–49
[106] Sayed A, Mahmoud G A, Said H and Diab A A 2022 Characterization and optimization of magnetic gum-PVP/SiO_2 nanocomposite hydrogel for removal of contaminated dyes *Mater. Chem. Phys.* **280** 125731
[107] Sarkar S, Chatterjee P K, Cumbal L H and SenGupta A K 2011 Hybrid ion exchanger supported nanocomposites: sorption and sensing for environmental applications *Chem. Eng. J.* **166** 923–31
[108] Turco A, Monteduro A G, Montagna F, Primiceri E, Frigione M and Maruccio G 2022 Nanoarchitectonics of highly sensitive and with large working range 3D piezoresistive microporous foam based on carbon nanotubes and elastomer *J. Colloid Interface Sci.* **607** 1436–45
[109] Tucek J, Kemp K C, Kim K S and Zboril R 2014 Iron-oxide-supported nanocarbon in lithium-ion batteries, medical, catalytic, and environmental applications *ACS Nano* **8** 7571–612
[110] Chen X M, Wu G H, Jiang Y Q, Wang Y R and Chen X 2011 Graphene and graphene-based nanomaterials: the promising materials for bright future of electroanalytical chemistry *Analyst* **136** 4631–40
[111] Norizan M N, Moklis M H, Demon S Z N, Halim N A, Samsuri A, Mohamad I S and Abdullah N 2020 Carbon nanotubes: functionalisation and their application in chemical sensors *RSC Adv.* **10** 43704–32
[112] Kausar A 2017 Pb (II) selective sensor of poly (vinyl chloride–vinyl acetate)/polyaniline/carbon black *Int. J. Instrum. Sci.* **6** 8–11
[113] Murr L E, Bang J J, Esquivel E V, Guerrero P A and Lopez D A 2004 Carbon nanotubes, nanocrystal forms and complex nanoparticle aggregates in common fuel–gas combustion source and ambient air *J. Nanoparticle Res.* **5** 1
[114] Dreher K L 2004 Health and environmental impact of nanotechnology: toxicological assessment of manufactured nanoparticles *Toxicol. Sci.* **77** 3
[115] Li Z J, Chapman R, Hulderman T, Salmen R, Shvedova A, Luster M I and Simeonova P P 2006 Relationship between pulmonary exposure to multiple doses of single wall carbon nanotubes and atherosclerosis in ApoE−/− mouse model *Toxicologist* **90** 213
[116] Jung J H, Hwang G B, Lee J E and Bae G N 2011 Preparation of airborne Ag/CNT hybrid nanoparticles using an aerosol process and their application to antimicrobial air filtration *Langmuir* **27** 10256–64
[117] Al-Hakami S M, Khalil A B, Laoui T and Atieh M A 2013 Fast disinfection of *Escherichia coli* bacteria using carbon nanotubes interaction with microwave radiation *Bioinorg. Chem. Appl.* **2013** 458943

[118] Chen H and Sholl D S 2004 Rapid diffusion of CH_4/H_2 mixtures in single-walled carbon nanotubes *J. Am. Chem. Soc.* **126** 7778–9
[119] Rahman A, Ali I, Al Zahrani S M and Eleithy R H 2011 A review of the applications of nanocarbon polymer composites *Nano* **6** 185–203
[120] Zurutuza A and Marinelli C 2014 Challenges and opportunities in graphene commercialization *Nat. Nanotechnol.* **9** 730–4
[121] Novoselov K S, Geim A K, Morozov S V, Jiang D, Katsnelson M I, Grigorieva I and Firsov A 2005 Two-dimensional gas of massless Dirac fermions in graphene *Nature* **438** 197–200
[122] Ciriminna R, Zhang N, Yang M Q, Meneguzzo F, Xu Y J and Pagliaro M 2015 Commercialization of graphene-based technologies: a critical insight *Chem. Commun.* **51** 7090–5
[123] Shearer C J, Cherevan A and Eder D 2014 Application and future challenges of functional nanocarbon hybrids *Adv. Mater.* **26** 2295–318
[124] Kawasumi K, Zhang Q, Segawa Y, Scott L T and Itami K 2013 A grossly warped nanographene and the consequences of multiple odd-membered-ring defects *Nat. Chem.* **5** 739–44
[125] Cheung K Y, Xu X and Miao Q 2015 Aromatic saddles containing two heptagons *J. Am. Chem. Soc.* **137** 3910–4
[126] Itami K and Maekawa T 2020 Molecular nanocarbon science: present and future *Nano Lett.* **20** 4718–20

www.ingramcontent.com/pod-product-compliance
Ingram Content Group UK Ltd.
Pitfield, Milton Keynes, MK11 3LW, UK
UKHW051903100926
8556IPUK00005B/744

9 780750 351751